SiC Technology

Maurizio Di Paolo Emilio
Editor

SiC Technology

Materials, Manufacturing, Devices and Design for Power Conversion

Springer

Editor
Maurizio Di Paolo Emilio
Power Electronics News
Pescara, Italy

ISBN 978-3-031-63417-8 ISBN 978-3-031-63418-5 (eBook)
https://doi.org/10.1007/978-3-031-63418-5

This Springer imprint is published by the registered company Springer Nature Switzerland AG
The registered company address is: Gewerbestrasse 11, 6330 Cham, Switzerland

Preface

In the ever-evolving landscape of semiconductor technology, the emergence of silicon carbide (SiC) stands as a transformative force, poised to revolutionize the field of power electronics. This comprehensive publication delves deep into the multifaceted realm of SiC, offering a detailed exploration of its properties, manufacturing processes, technological advancements, applications, market trends, and future prospects. Authored by leading experts and visionaries from academia and industry, this book is structured into nine chapters, each meticulously crafted to provide a holistic understanding of SiC's significance and potential impact.

Chapter 1 sets the stage with an insightful overview penned by esteemed contributors. Victor Veliadis, Ming Su, and Filippo Di Giovanni delve into the exceptional properties, evolutionary journey, and electron configurations of SiC, respectively, highlighting its transformative potential in power electronics.

Authored by Kevin Speer, Yogesh Sharma, Sergey Maximenko, and Filippo Di Giovanni, Chap. 2 delves into the essential material properties of SiC. From crystal structures to epitaxial growth techniques, this chapter underscores SiC's superiority for high-power and extreme environment applications. Ravi Bollina, William Gemmill, Terry Knight, Christian Jentgens, and Helge Willers of Pureon delve into the intricate process of preparing SiC wafer substrates, highlighting their importance for high-power and high-frequency applications.

Chapter 3, composed by John R. Silk, Dr. Victor Veliadis, Dario Tenaglia, Paolo Badalà, and Simone Rascuna, offers a detailed exploration of SiC device manufacturing processes. From bulk growth to ion implantation and device processing, this chapter unravels the intricacies of fabricating high-performance SiC semiconductor devices.

In Chap. 4, experts including Rudolf Elpelt, Andrea Piccioni, Jake Choi, Byeongyeop Lee, Sandol Lee, Paul Salmen, Olaf Zschieschang, Sebastian Fahlbusch, Ryo Takeda, Takamasa Arai, Llew Vaughan-Edmunds and Victor Veliadis discuss SiC device design, reliability, packaging, and dynamic characterization, highlighting its transformative impact on power electronics.

Authored by Sanbao Zheng, Jonathan Dodge, Andrea Colognese, and Levan Bidzishvili, Chap. 5 explores SiC's applications in power conversion technologies.

From motor drives to solid-state circuit breakers and traction inverters, this chapter showcases SiC's efficiency gains across diverse power conversion applications.

Chapter 6 presents real-world case studies illustrating SiC's widespread adoption. Contributions from industry experts like Suresh Thangavel, Aparna Kumar, Mike Zhu, Christoph Ihm, Tobias Asshauer, Rémi Comyn, Zan Huang, Didier Balocco, James Victory, and Stephen Oliver showcase SiC's transformative impact in renewable energy, data centers, electric vehicles, and more.

In Chap. 7, Peter Friedrichs and Victor Veliadis delve into the transformative impact of SiC technology on power devices. Friedrichs examines material innovations, device enhancements, and ecosystem developments driving performance and cost savings. Meanwhile, Veliadis tackles challenges such as production costs, defects affecting yield, and workforce needs. Together, they offer a comprehensive outlook on advancing SiC technology to meet global energy demands and environmental goals.

Aleksei Cherkasov provides a comprehensive overview of the SiC semiconductor market in Chap. 8, discussing key players, technological challenges, and investment initiatives driving innovation, particularly in the context of electric vehicle adoption and global semiconductor industry trends.

In the concluding chapter, I reflect on disruptive substrate and device innovations that promise to revolutionize SiC manufacturing. This chapter encapsulates the evolving landscape of semiconductor technology, poised to impact sectors like electric vehicles and renewable energy systems.

Together, these chapters offer a cohesive narrative, delving into the technological, industrial, and market dynamics surrounding SiC. They provide valuable insights for researchers, engineers, industry professionals, and enthusiasts keen on harnessing the transformative potential of silicon carbide in the realm of power electronics.

The semiconductor industry is witnessing a paradigm shift fueled by disruptive technologies across substrate innovation, wafering processes, and transition to larger wafer sizes. The industry's transition to 200 mm substrates is gaining momentum, driven by the substantial increase in die yield compared to traditional 150 mm wafers. This transition not only offers greater economies of scale but also aligns with the prevalent technology node in modern silicon power device manufacturing. These advancements hold immense promise for addressing critical challenges, unlocking new possibilities, and driving unprecedented progress in various application domains, from electric vehicles to renewables energy systems and beyond. As industry players continue to push the boundaries of technological innovation, the future of semiconductor devices appears brighter than ever before.

This publication serves as a definitive guide to understanding SiC's pivotal role in shaping the future of power electronics and its broader implications for sustainability, efficiency, and technological advancement.

Pescara, Italy Maurizio Di Paolo Emilio

Acknowledgments

Our goal with this book is to present a background to silicon carbide (SiC) technology, an explanation as to how and why its electrical characteristics are suited to power electronics systems in different vertical markets, and its practical implementation in devices. We also look at its market traction at the time of writing this book, plus its potential as a mainstream technology for the future.

The benefits of silicon carbide technology have become widely recognized in vehicle electrification, especially as the electric- and hybrid-vehicle market grows. This, in turn, has enabled the technology to mature and drive the commercialization of SiC devices into other sectors. Beyond automotive, SiC devices are becoming viable, cost-effective, and efficient replacements for silicon in other high-voltage applications such as electromobility, industrial, and renewable energy.

The book couldn't have come together without the help of keen observers, technology experts, and stakeholders in the power electronics industries. I extend my heartfelt gratitude to the multitude of individuals and organizations who have contributed to the fruition of this book. Without their invaluable insights, expertise, and support, this endeavor would not have been possible. Their dedication to advancing the field of silicon carbide is truly commendable.

Contributors

- Suresh Thangavel, Infineon Technologies
- Aparna Kumar, Qorvo
- Mike Zhu, Qorvo
- Christoph Ihm, Infineon Technologies
- Tobias Asshauer, Infineon Technologies
- Rémi Comyn, KnowMade
- Zan Huang, Wolfspeed
- Didier Balocco, onsemi
- James Victory, onsemi

- Stephen Oliver, Navitas Semiconductor
- Victor Veliadis, PowerAmerica
- Ming Su, Rohm Semiconductor
- Filippo Di Giovanni, Technical Writer
- Kevin Speer, Microchip
- Yogesh Sharma, Microchip
- Sergey Maximenko, Microchip
- Ravi Bollina, Pureon
- William Gemmill, Pureon
- Terry Knight, Pureon
- Christian Jentgens, Pureon
- Helge Willers, Pureon
- John R. Silk, SemiQ
- Dario Tenaglia, ST Microelectronics
- Paolo Badalà, ST Microelectronics
- Simone Rascuna, ST Microelectronics
- Rudolf Elpelt, Infineon Technologies
- Andrea Piccioni, Infineon Technologies
- Jake Choi, onsemi
- Byeongyeop Lee, onsemi
- Sandol Lee, Infineon Technologies
- Paul Salmen, Infineon Technologies
- Olaf Zschieschang, onsemi
- Sebastian Fahlbusch, Nexperia
- Ryo Takeda, Keysight Technologies
- Takamasa Arai, Keysight Technologies
- Llew Vaughan-Edmunds, Navitas Semiconductor
- Sanbao Zheng, Qorvo
- Jonathan Dodge, Qorvo
- Andrea Colognese, onsemi
- Levan Bidzishvili, onsemi
- Peter Friedrichs, Infineon Technologies
- Aleksei Cherkasov, Infineon Technologies

Many thanks to all of you!

Additionally, special thanks to the Springer team for their assistance and guidance throughout this journey.

Your combined contributions have significantly enhanced the content of this book, undoubtedly propelling the dialogue on cleaner, greener, and more efficient power conversion technologies, particularly those leveraging SiC.

About the Book

The book is a comprehensive exploration of silicon carbide (SiC) technology, highlighting its foundational principles, manufacturing processes, diverse applications across industries, challenges, and future trends. It features contributions from a multitude of experts in the field, each focusing on specific aspects of SiC technology:

Victor Veliadis, Ming Su, and Filippo Di Giovanni delve into the exceptional properties, evolutionary journey, and electron configurations of SiC, respectively, highlighting its transformative potential in power electronics.

Kevin Speer, Yogesh Sharma, Sergey Maximenko, and Filippo Di Giovanni and Pureon team delve into the essential material properties of SiC.

John R. Silk, Dario Tenaglia, Paolo Badalà, Simone Rascuna and Victor Veliadis offer a detailed exploration of SiC device manufacturing processes.

Rudolf Elpelt, Andrea Piccioni, Jake Choi, Byeongyeop Lee, Sandol Lee, Paul Salmen, Olaf Zschieschang, Sebastian Fahlbusch, Ryo Takeda, Takamasa Arai, Llew Vaughan-Edmunds and Victor Veliadis discuss SiC device design, reliability, packaging, and dynamic characterization, highlighting its transformative impact on power electronics.

Sanbao Zheng, Jonathan Dodge, Andrea Colognese, and Levan Bidzishvili explore SiC's applications in power conversion technologies.

Suresh Thangavel, Aparna Kumar, Mike Zhu, Christoph Ihm, Tobias Asshauer, Rémi Comyn, Zan Huang, Didier Balocco, James Victory, and Stephen Oliver showcase SiC's transformative impact in renewable energy, data centers, electric vehicles, and more.

Peter Friedrichs and Victor Veliadis delve into the transformative impact of SiC technology on power devices. Friedrichs examines material innovations, device enhancements, and ecosystem developments driving performance and cost savings. Meanwhile, Veliadis tackles challenges such as production costs, defects affecting yield, and workforce needs. Together, they offer a comprehensive outlook on advancing SiC technology to meet global energy demands and environmental goals.

Aleksei Cherkasov provides a comprehensive overview of the SiC semiconductor market.

Maurizio Di Paolo Emilio reflects on disruptive substrate and device innovations that promise to revolutionize SiC manufacturing.

Contents

Contributors

Takamasa Arai Keysight Technologies, Hachioji-shi, Tokyo, Japan

Tobias Asshauer Infineon Technologies, Warstein, Germany

Paolo Badalà STMicroelectronics, Catania, Italy

Didier Balocco onsemi, Vélizy, France

Levan Bidzishvili onsemi, Milan, Italy

Ravi Bollina Pureon, Lengwil, Switzerland

Aleksei Cherkasov Infineon Technologies, Villach, Austria

Jake Choi onsemi, South Portland, ME, USA

Andrea Colognese onsemi, Milan, Italy

Rémi Comyn KnowMade, Valbonne, France

Filippo Di Giovanni STMicroelectronics, San Gregorio, Italy

Maurizio Di Paolo Emilio Power Electronics News, Pescara, Italy

Jonathan Dodge Qorvo, Greensboro, NC, USA

Rudolf Elpelt Infineon Technologies, Erlangen, Germany

Sebastian Fahlbusch Nexperia, Hamburg, Germany

Peter Friedrichs Infineon Technologies, Neubiberg, Germany

William Gemmill Pureon, Monroe, NC, USA

Zan Huang Wolfspeed, Durham, NC, USA

Christoph Ihm Infineon Technologies, Neubiberg, Germany

Christian Jentgens Pureon, Lengwil, Switzerland

Terry Knight Pureon, Monroe, NC, USA

Aparna Kumar Qorvo, Greensboro, NC, USA

Byeongyeop Lee onsemi, Bucheon-si, Gyeonggi-do, Republic of Korea

Sandol Lee onsemi, Bucheon-si, Gyeonggi-do, Republic of Korea

Sergey Maximenko Microchip Technology, Chandler, AZ, USA

Stephen Oliver Navitas Semiconductor, Torrance, CA, USA

Andrea Piccioni Infineon Technologies, Berlin, Germany

Simone Rascuna STMicroelectronics, Catania, Italy

Paul Salmen Infineon Technologies, Ense, Germany

Yogesh Sharma Microchip Technology, Chandler, AZ, USA

John R. Silk SemiQ, Lake Forest, CA, USA

Kevin Speer Microchip Technology, Chandler, AZ, USA

Ming Su Rohm Semiconductor, Novi, MI, USA

Ryo Takeda Keysight Technologies, Hachioji-shi, Tokyo, Japan

Dario Tenaglia STMicroelectronics, Catania, Italy

Suresh Thangavel Infineon Technologies, Warstein, Germany

Llew Vaughan-Edmunds Navitas Semiconductor, Torrance, CA, USA

Victor Veliadis PowerAmerica, Raleigh, NC, USA

James Victory onsemi, Scottsdale, AZ, USA

Helge Willers Pureon, Willich, Germany

Sanbao Zheng Qorvo, Greensboro, NC, USA

Mike Zhu Qorvo, Greensboro, NC, USA

Olaf Zschieschang onsemi, Aschheim, Germany

Chapter 1
Overview of Silicon Carbide

Victor Veliadis, Ming Su, and Filippo Di Giovanni

Within the realm of power electronics, silicon carbide (SiC) stands out as a transformative material, offering a host of advantages over traditional silicon devices. Authored by Victor Veliadis, Executive Director and CTO of PowerAmerica and ECE Professor at NC State, paragraph 1.1 delves deep into the exceptional properties of SiC. Veliadis accentuates its wide bandgap (WBG) and high critical electric field, laying the groundwork for efficient power conditioning. Through meticulous graphical comparisons and comprehensive explanations, the potential of SiC emerges as a game-changer, promising reduced losses, heightened power density, and superior thermal performance.

Meanwhile, paragraph 1.2, penned by Ming Su, Ph.D., Senior Technical Marketing Manager at Rohm Semiconductor USA, delves into the evolutionary journey and profound significance of SiC devices in the electronics industry. Su's narrative navigates through the pivotal role SiC transistors play in revolutionizing high-voltage switching applications, setting a compelling stage for further exploration within subsequent sections.

In the concluding section, authored by Filippo Di Giovanni, the text furnishes a comprehensive examination of silicon carbide (SiC), with a particular emphasis on its electron configurations and crystal defects.

V. Veliadis
PowerAmerica, Raleigh, NC, USA

M. Su
Rohm Semiconductor, Novi, MI, USA

F. Di Giovanni (✉)
STMicroelectronics, San Gregorio, CT, Italy

M. Di Paolo Emilio (ed.), *SiC Technology*, https://doi.org/10.1007/978-3-031-63418-5_1

1.1 SiC Power Semiconductors

Victor Veliadis

Silicon devices dominate power electronics due to their superb starting material quality, streamlined fabrication, low-cost volume production, proven reliability and ruggedness, and design/circuit legacy. Although Si power devices continue to make progress, they are approaching their operational limits primarily due to their relatively low bandgap and critical electric field that result in high conduction and switching losses and poor high-temperature performance.

Wide bandgap semiconductors such as silicon carbide (SiC) and gallium nitride (GaN) are currently being developed for efficient high-power/temperature applications. Silicon carbide (SiC) is ideally suited for power conditioning applications due to its high saturated drift velocity, its mechanical strength, its excellent thermal conductivity, its wide bandgap, and its high critical electric field. For power devices, the ninefold increase in critical field strength of SiC, relative to Si, allows high-voltage blocking layers to be fabricated significantly thinner than those of comparably rated Si devices. This reduces device on-state resistance and the associated conduction losses while maintaining the same high-voltage blocking capability. Indeed, the specific on-state resistance of 4H-SiC is approximately 400 times lower than that of Si at a given breakdown voltage, enabling high-current operation at relatively low forward voltage drop. Thinner layers and low-specific on-resistance reduce capacitance, enabling low switching losses at high-frequency operation, which reduces the weight/volume of passive components and increases power density while lowering the cost of materials. In addition, the wide bandgap of SiC results in relatively low intrinsic carrier concentration, which combined with its high thermal conductivity, enables operation at high temperatures (where conventional Si devices fail), with low leakage currents and simplified thermal management. A graphical comparison of key Si, SiC, and GaN material properties is shown in Fig. 1.1.

Overall, SiC devices enable more efficient, lighter, smaller form-factor power electronics operating at high frequencies with reduced size of passive components and at elevated temperatures with reduced cooling. More efficient power conversion with lower losses will not only lead to significant energy savings but also reduce the use of fossil fuels and CO_2 emissions. Until recently, Si has been the dominating semiconductor for power electronics. Although the performance of Si devices has improved significantly over time, it is inevitably limited by the fundamental material properties of silicon.

There are also emerging "ultra-wide bandgap" semiconductors with excellent properties for power electronics. Those include diamond, gallium oxide, and aluminum nitride. However, their technological maturity is well below that of SiC and GaN, and thus, they are not expected to impact power electronics over the next decade.

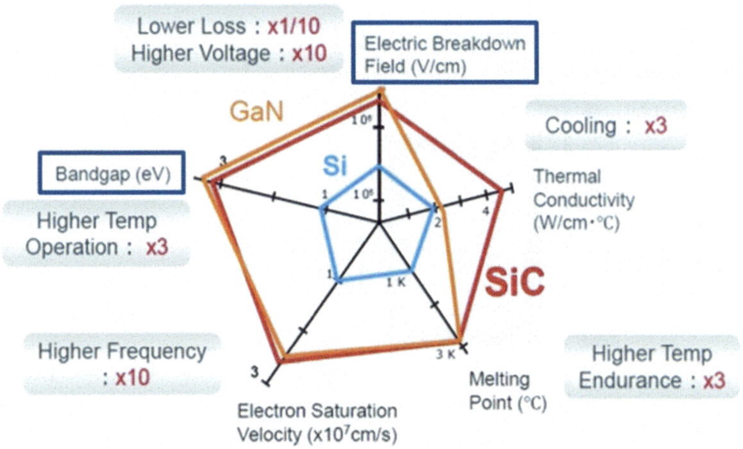

Fig. 1.1 Graphical comparison of key Si, SiC, and GaN material properties

1.1.1 Historical Development and State of the Art

To exploit SiC's compelling material properties in power devices, significant efforts started in the 1980s to develop high-quality, low-defect SiC substrates and epitaxy. Today, 150 mm SiC wafers are primarily used in the production of SiC devices. In 2015, 200 mm wafers were demonstrated and are now becoming commercially available from several vendors.

SiC substrate growth is more complex than that of Si, requiring the use of large seeds and high process temperatures. SiC is mainly grown by the seeded sublimation technique, as schematically shown in Fig. 1.2. The raw material, SiC powder, is placed at the bottom of a graphite crucible. A seed wafer is placed at the top of the crucible, which is heated by RF coils to a temperature of 2300–2500 °C. The seed wafer is kept at a lower temperature than that of the SiC powder, and the sublimed SiC species condensate and crystallize on the seed wafer. Reported sublimation growth rates are in the order of 0.5–2 mm/h. Increasing the growth rate can have the undesirable side effect of increased structural defect inclusion in the boule. Improvements in the growth process are being made by optimizing the growth zone.

The quality of the seed wafer is very important since structural defects in the seed can propagate into the growing crystal. The seed crystal determines the diameter of the grown crystal. SiC substrates have successively increased in diameter from 35 mm, 2″, 3″, 4″, 150 mm, to 200 mm (the latter in limited quantities) over the past 25 years. The 150 mm diameter substrates dominate the market, with good-quality wafers available from many suppliers. Several companies have demonstrated 200 mm substates and wide availability is expected in the next few years.

SiC hardness, comparable to that of diamond, and its chemical inertness pose a challenge for the preparation of high-quality wafer crystals. The grinding and slicing of SiC uses the same technology as that of Si, multiwafer wire slicing, but takes

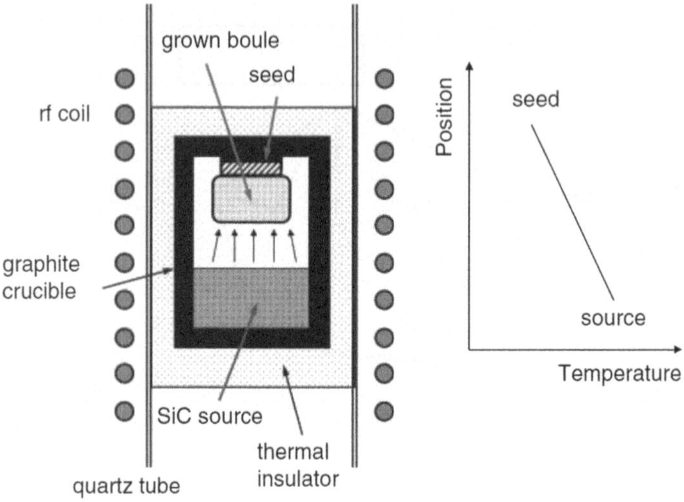

Fig. 1.2 Schematic description of crucible for seeded sublimation growth of SiC substrates

substantially longer time. Alternative techniques are under development, such as laser slicing, to reduce material losses and time—and hence cost.

The active SiC device layers are grown by chemical vapor deposition (CVD) epitaxy in horizontal or planetary reactors at about 1550 °C. Historically, epitaxial growth was performed on 8° off-cut substrates to maintain the polytype stability of the substrate, but it is now performed on 4° off-cut substrates. In addition to defects propagating from its underlying substrate, epitaxy can introduce morphological defects or ingrown stacking faults in the layer along the basal plane, which can be detrimental to device operation. Overall, SiC epitaxial growth is well established and strives to minimize defect generation, as well as defect propagation from the underlying substrate to the epitaxial layer.

Historically, killer defects limiting yield have been polytype inclusions and micropipes. These have practically been eliminated in commercial wafers. Micropipe density, which is detrimental to device operation, is typically below 0.1 cm^{-2}. The remaining structural defects are threading screw dislocations (300–600 cm^{-2}) that can increase reverse voltage leakage, threading edge dislocations (2000–5000 cm^{-2}) that are considered benign, and basal plane dislocations (BPDs) (500–3000 cm^{-2}), which can lead to catastrophic device degradation under bipolar current flow. Although threading dislocations do result in measurable disturbances of epitaxial layer surface morphology, the practical effects of these disturbances on device performance and reliability are minimal [1].

BPDs are the major remaining “killer” defect impacting bipolar SiC devices as well as unipolar devices that conduct bipolar current during portions of their operational cycle [2]. BPDs can be present in the starting SiC wafer and can also be generated during the high-temperature ion-implantation process. Under bipolar current flow, electron-hole pair recombination at basal plane dislocations induces

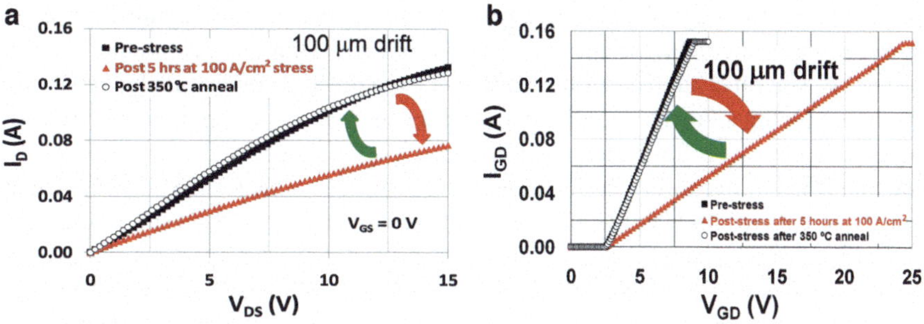

Fig. 1.3 BPD-related transistor on-state conduction degradation and full recovery by thermal annealing. (**a**, left) On-state current conduction and (**b**, right) gate-to-drain *pn* diode current conduction

stacking fault formation and expansion, which degrade on-state current conduction. In Fig. 1.3, degradation from the original state (black rectangles) for a SiC JFET that was subjected to bipolar current is shown by the red triangles. The BPD-related degradation is fully reversed by high-temperature annealing (350 °C), and the transistor returns to its original "undamaged" operational state, as shown by the white circles in Fig. 1.3. BPD-related degradation can be readily identified in post-fabrication production testing and is thus a yield issue, not a reliability one.

Numerous well-established processes from silicon technology have been successfully transferred to SiC. However, SiC material properties necessitate the development of specific processes, the parameters of which must be optimized and qualified [3, 4]. SiC is inert against chemical solvents and only dry etching is practical. Conventional thermal diffusion is not realistic in doping SiC due to its high melting point and the low diffusion constant of dopants within SiC; heated ion implantation must be performed, followed by a 1600–1800 °C anneal for SiC recrystallization and implant activation. The high value of the SiC/metal barrier results in rectifying metal contacts. Post metal-deposition anneal is required for Ohmic contact formation. Unlike Si wafers, those of SiC are transparent. This complicates CD-SEM and metrology measurements, as the focal plane is determined with the use of an optical microscope. SiC-specific metrology/inspection tools are now available from multiple vendors. SiC wafers' relative lack of flatness, compared to that of Si, can complicate photolithography and other processing, particularly as high-temperature processes can further degrade wafer flatness and render them unusable. To elevate high-voltage performance close to the device's theoretical limit, specialized edge-termination structures like multiple-junction termination extensions and floating guard rings have been successfully implemented [5, 6].

The poor SiC/SiO_2 interface quality reduces inversion-layer mobility in metal-oxide-semiconductor field-effect transistors (MOSFETs), and passivation techniques are utilized to improve the SiC/SiO_2 interface quality. MOSFETs are normally-off gate voltage-controlled power switches. In MOSFETs, injection of semiconductor electrons into the gate oxide occurs when the electrons gain sufficient energy to overcome the potential barrier between the semiconductor and the oxide. Due to the large 4H-SiC bandgap, the offset between the conduction band edges of the semiconductor and the silicon dioxide is significantly smaller than in the case of silicon. Thus, electrons can tunnel from the SiC conduction band into the oxide, producing a Fowler-Nordheim tunneling current. The trapping of electrons in the gate oxide sets a fundamental limit on high-temperature SiC MOSFET operation and leads to instability in threshold voltage and unwanted MOSFET turn-on at temperatures over 200 °C [7]. Consequently, commercial SiC MOSFET offerings typically have a maximum temperature rating of 200 °C. SiC process-integration technology has made significant advancements, and optimizations are continuing, particularly in improving the SiC/gate-oxide interface, as well as in streamlining the labor-intensive heated-implantation process and subsequent anneal.

WBG lateral power devices have an advantage from a system integration perspective but can be challenging to realize at blocking voltages exceeding 900 V. To support very high blocking voltages, lateral devices necessitate impractically large wafer areas because they require large gate-to-drain spacings. In addition, the high voltage must be supported by the vertical path from drain to substate. Consequently, +600 V SiC power devices are almost exclusively of vertical configuration, in which the blocking voltage capability is primarily determined by the thickness and doping of the vertical drift layer (assuming a well-performing edge termination).

The junction barrier Schottky (JBS) diode was the first SiC device to be commercialized, in 2001. The JBS diode, shown schematically in Fig. 1.4, relies on the Schottky contact to achieve rectifying action. In reverse mode, it uses closely spaced implanted *p*-regions to generate a potential barrier that shields the Schottky contact from the large SiC electric field in the drift layer, thereby minimizing reverse leakage current and maximizing blocking voltage capability. In forward mode, the undepleted spacing between the *p*-regions allows for unipolar on-state current flow. The JBS diode exhibits unipolar Schottky behavior at rated current conduction, while minority carrier injection (bipolar current flow) occurs at high currents enabling its excellent surge current capability. The "large extent" *p* + regions at the periphery of the JBS diodes in Fig. 1.4 denote the edge-termination structure that maximizes blocking voltage.

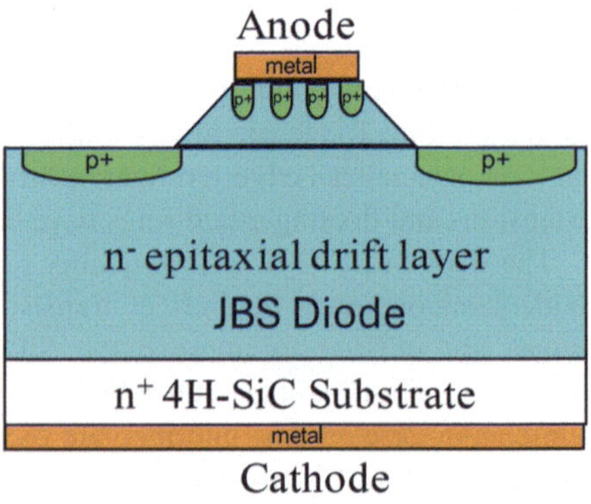

Fig. 1.4 SiC junction barrier Schottky diode schematic. In blocking voltage mode, the closely spaced *p*-regions underneath the Schottky contact generate a potential barrier that shields it from the large electric field in the drift layer

Unipolar transistors such as MOSFETs and JFETs, as well as bipolar devices such as BJTs, thyristors, IGBTs, and PiN diodes, have demonstrated their viability in SiC. Commercially, the dominant switch used in SiC power electronics is the MOSFET, which was released as a product in 2011. Different MOSFET suppliers use different design configurations such as planar and trench MOSFETs, schematically illustrated in Fig. 1.5.

Each design has had several generations of maturity, with increasingly better performance and lower on-state resistance. The JFET region contribution to the on-state resistance is eliminated in the trench MOSFET, and the MOS mobility is higher in the vertical sidewall, both of which lower channel resistance. In addition, the cell pitch can be made smaller in the trench architecture, which yields more devices per wafer. However, the trench MOSFET blocking-voltage capability can be lower than that of an "equivalent rated" planar MOSFET (same drift layer thickness and doping) due to electric field crowding at the trench corners of the gate oxide. Schemes to overcome this drawback have been implemented, making trench MOSFET fabrication more complex than that of its planar counterpart. Additionally, the need to have adequately large chip area for satisfactory short-circuit robustness sets a practical limit on the area reduction in trench architectures. Large suppliers such as Microchip, onsemi, STMicroelectronics, and Wolfspeed offer planar MOSFETs, while those of Infineon and Rohm are of the trench architecture. The reduced channel-resistance advantage of trench MOSFETs diminishes as the voltage rating exceeds 1700 V. At these higher voltages, the drift layer resistance

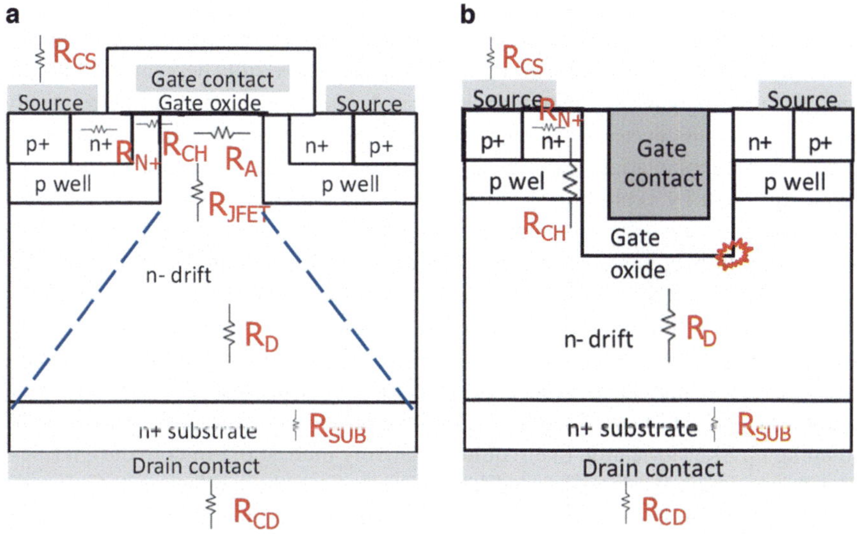

Fig. 1.5 (**a**, left) Planar MOSFET configuration and its various resistance elements and (**b**, right) trench MOSFET configuration and its various resistance elements. The JFET resistance of the planar MOSET is not present in the trench architecture. However, the trench MOSFET requires an "etch" fabrication step and is prone to premature voltage breakdown due to electric field crowding at the corners of the trench

becomes dominant, and the trench MOSFET's fabrication complexity makes it a less attractive option. Early MOSFETs suffered from low channel mobility and threshold voltage instability due to the poor quality of the SiC/gate-oxide interface. Material quality improvements have alleviated but not fully resolved these issues. As a result, enhancing MOSFET electron mobility and oxide stability are active areas of research and development. SiC MOSFETs are commercially available in the 600 V–3.3 kV range from multiple large vendors.

Junction field-effect transistors (JFETs) are commercially available in the blocking voltage range of 600–1700 V, but they have not been adopted in large volumes. JFETs use *pn* junctions in the active device area, where the high electric fields occur, and can therefore fully exploit the high-temperature properties of SiC in a gate voltage-controlled switching device. Provided the gate-to-source junction of the JFET is biased below its built-in potential, a negligible gate leakage current is needed to drive the device and voltage-controlled switching is realized. JFETs are free of MOS native oxide issues such as low channel mobility, threshold voltage instability, and elevated temperature operation limitations. They have demonstrated electrostatic discharge immunity to 16 kV [8]. However, practical JFETs are native normally-on in the +600 V range and this has limited their adoption [9]. The cascode configuration allows for normally-off JFET-based switching and has commercial offerings in the 600–1700 V range. Reviews of SiC JFET technology for RF and power switching applications are given in references [4, 10–14].

The SiC industry is successfully leveraging the fully depreciated legacy silicon fab infrastructure, and it is making the relatively small financial investments needed to allow mature silicon fabs to process SiC. Consequently, SiC chip fabrication alongside silicon has emerged as a cost-effective model that exploits silicon manufacturing economies of scale. Through re-purposing older, fully depreciated 150 mm and 200 mm Si fabs, SiC power devices can be manufactured with the relatively small investments necessary to support unique SiC processing steps such as high-temperature implantation and anneal and Ohmic contact formation. Minimizing fabrication cost by exploiting the mature Si volume production assumes the fab is loaded close to capacity with standard Si and SiC processes running on the same line. In addition, aggregating the demand for SiC substrates and epilayers in a few volume fabs contributes to lower material costs. Lower fabrication costs in a fully depreciated Si + SiC "capacity" loaded fab, coupled with decreased material costs, leads to significant price reductions for SiC devices. This approach offers a new opportunity for mature Si fabs, which have not kept up with the channel length reductions of the last two decades, to continue manufacturing legacy Si parts while ramping up SiC fabrication that requires relatively modest 0.3-micron design rules. Today, numerous IDMs and foundries across the world manufacture SiC devices and are expanding their capacity to meet the increasing demand.

SiC power devices are commercially available and are steadily gaining market share. However, high device cost, reliability/ruggedness concerns, and the need for a workforce trained in their use are barriers to their mass adoption. In several power applications, such as PV systems, insertion of SiC reduces overall system cost compared to Si, even though SiC devices cost three times more than their Si counterparts. This is due to the passive component and cooling system simplifications enabled by SiC high-frequency operation. Nevertheless, reducing SiC device cost is highly desirable. The SiC wafer represents over 50% of the overall transistor cost, a consequence of the complex substrate fabrication technology outlined in the discussion of Fig. 1.2. A device cost reduction of over 20% is expected in the initial transition from 150 to 200 mm SiC substrates.

Voltage- and switching-frequency considerations drive unipolar versus bipolar SiC device selection. Current flow in unipolar devices such as the MOSFET and JFET comes from only one type of charge carrier, an electron or a hole. Since only one type of charge carrier flows during conduction, the on-state resistance is relatively high. Unipolar conduction, however, allows for faster switching and relatively lower switching losses. In bipolar devices such as IGBTs, current flow comes from both electron and hole charge carriers, which lowers conduction losses. However, bipolar conduction results in slower switching, as minority carriers also need to be swept during switching transitions, which increases switching losses. Typically, unipolar SiC devices operate efficiently in applications of up to 10 kV, and bipolar devices are better candidates for voltages above that level. SiC MOSFETs rated at 6.5 and 10 kV, as well as IGBTs with a 15 kV rating, have been fabricated and demonstrated in power electronic circuits throughout the past 20 years. They have yet to be commercialized.

1.2 SiC Devices

Ming Su

Even though silicon carbide (SiC) has been used in the electronics industry for over a century, it is only in the last 20 years that the manufacturing technology and market applications have brought it to fruition as a mainstay of high-power design. The earliest applications in the 1900s were limited to basic voltage-dependent resistors for arrestors and surge protection. Diode structures soon followed for simple RF circuitry. Eventually, by the late twentieth century, SiC devices were designed for Schottky barrier diodes, JFETs, and MOSFET transistors. For example, the figure below shows ROHM's several major milestones in the development of prime-time SiC devices in the twenty-first century, beginning with the 2-in. SiC wafers used for research purposes and ending with mass production of the fourth-generation MOSFETs (Fig. 1.6).

At each stage of the journey, SiC material quality and device structure complexity have improved the electrical and thermal performance by leaps and bounds. Today, SiC MOSFETs and diodes are some of the fastest, most efficient, and smallest power switches available on the market.

Wide bandgap (WBG) semiconductors are the critical building blocks of high-power electronic switches. Compared to common semiconductors such as silicon, whose bandgap lies in the range of 0.6–1.5 eV, WBG materials exhibit bandgaps

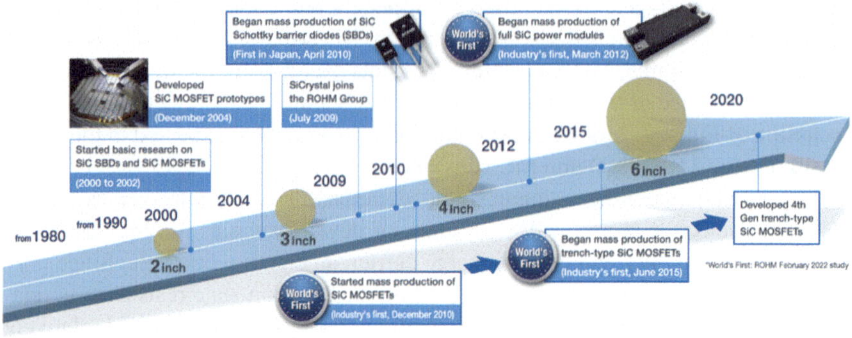

Fig. 1.6 Evolution of ROHM SiC technology since 2000

beyond 2.0 eV, enabling operation at much higher voltage, frequency, and temperature. Most common WBG materials are composed of semiconductor alloys of III–V and II–VI elements—for example, gallium nitride and zinc oxide. Interestingly, there exist two materials composed of group IV elements that make excellent WBG semiconductors: diamond (carbon) and silicon carbide (SiC).

Diamond is one of the highest performing semiconductors, with a bandgap of 5.47 eV, extremely high resistivity, and excellent carrier mobility. In addition, diamond exhibits unusually high thermal conductivity and stability. Unfortunately, aside from very niche experimental devices, the cost and difficulty in using diamond to mass produce electronic components is prohibitive.

Diamond's group IV cousin, silicon carbide, exhibits a lower bandgap, typically around 3 eV depending on the specific crystalline form, thermal conductivity greater than 300 W/mK, and electron mobility in excess of 900 cm^2/Vs. SiC crystals can exist in over 250 different forms, each with a specific recipe for construction and slightly different physical properties. The most common are the hexagonal alpha structure (6H), the hexagonal 4H structure, and the zinc blende beta structure (3C). Most devices today make use of the 4H structure due to its manufacturability and higher thermal conductivity. While SiC has been manufacturable since the late 1800s, its role has been limited primarily as an abrasive due to the difficulties in maintaining purity of the SiC crystal. Since the early 2000s, advanced manufacturing techniques have solved the defect issues and enabled the development of commercial SiC devices including JFETs, MOSFETs, and Schottky barrier diodes at 600 V and higher rated voltages.

1.2.1 SiC Transistors

As with most power transistors, SiC MOSFETs are constructed with a vertical device design where the drain terminal is coincident with the SiC substrate and the source terminal is located above it. Voltage isolation in the off state depends on the depth and doping of the epitaxial layer, while current capacity in the on state depends on the lateral width of the device. Since SiC has a nearly 10× higher critical breakdown electric field compared to silicon, the voltage blocking layer of the device can be made proportionally thinner, as shown in the figure below. As a result, on-state resistance is dramatically reduced—making SiC MOSFETs more power efficient and easier to manage thermally (Fig. 1.7).

Over time, the vertical SiC structure has evolved to further reduce on-state resistance and minimize switching losses. As shown in the figures below, this evolution has largely focused on the application of trench structures in ROHM's technology development initiatives. With each generation, the on resistance is further improved by a factor of 40–50%, and the switching losses are minimized through a reduction in gate capacitance (Cgd) (Fig. 1.8).

A key figure of merit for any power transistor is the relationship between on-state resistance and off-state blocking voltage. As alluded to previously, traditional silicon-based devices trade off these two parameters as a function of the device thickness. As shown in Fig. 1.9, the inherent benefit of silicon carbide as a material almost eliminates this trade off, and the resulting devices are extremely low resistance, even at very high-voltage applications (Fig. 1.9).

The low on-state resistance of SiC MOSFETs directly results in higher efficiency and better thermal performance at a given drain current. These advantages persist across the full operating temperature range, due to the smaller temperature dependence when compared to silicon MOSFETs. As shown in Fig. 1.10 comparing SiC devices to traditional silicon devices, the Vds-Id curve of the SiC device is linear and highly efficient across the entire power and temperature range (Figs. 1.6, 1.7, and 1.8).

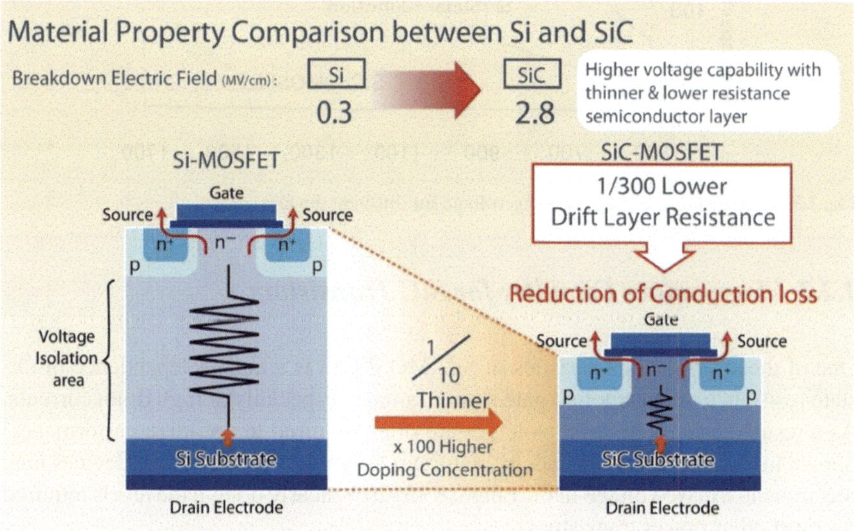

Fig. 1.7 Comparison between silicon and SiC power transistors

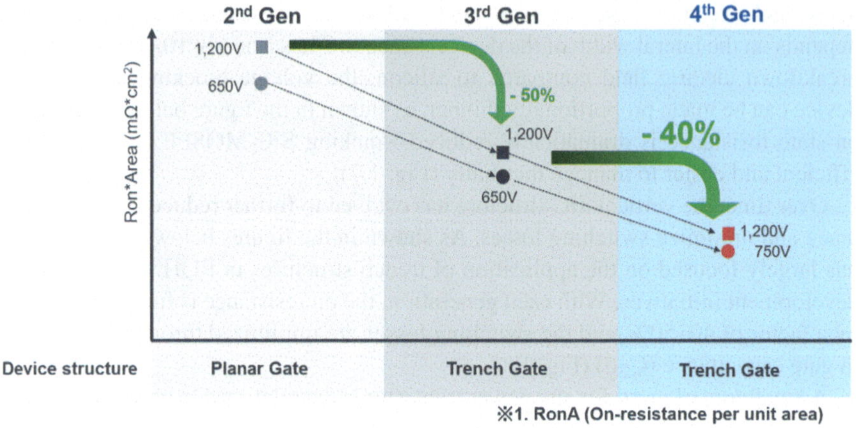

Fig. 1.8 Generational improvements in SiC MOSFET performance

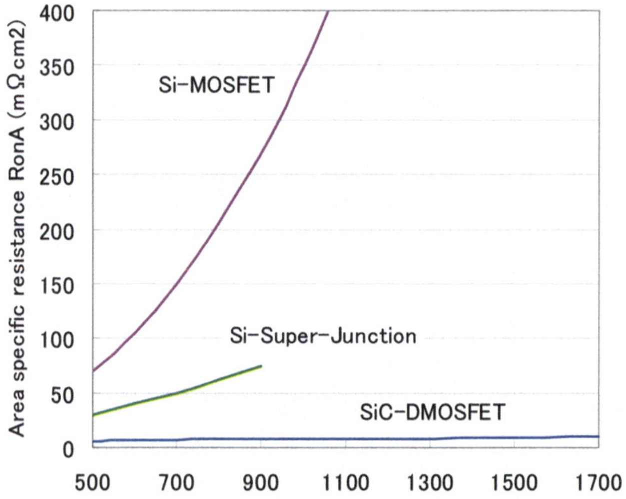

Fig. 1.9 On resistance versus blocking voltage for different devices

1.2.2 Supporting Circuitry for SiC Transistors

One of the special characteristics of SiC MOSFETs is a higher dependence of on-state resistance on the applied gate source voltage, especially at high drain currents. As a result, higher gate drive voltages are often required to optimize performance. Shown in Fig. 1.11, this dependence saturates around 20 V, and typical devices may recommend a drive voltage in the range of 15–20 V, nearly double the levels required for most other power transistors.

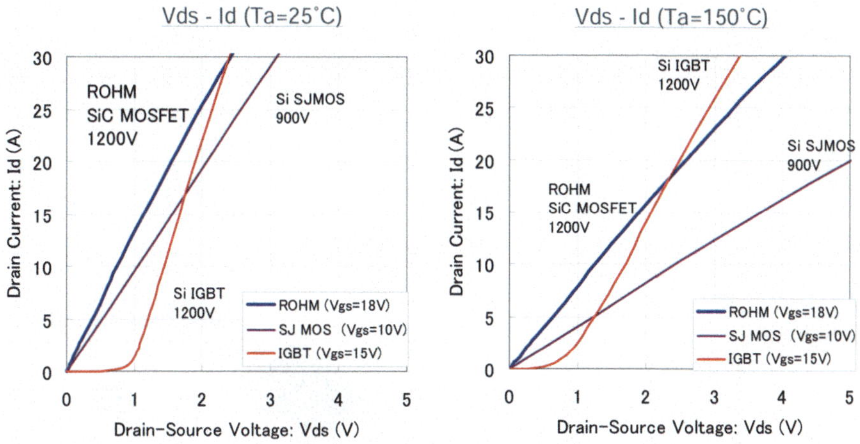

Fig. 1.10 Comparison of Id-Vds curves at different temperatures

Fig. 1.11 SiC on resistance as a function of gate drive voltage

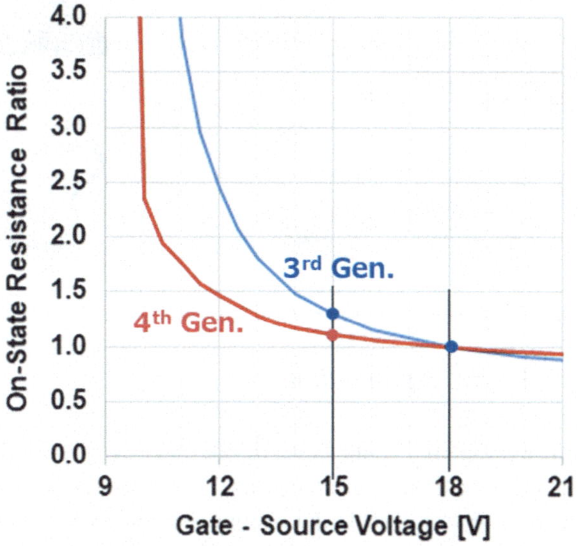

For certain applications, negative gate drive voltages may be optimal for ensuring off-state switching speed. Combining this with the high on-state voltage requires a driver circuit that operates with supply rails spanning 22–28 V. Specialized gate driver ICs are available to simplify the task of efficiently controlling SiC MOSFETs and include features such as galvanic isolation, active Miller clamping, and undervoltage lockout protection. An example is shown in Fig. 1.12, where a SiC MOSFET is controlled with a 24 V supply voltage from a 3.3 V logic input.

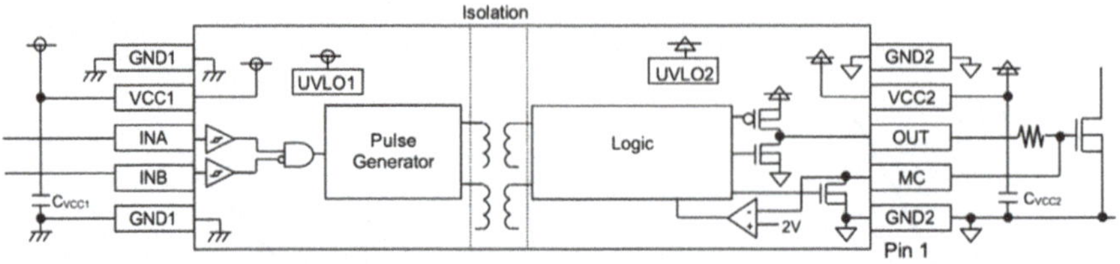

Fig. 1.12 ROHM's BM61S41RFV-C SiC MOSFET driver IC

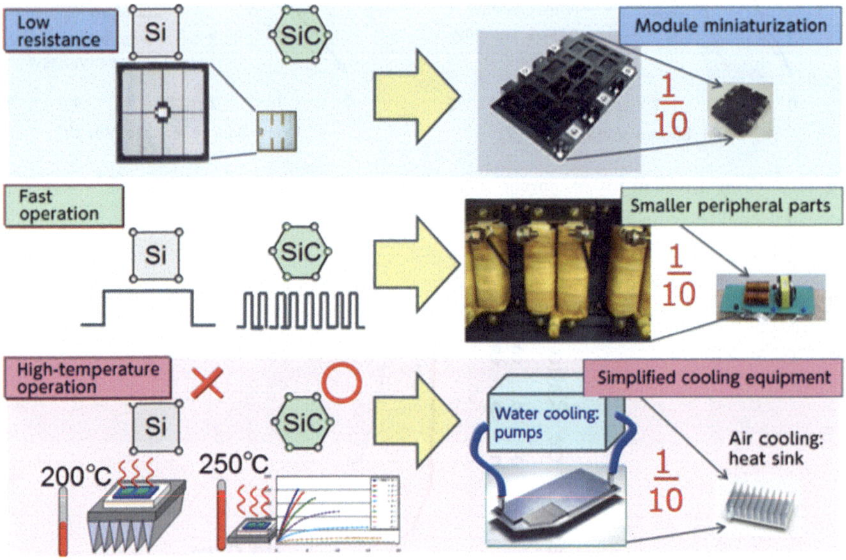

Fig. 1.13 Summary of three key SiC advantages

1.2.3 SiC Applications

When discussing applications for SiC devices, it is important to do so in the context of SiC advantages and trade-offs. In general, as shown below, SiC devices offer benefits in size, speed, and cooling simplicity. This is largely due to the inherent material properties of silicon carbide itself. The trade-offs include higher device cost and specialized gate driver requirements and must be considered as part of the entire system cost and complexity. For example, the simpler cooling requirements of SiC MOSFETs may reduce system cost more than the individual increases in device cost (Fig. 1.13).

While SiC devices are gaining in popularity across the power electronics industry, three areas in particular have found them to be an excellent driver for growth: electric vehicles, data centers, and renewable energy.

The rise of electric vehicles (EVs) has created a demand for high-performance, high-voltage switching devices. With bus voltages of 400–900 V and an intense focus on energy efficiency, SiC MOSFETs and SBDs are ideal candidates for EV designs.

In a similar vein, data centers and the associated uninterruptible power supplies are constrained by energy efficiency and cooling. Two percent of all electrical energy in the USA is dedicated to data centers and that figure is growing every year. From the AC/DC input converters to the DC/DC domain converters, any advantage in energy efficiency can translate into huge cost savings and lower environmental impact. It is estimated that transitioning to SiC switching devices can yield a 40% reduction in the cooling costs of a typical data center.

Solar inverter systems that use maximum power point tracking (MPPT) to efficiently convert solar energy into residential or industrial AC power also benefit tremendously from SiC switching technology. As seen in the figure below, a typical system includes a front-end diode along with several switching blocks to convert high-voltage DC into the AC grid output. At each of these stages, any reduction in switching loss, on-state conduction loss, and cooling support yields significant system level gains. For alternative energy applications where efficiency is the fundamental figure of merit, the gains from SiC can realize systems with greater than 98% efficiency from outputs that range from kW to MW (Figs. 1.13 and 1.14).

1.2.4 Conclusion

Silicon carbide's unique material properties make it one of the most promising wide bandgap semiconductors for high-power, high-speed switching applications. The vertical transistor construction with extreme voltage tolerance allows for a much smaller die volume to handle the same power as its silicon counterparts. When

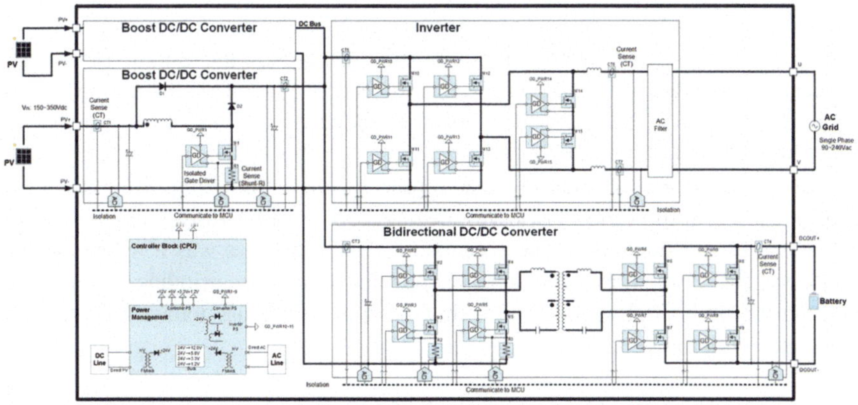

Fig. 1.14 Block diagram of a photovoltaic inverter and battery energy storage system

leveraged with advanced trench fabrication techniques, the resulting on-state resistance and reduced gate capacitance put SiC transistors in a league of their own for switching speed, power efficiency, and thermal capability. Adoption of these devices into electric vehicles, data centers, and renewable energy systems has further accelerated their development, and it is expected that SiC transistors will become the dominant solution for high-voltage devices across the power electronics industry.

1.3 Electronic Configurations and Defects

Filippo Di Giovanni

Silicon carbide keeps drawing the attention of a growing number of power electronics engineers as the material opens up good prospects for reducing carbon footprint globally. The use of this versatile semiconductor cannot be decoupled from a sufficiently deep understanding of its crystalline morphology. This paragraph is aimed to bridge this gap by discussing the material's physical structure along with an overview of inherent defects that could pose manufacturing challenges or limit its use in applications. To this end, some space is devoted to the analysis of the main crystal defects whose handling is of paramount importance for improving manufacturing yields and putting in place effective screening techniques to neutralize their effects.

1.3.1 Brief History of SiC

SiC is not native to earth. In fact, fragments of this material (also called *carborundum*), the second-hardest natural mineral after diamond, were discovered in 1893 by the French chemist Henri Moissan in a crater at Canyon Diablo, Arizona, caused by the impact of a meteorite almost 50,000 years ago. Today, the alien origin of SiC is demonstrated by its carbon isotopic composition, defined as the ratio of ^{12}C isotope concentration to that of ^{13}C. On earth such ratio is approximately 93:1 as opposed to 64:1 of the crater specimen. The relative abundance of the two isotopes is $^{12}C \approx 98.89\%$ and $^{13}C \approx 1.10\%$.

Edward Goodrich Acheson invented a process for synthesizing SiC from silica (SiO_2), carbon in the form of powder coke, and some salt additives in an effort to produce artificial diamonds. The so-called Acheson process was a technique for producing SiC in the form of an abrasive powder for cutting, grinding, and polishing, the first industrial application of this material still in use today. The reaction associated with this process is:

$$SiO_2(s) + 3C(s) \rightarrow SiC(s) + 2CO(g)$$

1.3.2 Electron Configurations

SiC is an IV–IV wide bandgap compound material with rigid 50% Si, 50% C stoichiometry used in power electronics, LEDs, and particle detectors. Its physical properties—a high breakdown field, high saturation drift velocity, and high thermal conductivity—make it ideal for applications where it is operated at high temperature, high intensity of radiation, high voltage, or high-power dissipation without compromising efficiency. The strong chemical bonding between Si and C atoms confers this material very high hardness, chemical inertness, and high thermal conductivity.

Let us recap the distribution of electrons in an atom according to the rules of quantum physics. An electron in an atom is completely described by three quantum numbers: n, l, and m_l. The fourth quantum number, m_s (of values ±1/2), is the spin quantum number and was introduced later to explain small energy anomalies in some atomic spectra.

The electron energy levels are labeled with n, where n is an integer and is referred to as the principal quantum number. The higher the value of n, the higher the energy of an electron. An atomic *orbital*, which is distinct from an orbit, is a general region of space around an atom within which an electron can be found with some probability and is determined by solving Schrödinger's wave equation. Orbitals having the same value of n are said to be in the same *shell*. Another quantum number is l, the angular momentum quantum number (or azimuthal quantum number). It is an integer that defines the shape of the orbital and takes on the values l = 0, 1, 2, …, $(n - 1)$. This means that an orbital with $n = 1$ can have only one value of l, l = 0, whereas $n = 2$ permits l = 0 and l = 1, and so on. Orbitals with the same value of l form a subshell. In addition, the greater the angular momentum quantum number, the greater is the angular momentum of an electron at this orbital. Orbitals with l = 0 are called s orbitals (or the s *subshells*). The value l = 1 corresponds to the p orbitals. For a given n, p orbitals constitute a p subshell (e.g., 3p if $n = 3$). The orbitals with l = 2 are called the d orbitals, followed by the f-, g-, and h-orbitals for l = 3, 4, and 5. The *magnetic quantum number*, called m_l, specifies the z-component of the angular momentum for a particular orbital and specifies orientation of the orbital in space. Its values vary between −l to +l.

The SiC material can be synthesized in different variants called polytypes; "polytypism" is akin to polymorphism, which refers to the possibility of an element or compound crystallizing in a range of structures, while polytypes differ by virtue of the stacking sequence of atomic layers along a given crystalline direction without change in the chemical composition. There exist more than 250 total polytypes; the common polytypes are 6H, 15R, 4H, and 3C. The letters H, C, and R denote the overall lattice type as being hexagonal, cubic, or rhombohedral, respectively (Table 1.1).

Table 1.1 Physical properties of SiC polytypes

		4H-SiC	6H-SiC	15R-SiC	3C-SiC
Band gap (eV)		3.26	3.03	2.986	2.39
Effective mass	m_e	0.37	0.69	0.53–0.28	0.68–0.25
Effective mass	m_h	0.94	0.92		
Lattice constant	a	3.08	3.08	3.08	4.36
Lattice constant	c	10.05	15.12	37.7	
Mobility @300 K (cm^2/Vs)	μ_e	500	300	400	900
Mobility @300 K (cm^2/Vs)	μ_h	50	50		20
Thermal conductivity (W/cm K)		3.0–3.8	3.0–3.8	4.9	3.2

1.3.3 Defects in SiC Crystals

Defects in SiC wafers are typically classified into two major groups: (a) crystallographic defects within the wafer and (b) surface defects at or in the vicinity of the wafer surface. Crystallographic defects include basal plane dislocations (BPDs), stacking faults (SFs), threading edge dislocations (TEDs), threading screw dislocations (TSDs), micropipes, and grain boundaries. Epitaxial growth parameters are very critical to the quality of the SiC wafer. In fact, crystallographic defects and contaminations during growth processes may extend into the epitaxial layer and wafer surface to form various surface defects, including carrot defects, polytype inclusions, and scratches, or even convert to produce other defects.

The performance of SiC-based devices is greatly influenced by the presence of killer defects, formed during crystal growth. Silicon carbide boules and substrates contain various crystal imperfections, both extended and point-like. The density of point (i.e., the size of an atom) defects in SiC boules is quite high, in the 10^{14}–10^{16} cm^{-3} range. The most common defects in SiC crystals are micropipes, dislocations, and stacking faults. Micropipes are hollow tubular structures that cross the crystal and can lead to a significant reduction in the electrical conductivity of the material. Dislocations are linear defects that occur when the crystal lattice is distorted, and they can affect the material's mechanical and electrical properties. Stacking faults are planar defects that build up when there is an interruption in the perfect stacking sequence of the crystal lattice, and they can affect the electronic properties of the material. Stacking faults are common defects because of the low stacking-fault energy (14 mJ m^{-2} for 4H-SiC and 2.9 mJ m^{-2} for 6H-SiC) and the occurrence of many polytypes in SiC. For example, in face-centered cubic structures b-SiC, the normal stacking sequence is ABCABC…, but if a stacking fault is introduced, an irregular sequence such as ABCBCABC is generated into the normal stacking order. These irregularities carry a certain energy that is called stacking-fault energy that is measured in energy per unit area.

All such types of defects cause different degrees of deterioration to device performance, if not catastrophic failure. Therefore, tailored techniques enabling nondestructive defect inspection prior to device fabrication become vital.

Table 1.2 Crystal imperfections

Extended defect (dislocation)	Typical density (cm^{-2})	Impact on unipolar device	Impact on bipolar device
Micropipe	0–0.1	N	Y
Threading screw dislocation	300–600	N	N
Threading edge dislocation (TED)	2000–5000	N	N
Basal plane dislocation (BPD)	5000–3000	N	Y
In-grow SF	0.1	Y	Y
Carrot, triangular	0.3	Y	Y
Downfall	0.3	Y	Y

Table 1.2 shows the major extended defects observed in SiC boules and wafers with respective densities for n-type 4H-SiC. Note that recent improvement efforts have eliminated three-dimensional defects such as large carbon inclusions and voids.

As said, dislocations are one-dimensional or line imperfections and can be considered as abrupt changes in the periodic arrangement of atoms along a line (dislocation line). Dislocations can be an edge type, screw type, or mixed type, depending on how they distort the lattice. It is important to note that dislocations cannot end inside a crystal, but they must end at a crystal edge or other dislocations. Otherwise, they must close back on themselves.

Geometrically, they are characterized by dislocation line l and the Burgers vector b (named after Dutch physicist Jan Burgers). The Burgers direction and magnitude can be determined by drawing a loop around the disrupted region and noticing the extra interatomic spacing needed to close the loop in the start-to-finish direction. In other words, the Burgers vector will be the vector to complete the circuit (or loop), that is, from the end to the start of the circuit. Dislocations occur when an extra incomplete plane is inserted, and the dislocation line is at the end of the plane. In edge dislocations, the Burgers vector and dislocation line are perpendicular to one another. In screw dislocations, they are parallel.

A micropipe, also called a microtube or pinhole defect, is a crystallographic defect similar to a hollow core associated with a super-screw dislocation that possesses a large Burgers vector. The high stress along the center core of this super-screw dislocation causes preferential sublimation during the growth process and, therefore, the hollow core nature of the defect. These hollow core screw dislocations typically run parallel to the growth direction through the entire SiC boule. In other words, a screw dislocation transforms successive atomic planes within a crystal lattice into the shape of a helix. Once a screw dislocation propagates through the bulk of a sample during wafer growth, a micropipe is formed with the diameter of the pinhole ranging from 0.5 mm to few micrometers. Micropipes and screw dislocations in epitaxial layers are normally derived from the substrates on which the epitaxy is performed. Micropipes are considered to be empty-core screw dislocations with large strain energy; they follow the growth direction (c-axis) in SiC boules and substrates propagating into the deposited epitaxial layers.

Factors that influence formation of micropipes and other defects include growth parameters such as temperature, supersaturation, vapor phase stoichiometry, impurities, and the polarity of the seed crystal surface. Because a micropipe is a pinhole extending along the c-axis ([0001]) direction, it is no surprise that SiC devices containing a micropipe exhibit severely impaired performance, with effects such as high leakage current and premature breakdown. Micropipes also act as a source of impurity contamination in epitaxial growth and device processing. Thus, micropipes were identified as the most important killer defects, and growth technology has now been developed to eliminate them.

Currently, commercially available SiC bulk crystals are almost exclusively grown by physical vapor transport (PVT), a modification of the Lely method. However, dislocation formation during the PVT growth of bulk SiC crystals is a real challenge to realizing high-performance SiC power devices, including yield and reliability issues. Dislocations in PVT-grown SiC crystals are classified into (i) threading dislocations extending along the c-axis and (ii) basal plane dislocations (BPDs) lying in the basal plane. Most TSDs are inherited from the seed crystal and often form in the early stage of crystal growth. BPDs can also be inherited from the seed crystal and form during the initial seeding process, but they do not propagate throughout the entire crystal because their propagation or extension directions are restricted solely to the basal plane, almost perpendicular to the direction of growth. TSDs are harmful because they degrade the blocking capabilities of SiC diodes. BPDs, on the other hand, have a serious impact on the reliability of unipolar devices such as SiC MOSFETs and JFETs as well as SiC bipolar transistors. Therefore, over the decades, considerable effort has been devoted into reducing the TSD and BPD densities in SiC crystals. Today, threading dislocations (TDs) are no longer considered killer defects in state-of-the-art wafers. For example, significant leakage spots show up at TDs in SiC Schottky barrier diodes (SBDs) only when a surface pit is formed at the location where dislocation intersects the surface of the epitaxial layer; after removing such pits, the SBD exhibits an ideal characteristic in spite of the TDs.

Another concern is associated with the poor quality of SiO_2/SiC interface in a vertical MOSFET. The performance-to-reliability trade-off is severely impacted by traps and carrier energy states at the SiC-dielectric interface. The origin of these traps is linked to the presence of carbon clusters and/or dangling Si- and C-bonds. A dangling bond is a chemical bond associated with an atom on the surface of a solid that does not join with a second atom but extends in the direction of the solid's exterior. In an energy band diagram, these traps are located between the Fermi level and the conduction band of the MOSFET SiC polytype. A large number of electrons, whose flow is expected to contribute to the low on-resistance of the inversion layer, get trapped in these energy states instead and scatter mobile electrons, further increasing the resistance in the channel region. These interface states also influence the Fowler-Nordheim (FN) tunneling currents at the SiC-dielectric interface because they lower the theoretical tunneling barrier height. If Φ_F is the theoretical barrier height and Φ_C is the conduction band offset due to the interface states, then the effective height is $\Phi_{eff} = \Phi_F - \Phi_C$.

The lower mobility in the inversion layer of a MOSFET may be acceptable for voltages above 2000 V since the channel resistance becomes negligible with respect to the drift component. However, if the MOSFET interface state density in these devices is high, their practical use will be determined mainly by FN tunneling. Furthermore, charging and discharging the oxide traps lead to a threshold voltage instability where an oxygen vacancy has been hypothesized as the origin of the primary oxide traps. Recent advances in technology have removed such instability.

One of the most successful techniques to reduce extended defects in SiC boules is the so-called "repeated a-face growth (RAF)" method. The main concept is the preparation of an almost dislocation-free seed and subsequent sublimation growth on the high-quality seed under stabilized conditions. This means that the boule is sliced parallel to the growth direction; because most dislocations are propagating along the growth direction, only a limited number of dislocations (mainly BPDs) appear on the surface of the resulting SiC(1120). By repeating the process, the density of elementary screw dislocations and TEDs can be lowered. Once this RAF process is done successfully, high-quality SiC(0001) seed crystals can be directly obtained by slicing the high-quality boule eliminating the need to repeat these complicated processes.

Let us recall that in the Miller-Bravais notation, a four-axis coordinate system for three-dimensional crystals is used, based on the unit cell. This coordinate system is based on three indices (h,k,i) referring to axes a_1, a_2, and a_3. An extra axis (c) is used for hexagonal crystals which gives rise to index l. The system can indicate directions or planes and are often written as (h,k,i,l). Simple equations allow conversion from one system to the other.

The three Miller indices specify the orientation of a surface or a crystal plane by considering how the plane (or indeed any parallel plane) intersects the main crystallographic axes of the solid. Such integers are the reciprocals of the fractional intercepts along the three axes. If fractional intercept is ∞, the reciprocal is 0. For example, if a plane is parallel to the yz plane at a distance a along the x-axis, then intercepts are (a, ∞,∞). By dividing each number by a and taking the reciprocals, we get the three indices (1,0,0). If a plane intercepts the axes at (1/2a,a,∞), the Miller indices become (2,1,0).

References

1. E. Van Brunt, A. Burk, D.J. Lichtenwalner, R. Leonard, S. Sabri, D.A. Gajewski, A. Mackenzie, B. Hull, S. Allen, J.W. Palmour, Performance and reliability impacts of extended epitaxial defects on 4H-SiC power devices. Mater. Sci. Forum **924**, 137
2. V. Veliadis, H. Hearne, E.J. Stewart, M. Snook, W. Chang, J.D. Caldwell, H.C. Ha, N. El-Hinnawy, P. Borodulin, R.S. Howell, D. Urciuoli, C. Scozzie, Degradation and full recovery in high-voltage implanted-gate SiC JFETs subjected to bipolar-current stress. IEEE Electron. Dev. Lett. **33**(7), 952–954 (2012)
3. V. Veliadis, Chapter 13: Silicon carbide vertical junction field effect transistors and cascode switches for 1200 V power conditioning applications, in *Advanced Semiconductor Materials*

and Devices Research—SiC and III-Nitrides, ed. by H.-Y. Cha. ISBN: 978–81–7895-371-7 (Research Signpost, 2010), pp. 407–446
4. V. Veliadis, Silicon carbide junction field effect transistors (SiC–JFETs), in *Wiley Encyclopedia of Electrical and Electronics Engineering*, pp. 1–37, Online ISBN: 9780471346081, 10.1002/047134608X.W8232 (2014)
5. V. Veliadis, M. Snook, T. McNutt, H. Hearne, P. Potyraj, A. Lelis, C. Scozzie, A 2055-V (at 0.7 mA/cm^2), 24-A (at 706 W/cm^2) normally-ON 4H-SiC JFET with 0.068-cm^2 active area and blocking voltage capability of 94% of the SiC material limit. IEEE Electron. Dev. Lett. **29**(12), 1325–1327 (2008)
6. V. Veliadis, M. Snook, H. Hearne, B. Nechay, S. Woodruff, C. Lavoie, C. Kirby, E. Imhoff, J. White, S. Davis, Process tolerant single photolithography/implantation 120-zone junction termination extension. Mater. Sci. Forum **740–742**, 855–858 (2013)
7. A. Agarwal, S.H. Ryu, Status of SiC power devices and manufacturing issues. *CS MANTECH Technical Digest*, pp. 215–218 (April 2006)
8. V. Veliadis, T. McNutt, E. Stewart, M. McCoy, H. Hearne, C. Clarke, Recent progress in 300°C silicon carbide JFET technology for power conditioning in electric vehicles, in 7th International All Electric Combat Vehicle Conference, paper FCXST-07060711-545141-1 (June 2007)
9. V. Veliadis, H. Hearne, T. McNutt, M. Snook, R. Howell, A. Lelis, C. Scozzie, Investigation of the suitability of 1200-V normally-off recessed-implanted-gate SiC JFETs for efficient power-switching applications. IEEE Electron. Dev. Lett. **30**(7), 736–738 (2009)
10. W. Shockley, A unipolar field-effect transistor. Proc. IRE **40**, 1365 (1952)
11. G.C. Dacey, I.M. Ross, Unipolar field-effect transistor. Proc. IRE **41**, 970 (1953)
12. D. Stephani, P. Friedrichs, Silicon carbide junction field effect transistors. Int. J. High Speed Electron. Syst. **16**(3), 825–854 (2006)
13. G. DeSalvo, Silicon carbide static induction transistors. Int. J. High Speed Electron. Syst. **15**(4), 997–1033 (2005)
14. V. Veliadis, Chapter 7: 1200-V SiC vertical-channel-JFETs and cascode switches, in *Silicon Carbide: Volume 2: Power Devices and Sensors*, ed. by P. Friedrichs, T. Kimoto, L. Ley, G. Pensl. ISBN: 978-3-527-41002-6 (Wiley-VCH, 2009), pp. 157–191

Dr. Victor Veliadis is the executive director and CTO of PowerAmerica, a member-driven consortium of industry, universities, and national labs accelerating the commercialization of energy-efficient silicon carbide and gallium nitride power semiconductor chips and electronics. At PowerAmerica, he has managed a budget of $156 million that he strategically allocated to over 210 industrial and university projects to catalyze SiC and GaN semiconductor and power electronics manufacturing, workforce development, and job creation. His PowerAmerica educational activities have trained 430 full-time students in collaborative industry/university WBG projects and engaged over 7000 attendees in tutorials, short courses, and webinars. Dr. Veliadis is currently negotiating a $64 M US Department of Energy PowerAmerica renewal to further catalyze WBG power technologies.

Dr. Veliadis is an ECE professor at NC State and an IEEE fellow and EDS distinguished lecturer. He has 27 issued US patents, 6 book chapters, and over 150 peer-reviewed publications to his credit. He has given over 150 keynote/tutorial/invited presentations including keynotes at ICSCRM, APEC, ECCE, and WiPDA. Prior to entering academia and taking an executive position at PowerAmerica in 2016, Dr. Veliadis spent 21 years post-PhD in the semiconductor industry where his work included design, fabrication, and testing of SiC devices, GaN devices for military radar systems, and financial and operations management of a commercial semiconductor fab. Dr. Veliadis has a Ph.D. degree in electrical engineering from John Hopkins University (1995).

Ming Su is the technical marketing manager responsible for Rohm's SiC and high-power device products for the Americas market. He received the bachelor's degree in electronic engineering from Tsinghua University and the M.S. and Ph.D. degrees in electrical engineering from Rutgers University. During his doctoral research, he developed normally-off SiC JFET transistors for power switching applications and power IC. Prior to joining Rohm, Dr. Su worked as a research engineer at Ford Motor Company, responsible for discovering and evaluating emerging power semiconductor technologies in support of the company's vehicle electrification strategies. He has contributed over 35 scientific journal articles and conference presentations in the area of silicon and wide bandgap semiconductors and holds two US patents in power devices and modules.

Filippo Di Giovanni Filippo Di Giovanni has previously been with STMicroelectronics as marketing manager for power transistors and, more recently, also for wide bandgap (WBG) products such as silicon carbide and gallium nitride, plus power modules. Doctor Di Giovanni's current interests span from power technologies to their automotive and industrial applications. His working experience also includes attendance to conferences and workshops dedicated to power conversion and coordination activities for European-funded projects. At the end of 2023, after retirement, Filippo is a full-time technical writer about semiconductor technology and market analysis for professional magazines.

Chapter 2
SiC Material Properties

Kevin Speer, Yogesh Sharma, Sergey Maximenko, Filippo Di Giovanni, Ravi Bollina, William Gemmill, Terry Knight, Christian Jentgens, and Helge Willers

This chapter of this publication is a comprehensive exploration of the essential properties related to silicon carbide (SiC) semiconductor materials. Authored by esteemed experts in the field, the chapter is divided into several sections, each focusing on distinct aspects crucial for understanding and utilizing SiC in semiconductor devices.

In Sects. 2.1 and 2.2, Dr. Kevin Speer, Dr. Yogesh Sharma, and Dr. Sergey Maximenko from Microchip Technology delve into the intricate crystal structure of SiC. Following this, the authors further explore the electronic properties of SiC, particularly focusing on its wide bandgap, breakdown electric fields, carrier mobilities, and thermal properties. Their analysis underscores the exceptional attributes of

---

This paragraph represents a collaborative effort by Dr. Ravi Bollina (CSO and Executive Board Member), Ph.D. William Gemmill (Chief Innovation Officer and Executive Board Member), Terry Knight (Technical Services Manager), Dr. Christian Jentgens (Head of Research and Development Industrial), and Dipl. Ing. Helge Willers (Sales Director) from Pureon, the leading provider of high-quality surface finishing solutions. Their combined expertise reflects Pureon's commitment to driving innovation and developing next-generation solutions.

K. Speer · Y. Sharma · S. Maximenko
Microchip, Chandler, AZ, USA

F. Di Giovanni (✉)
STMicroelectronics, San Gregorio, CT, Italy

R. Bollina · C. Jentgens
Pureon Inc., Lengwil, Switzerland

W. Gemmill · T. Knight
Pureon Inc, Monroe, NC, USA

H. Willers
Pureon GmbH, Willich, Germany

M. Di Paolo Emilio (ed.), *SiC Technology*,
https://doi.org/10.1007/978-3-031-63418-5_2

SiC, positioning it as a standout material for high-power, high-frequency, and extreme environment applications.

In Sect. 2.3, Filippo Di Giovanni discusses substrates and epitaxial growth techniques essential for SiC semiconductor fabrication. He elaborates on methodologies such as physical vapor transport (PVT) and high-temperature chemical vapor deposition (HTCVD), crucial for producing monocrystalline wafers and epitaxial layers, respectively.

Lastly, in Sect. 2.4, Ravi Bollina, William Gemmill, Terry Knight, Christian Jentgens, and Helge Willers of Pureon delve into the intricate process of preparing SiC wafer substrates, highlighting their importance for high-power and high-frequency applications.

2.1 Crystal Structure of SiC

Kevin Speer, Yogesh Sharma, and Sergey Maximenko

Appreciating a material's crystal structure is essential for unraveling its electronic, mechanical, thermal, and optical properties; these properties then determine the material's suitability for a variety of end uses, such as in high-voltage semiconductor devices. In this chapter, we explore the crystallography of silicon carbide (SiC), beginning with an examination of the most basic Si-C interatomic bond. Next, we take multiple Si-C bonds and construct the next most basic building block, known as the Si-C tetrahedron (plural: tetrahedra). With these two basics established, the chapter moves into stacking the tetrahedra atop one another to build a crystal; the stacking must follow certain rules discussed herein, resulting in a variety of SiC crystal structures, also known as polytypes.

To understand the nature of the bond between Si and C atoms, it is helpful to first write the electronic configuration of SiC. The configuration for Si, having an atomic number of $Z = 14$, is written as $1s^2 2s^2 2p_x^2 2p_y^2 2p_z^2 3s^1 3p_x^1 3p_y^1 3p_z^1$. The configuration for C ($Z = 6$) is written as $1s^2 2s^1 2p_x^1 2p_y^1 2p_z^1$. Looking at the outermost shells, these configurations tell us that every Si and C atom will have four unpaired electrons available for participation in bonding: each with one *s*-orbital electron and three *p*-orbital electrons. To help with visualization, Fig. 2.1 shows the characteristic shapes of the atomic orbitals within the different subshells. The *s* orbital takes the shape of a sphere, while the *p* orbitals take the shape of dumbbells aligned along three mutually orthogonal axes; as in our case, the three *p* orbitals are often distinguished by subscripts denoting their axial alignment (i.e., p_x, p_y, and p_z). These shapes can be thought of as *directed* distributions of the spatial probability density functions used to evaluate the location of the electrons; that is, the axial diameter of these must exist within these volumes; however, quantum mechanics dictates that there must be a finite probability that the electrons may exist outside of these volumes.

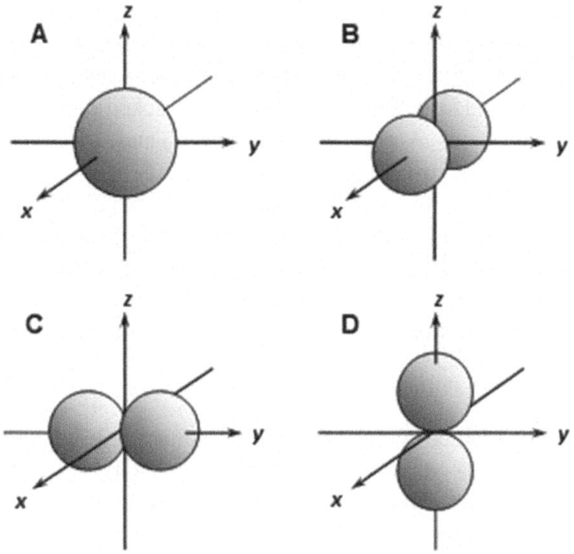

Fig. 2.1 Characteristic shapes of the (**a**) s orbital, (**b**) p_x orbital, (**c**) p_y orbital, and (**d**) p_z orbital

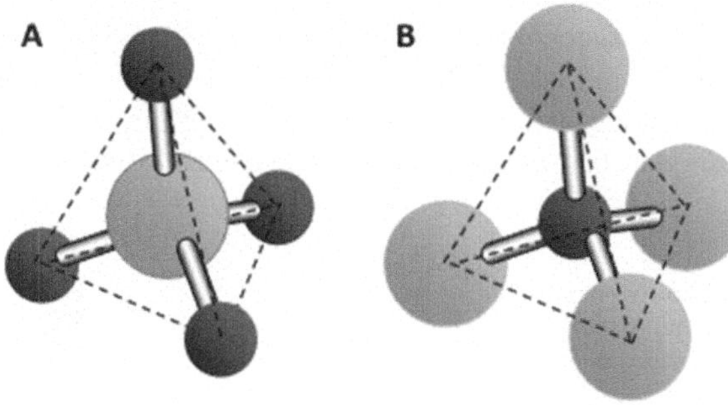

Fig. 2.2 (**a**) A SiC_4 tetrahedron with a Si centroid and (**b**) a Si_4C tetrahedron with a C centroid

Due to the differently shaped orbitals, the bond between a Si atom and a C atom forms through *hybridization*, a process through which a number, *x*, of bondable electrons from different orbitals combine to form *x* orbitals that are equivalent in size and shape. In the case of SiC, the hybridized bonds, classified as sp^3 bonds, are directed toward four corners of a regular tetrahedron, as illustrated in Fig. 2.2. The angle scribed between any two adjacent bonds is approximately 109°. Hybridized bonds possess greater strength than non-hybridized bonds; evidence of SiC's superior bond strength is found in its widespread use as an abrasive material.

To approximate the length of a covalent sp^3 bond, one sums the so-called tetrahedral radii of constituent atoms; these tetrahedral radii are applicable when sp^3 hybridization takes place, as in the case of SiC. Tetrahedral radii may be routinely found in tabulated form. From Evans [1], the tetrahedral radius for Si is 0.117 nm, and for C it is 0.077 nm; it follows, then, that the length of the sp^3-hybridized Si-C bond is 0.194 nm. As examples for comparison, the same sp^3-hybridized bond between atoms is 0.234 nm for pure Si, 0. 244 nm for GaAs, and 0.196 nm for

GaN. The tetrahedra are then arranged to form various crystal structures. In contrast to the majority of semiconductors, which typically exhibit only one or two crystal structures, SiC has more than 200 distinct polytypes. These polytypes feature various crystal structures including close-packed cubic, hexagonal, and rhombohedral arrangements. Among these, cubic (3C) and hexagonal (4H) are the most common forms of SiC used in electronic devices.

The cubic form of SiC crystallizes in the zincblende structure, which is physically identical to the diamond structure with the exception that there are two elemental constituents instead of one—namely, silicon and carbon. Schematically illustrated in Fig. 2.3, the zincblende structure can further be thought of as two interpenetrating face-centered cubic (FCC) lattices, one of which is offset from the other by a lattice-parameter distance of $\left[\frac{a}{4},\frac{a}{4},\frac{a}{4}\right]$. The conventional FCC lattice has four lattice points; as such, the interpenetrative dual-FCC zincblende unit cell has eight lattice points, four of which are occupied by Si atoms and the other four by C atoms. Symmetrically, 3C-SiC has a fourfold rotation inversion axis about the $\langle 100 \rangle$ family of directions, a threefold rotation axis about the $\langle 111 \rangle$ family of directions, and a mirror reflection across the $\{110\}$ planes; thus, its space group is $F\overline{4}3m$, F denoting its *all*-face-centered Bravais lattice. Finally, its lattice parameter is 0.436 nm.

The hexagonal form of SiC crystallizes in the wurtzite structure, illustrated schematically in Fig. 2.4. Note that the dashed gray lines outline the frame of the hexagonal prism, while the solid gray lines outline the frame of the unit cell. In a similar fashion as that in which zincblende can be visualized as two interpenetrating FCC

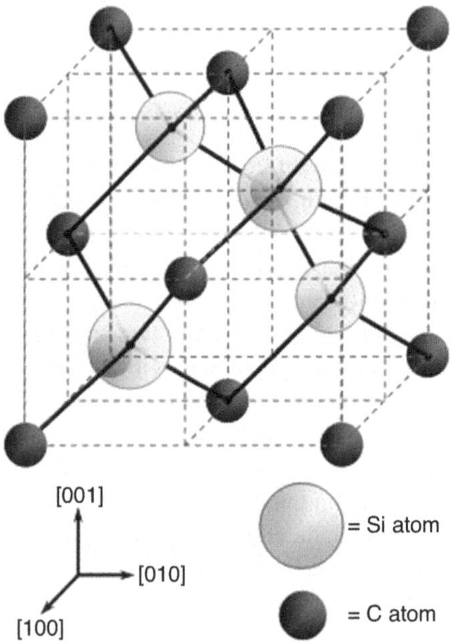

Fig. 2.3 Schematic illustration of the zincblende structure of 3C-SiC

lattices, wurtzite may be visualized as two interpenetrating hexagonal close-packed (HCP) lattices.

Although wurtzite is similar to the zincblende structure, the fundamental difference has to do with the stacking of tetrahedra along the [111] direction (or the crystallographically equivalent [0001] direction using the Bravais-Miller indexing system). This subtle difference can be imagined using the following adaptation from Burns and Glazer (2): Looking down the [0001] direction in an HCP crystal, one will see that by taking a single lattice point as the origin, the point will have 12 nearest neighbors: six in the shared (0001) plane, three in an (0001) plane *above* the shared plane, and three in an (0001) plane *below* the shared plane. One will also see that the three above-plane neighbors are exactly aligned with the three below-plane neighbors along [0001]. Now consider a single lattice point in an FCC structure while looking along the [111] direction. It is seen that the three above-plane neighbors are no longer aligned with the three below-plane neighbors; in fact, they are rotated about the [111] direction by 60° with respect to one another. This is a manifestation of the 6_3 screw axis of the wurtzite structure, wherein a proper sixfold rotation is coupled with a translation parallel to the screw axis by a distance equal to three-sixth, or half, the lattice parameter along the [0001] axis. This is also why the stacking sequence of cubic structures is …ABCABC…, while the stacking sequences of hexagonal structures are of the form …ABABAB… for 2H-SiC, …ABAC… for 4H-SiC, or …ABCACB… for 6H-SiC.

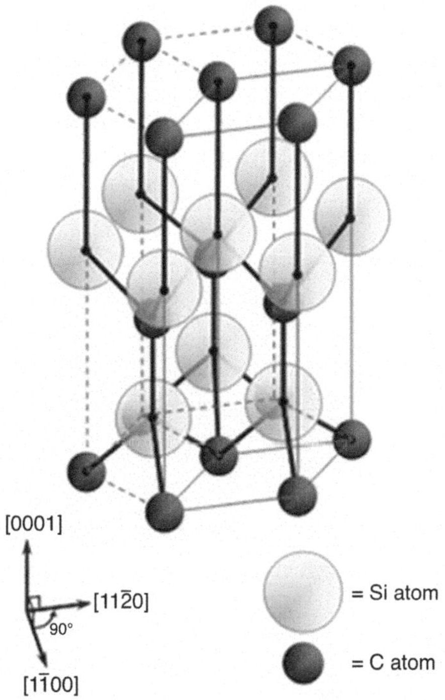

Fig. 2.4 Schematic illustration of the wurtzite structure of the hexagonal SiC

Symmetrically—and in addition to the 6_3 screw axis—4H-SiC has a mirror reflection across $\left(10\bar{1}0\right)$ and an axial glide plane along the c direction across $\left(11\bar{2}0\right)$; thus, its space group is $P6_3mc$, P denoting its primitive Bravais lattice. Finally, its lattice parameters are $a = 0.3073$ nm and $c = 1.005$ nm. It is worth noting here that each Si-C bilayer contributes approximately 0.25 nm to the [0001] direction (or $\langle 111 \rangle$ direction) in all polytypes. This makes an approximation of the c lattice parameter for hexagonal polytypes simple; 4H is 1.0 nm, 6H is 1.5 nm, 8H is 2.0 nm, and so forth.

When beginning to contemplate the structure of a bulk crystal of SiC—cubic, hexagonal, or otherwise—it is often convenient to visualize the tetrahedra of Fig. 2.2 as hard spheres. The way in which crystals of SiC are constructed, then, is to stack these spheres on top of one another, sheet by sheet. The SiC close-packed structure can be presented as sheets of the close-packed array of spheres forming a two-dimensional pattern with sixfold symmetry. Each sheet of spheres represents a bilayer composed of one layer of Si atoms and one layer of C atoms. The most stable way to stack an identical second sheet of close-packed spheres is to place the spheres atop the "valleys" in the first sheet. There are two possibilities "B" or "C" for arranging the second sheet relative to the first sheet as presented in Fig. 2.5. Thus, a sheet can be denoted A, B, or C, depending on the positions of its spheres.

Only two things determine the polytype of a crystal: the sequence with which the sheets are stacked and the periodicity of the sequence. The crystal structure (i.e., cubic, hexagonal, and rhombohedral) is merely a consequence of these two factors. Periodicity refers to the number of sheets in the sequence that are repeated periodically in the bulk crystal. In the case of our "ABC" structure, if one were to stack many sets of "ABC" crystals on top of one another to make a bulk crystal, the crystal's sequence would be …ABCABCABCABC…. Since the sequence "ABC" is repeated over and over again, the periodicity is 3—one each for the "A," "B," and "C." For a bulk crystal based on the …ABCACB… structure, the periodicity would

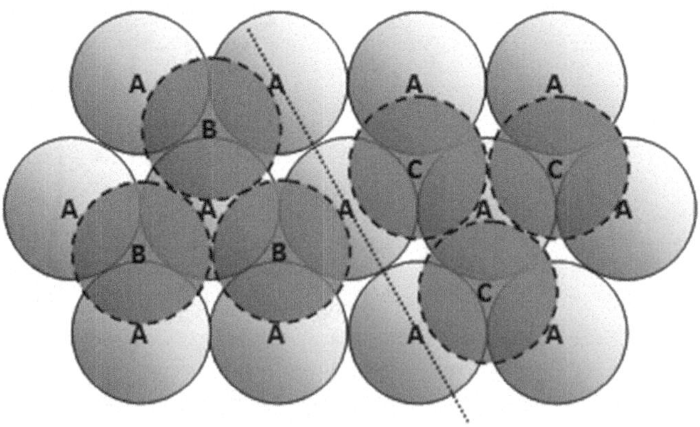

Fig. 2.5 (Left) An upward-pointing tetrahedron resulting from atoms deposited on "B" sites and (right) a downward-pointing tetrahedron resulting from atoms deposited on "C" sites

be 6—one each for the “A,” “B,” “C,” “A,” “C,” and “B” and so on. The number of times these sequences are stacked is irrelevant; the only thing that matters is that the number of sheets in the sequence is indicated by the periodicity number. The notation for each polytype is written based on (1) the crystal’s periodicity and (2) the first letter of the resulting crystal’s crystal system (e.g., “C” for cubic, “H” for hexagonal, and “R” for rhombohedral). For example, our …ABC… structure, with periodicity of 3, has a cubic structure; thus, the notation for this polytype would be 3C-SiC. For …ABCACB… that takes a hexagonal structure, the polytype’s notation is 6H-SiC and so on. This nomenclature, now broadly used in SiC, was suggested by Ramsdell (3). Also, the cubic polytype is named as *b*-SiC, whereas the hexagonal and rhombohedral polytypes are referred to as *a*-SiC (4).

To construct ball-and-stick models for each polytype in an effort to visualize them in three dimensions would be a timely and laborious undertaking; however, it can be done more simply by projecting the tetrahedra in two dimensions onto a $\{11\bar{2}0\}$, or equivalently, a $\{1\bar{1}0\}$, plane, resulting in a tetrahedron and its so-called twin variant (the twin variant of a tetrahedron is the tetrahedron rotated 180° about the [0001], or [111], direction) appearing as they do in Fig. 2.6.

Figure 2.7 shows the grid and the six possible tetrahedral positions corresponding to the normal and twin variants for positions A, B, and C. A grid with vertical lines 1, 2, and 3 represents the three possible centroid sites A, B, and C, respectively. Horizontal lines whose intersections with the vertical lines represent the atomic centroids of the tetrahedra and are labeled α, β, and γ in correspondence with their positions A, B, and C, respectively. The positions’ normal (twin) tetrahedra can be named according to the following convention: position A will be T_1 (T_1'), position B will be T_2 (T_2'), and position C will be T_3 (T_3'). Further, the stacking constraints

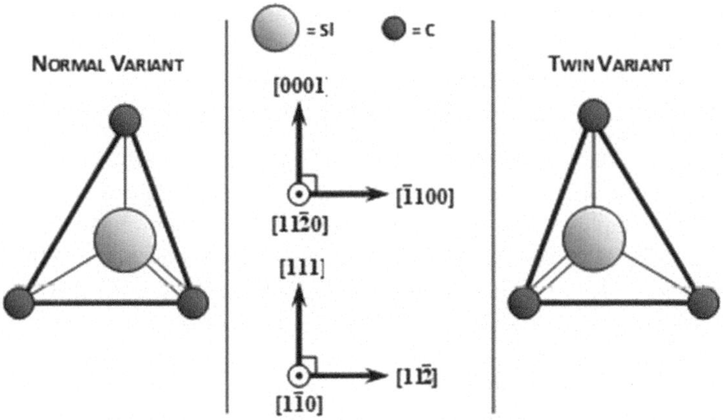

Fig. 2.6 SiC4 tetrahedra projected along the [“11” “2”‾0] direction. (Left) Orientation of tetrahedra arbitrarily defined as the “normal variant” and (right) orientation of tetrahedra subsequently defined as the “twin variant.” Note that one is rotated 180° about the [0001] direction with respect to the other

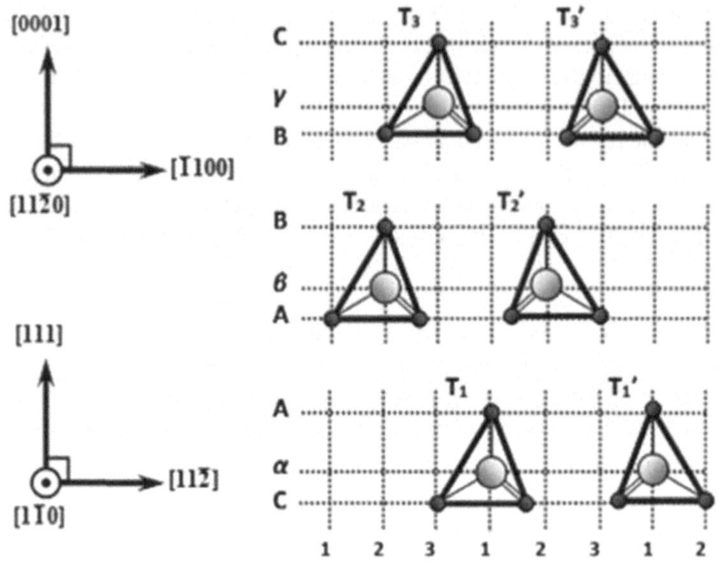

Fig. 2.7 The "stacking grid" and the six possible tetrahedra, corresponding to the normal and twin variants for each of the three positions, A (T1 and T1′), B (T2 and T2′), and C (T3 and T3′)

may be verbalized by the following four rules, which are illustrated on the grid in Fig. 2.8.

(i) Normal variants may be stacked atop one another if the subscript is increased cyclically (T_1 T_2, T_2 T_3, and T_3 T_1). See Fig. 2.8 (1–3).
(ii) Twin variants may be stacked atop one another if the subscript is decreased cyclically (T_2' T_1', T_1' T_3', and T_3' T_2'). See Fig. 2.8 (4–6).
(iii) Normal variants may be stacked atop twin variants if the subscript is increased cyclically (T_3' T_1, T_1' T_2, and T_2' T_3). See Fig. 2.8 (7–9).
(iv) Twin variants may be stacked atop normal variants if the subscript is decreased cyclically (T_1 T_3', T_3 T_2', and T_2 T_1'). See Fig. 2.8 (10–12).

Now that the rules have been formalized, the grid will further assist with the ultimate objective of building the polytypes from periodic sequences of Si-C tetrahedra. The simplest SiC polytype is 2H-SiC, which has the repeated sequence …ABAB…; next is 3C-SiC, with the repeated sequence …ABC…; next is 4H-SiC, with the repeated sequence …ABAC…; next is 6H-SiC, with the repeated sequence …ABCACB…; etc. These structures are illustrated in Figs. 2.9 and 2.10.

A point should be made about SiC and crystallographic polarity. As a compound semiconductor, the valence electrons in SiC tend to localize near the carbon (C) atoms, leading to ionicity and polarity within the material. A silicon (Si) atom, bonded tetrahedrally and aligned along the c-axis (<0001>), is termed the "Si-face," while the (0001) face, where a bond from a tetrahedrally bonded carbon (C) atom

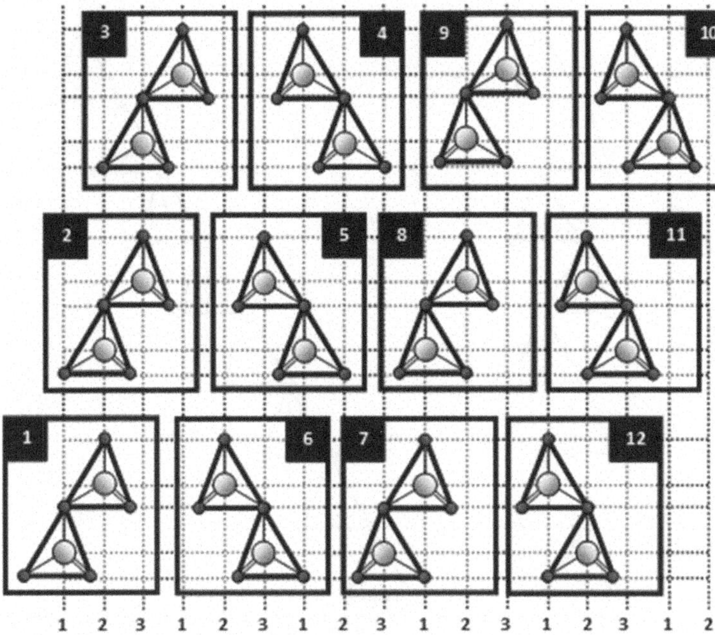

Fig. 2.8 Schematic illustrating all 12 possibilities for tetrahedral stacking. Refer to the text for stacking rules

aligns along the c-axis, is known as the "C-face" as shown in Fig. 2.11 In the context of 3C-SiC, the (111) and $(\bar{1}\bar{1}\bar{1})$ faces correspond, respectively, to the Si-face and C-face.

Finally, it is worth mentioning one more naming convention that is commonly used to discuss differences in SiC metal-oxide-semiconductor field-effect transistor (MOSFET) channel mobilities. Figure 2.12 highlights major planes within cubic and hexagonal SiC polytypes; the $\{11\bar{2}0\}$ face is referred to as the "A-face (or a-face)," and the $\{1\bar{1}00\}$ face is called the "M-face (or m-face)." These crystal planes are considered major due to their substantial differences in surface energy, chemical reactivity, and electronic properties.

2.2 Bandgap and Electrothermal Properties

Sergey Maximenko, Yogesh Sharma, and Kevin Speer

The optimal semiconductor for advanced electronic devices requires a very wide bandgap to ensure stable operation at high temperatures, a large breakdown electric field to minimize on-state resistance and power losses, swift carrier mobilities for rapid switching, and an electronic structure that supports the incorporation of

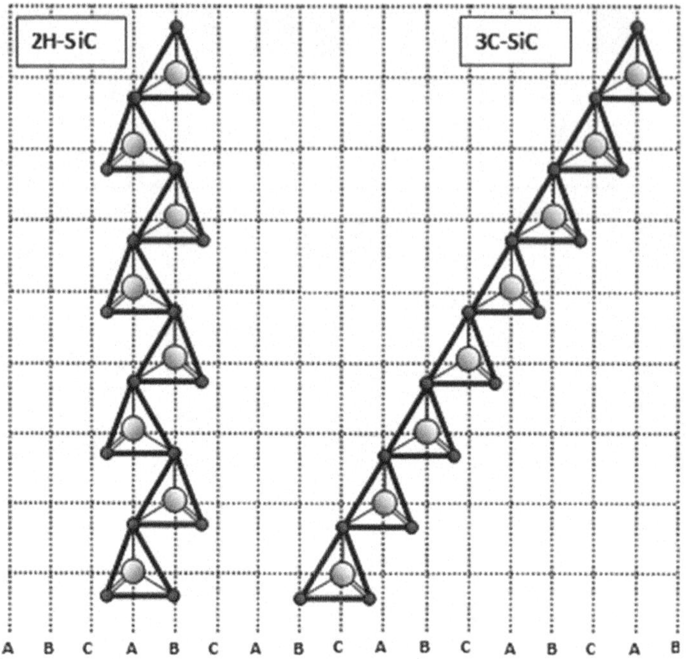

Fig. 2.9 (Left) 2H-SiC with repeated sequence …ABAB… and (right) 3C-SiC with repeated sequence …ABC…

dopants with relatively low ionization energies. Silicon carbide emerges as a standout material, possessing most of these desired attributes. Furthermore, the robustness of the sp^3 bond endows SiC with exceptional toughness, thermal dissipation capabilities, and resilience to radiation damage, qualities coveted in the electronics sector. These traits position SiC to surpass traditional semiconductors in applications involving high power, high frequency, and extreme conditions.

Table 2.1 lists the key electronic properties of various semiconductor materials. These properties, listed from left to right at the top of the table, include the bandgap (E_{gap}), breakdown electric field (E_b), electron and hole mobilities (μ_n and μ_p), saturated electron drift velocity (υ_{sat}), dielectric constant (ε) (also known as relative permittivity), and thermal conductivity (κ). The properties for three predominant SiC polytypes, 3C, 4H, and 6H, are displayed. For use in electronic devices, 4H-SiC is the best polytype due to its relatively high E_{gap}, E_b, υ_{sat}, and κ, despite its modest carrier mobilities and decent thermal conductivity. Consequently, this chapter will discuss the properties of SiC specifically through the lens of 4H-SiC.

For many years, there has been a hot debate in the electronic materials community about which of the two wide bandgap semiconductors—4H-SiC or 2H-GaN—is the better choice for use in so-called "high-voltage" diodes and transistors. (In this context, high voltage is assumed to be greater than or equal to 100 V.) Referring to Table 2.1, one can see that SiC has much higher thermal conductivity (κ) and

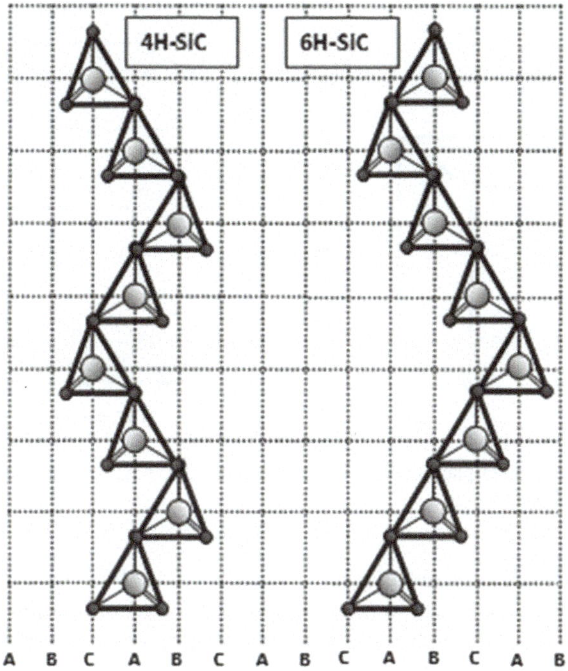

Fig. 2.10 (Left) 4H-SiC with repeated sequence ...ABAC... and (right) 6H-SiC with repeated sequence ...ABCACB...

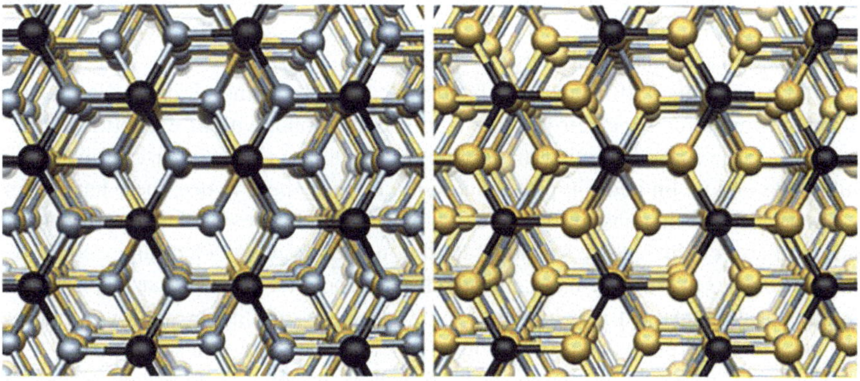

Fig. 2.11 Schematic drawing of the SiC polar structures. (Adapted from [2]). Plan views of (left) the (0001) Si-terminated and (right) the (000$\bar{1}$) C-terminated 4H-SiC surfaces. Yellow, gray, and black spheres represent Si, C, and surface sites, respectively

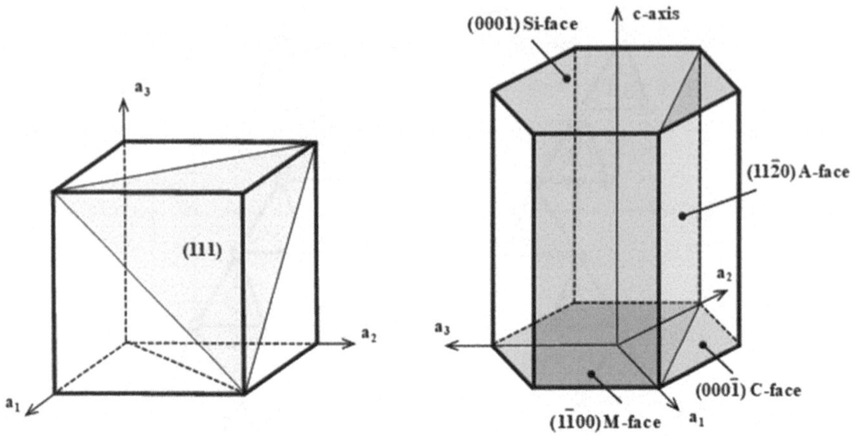

Fig. 2.12 Identification of planes in the (left) cubic and (right) hexagonal SiC structure using the Bravais-Miller indexing system

exhibits similar saturation velocity (υ_{sat}). However, GaN boasts a larger breakdown field strength and, when used to make a high electron mobility transistor (HEMT), has near-ballistic carrier transport due to the formation of a two-dimensional electron gas (2DEG).

The primary limitation of high-voltage GaN electronics is the scarcity of native substrates. There is also the option of heteroepitaxial GaN, which is grown on larger silicon wafers, but these devices are limited in their voltage-blocking capability due to a large defect density in the epilayers. Moreover, high-voltage GaN devices must be lateral in the absence of GaN substrates; and even if using GaN substrates, the large amounts of dissipated heat cannot be removed as efficiently as in SiC due to GaN's lower thermal conductivity.

Other materials with ultra-wide bandgaps, such as diamond and β-Ga_2O_3, are also being explored for electronic applications. Diamond, with its exceptional electronic properties, faces significant challenges due to the restricted size of available substrates and its limited n-type conductivity. Despite these issues, there have been a few successful demonstrations of working bipolar devices made from diamond. Should these challenges be surmounted, the impact on power electronics could be revolutionary. On the other hand, the novel semiconductor β-Ga_2O_3 can be produced in larger substrates, but it is hindered by the absence of suitable shallow acceptor dopants, which presents a hurdle for its practical use in electronics.

Returning to SiC, one of the most frequently mentioned beneficial electrical characteristics is its wide bandgap. The relationship between bandgap size and temperature can be expressed using the following formula [3]:

$$E_{gap}(T) = E_{gap}(300\,\mathrm{K}) - \eta(T - 300) \tag{2.1}$$

Table 2.1 Electronic properties of semiconductors at room temperature [3–12]

	E_{gap} (type) [eV]	E_b [MV/cm]	$\mu_{n,p}$ [cm^2/V·s]	v_{sat} [x10^7 cm/s]	ε	k [W/m·k]
3C-SiC	2.36 (I)	1.4 (// c-axis)	$\mu_n = 1000$ (1000) (//(⊥) c-axis)	2	9.72 (// c-axis)	360
		1.4 (⊥ c-axis)	$\mu_p = 100$ (100) (//(⊥) c-axis)		9.72 (//(⊥) c-axis)	
4H-SiC	3.23 (I)	2.8 (// c-axis)	$\mu_n = 1200$ (1020) (//(⊥) c-axis)	2.2	10.32 (// c-axis)	280–390
		2.2 (⊥ c-axis)	$\mu_p = 120$ (120) (//(⊥) c-axis)		9.76 (⊥ c-axis)	
6H-SiC	3.00 (I)	3.0 (// c-axis)	$\mu_n = 100$ (450) (//(⊥) c-axis)	1.9	10.03 (// c-axis)	490
		1.7 (⊥ c-axis)	$\mu_p = 100$ (100)		9.66 (⊥ c-axis)	
Si	1.12 (I)	0.3	$\mu_n = 1450$	1.0	11.7	156
			$\mu_p = 370$			
GaAs	1.42 (D)	0.4	$\mu_n = 2850$	1.0	12.9	45.5
			$\mu_p = 400$			
GaN (wurtzite)	3.44 (D)	~ 5	$\mu_n = 900$	2.5	8.9	130
			$\mu_p = 200$			
β-Ga_2O_3	4.5–4.87 (D)	~ 8	$\mu_n = 300$	1.8–2	10	10–30
Diamond	5.47 (I)	5.5	$\mu_n = 2000$	2.7	5.7	2000
			$\mu_p = 2100$			
AlN (wurtzite)	6	1.8	$\mu_n = 300$	1.6	8.5	285
			$\mu_p = 14$			
c-BN	6.2–6.4 (I)	4	$\mu_n = 200$	4.3	7.1	1300

where E_{gap}(300 K) is the bandgap at 300 °K, T is the temperature, and η is the material-specific constant selected to fit experimental data. For 4H-SiC, $E_{gap}(300) = 3.23$ eV and $\eta = 3.3 \times 10^{-4}$. The benefit of possessing a wide bandgap is particularly attributed to the bandgap's responsiveness to fluctuations in temperature.

The bandgap energy value has direct impact on the intrinsic carrier concentration n_i in semiconductor and expressed as [26].

$$n_i(T) = (N_C N_V)^{0.5} \exp\left(-\frac{E_{gap}(T)}{2kT}\right) \tag{2.2}$$

where effective densities of states in valence N_V and conduction (N_C) bands are evaluated as $N_V = 2\left(\frac{2\pi m_h^* kT}{h^3}\right)^{\frac{3}{2}}$ and $N_C = M_C 2\left(\frac{2\pi m_e^* kT}{h^3}\right)^{\frac{3}{2}}$, h is Planck's

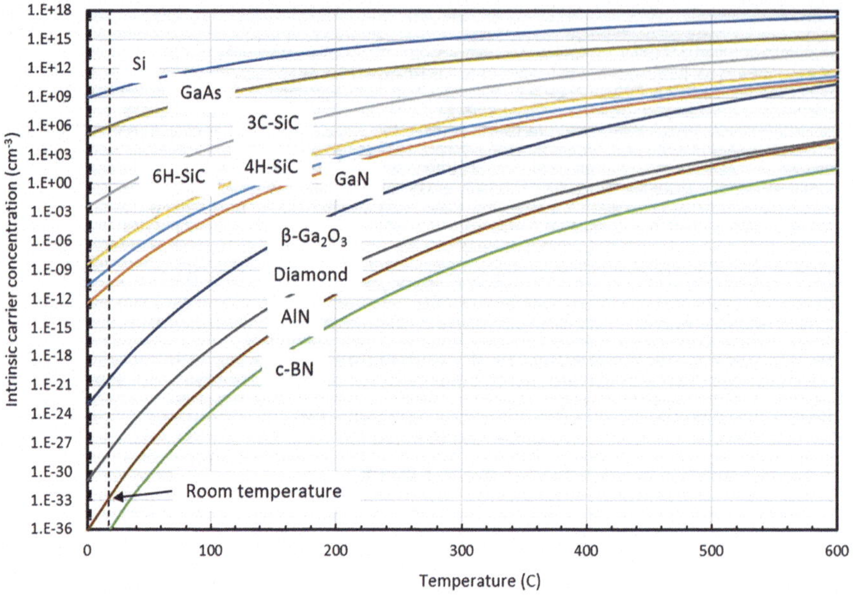

Fig. 2.13 Intrinsic carrier concentration versus temperature for various semiconductor materials

constant, and k is the Boltzmann constant with effective masses of holes ($m_h = 0.82\ m_0$) and electrons ($m_e = 0.39\ m_0$) for 4H-SiC and the number of equivalent band minima (M_C) 3 for 4H-SiC [12]. It could be seen that a wide-bandgap material is less susceptible to temperature-based conductivity modulations. Thus, 4H-SiC could reliably operate even for temperatures above 600 °C [19]. In the case of conventional Si semiconductor, Si-based devices are excluded from applications if junction temperatures reach 125 °C [20]. Figure 2.13 demonstrates the influence of bandgap on intrinsic carrier concentration for the semiconductors listed in Table 2.1. One can see that, at room temperature, $n_{i,Si}$ is ~1 × 10^{10} cm^{-3} and $n_{i,4H\text{-}Sic}$ is ~1 × 10^{-8} cm^{-3}—a difference of nearly 19 orders of magnitude.

Doping of SiC is typically accomplished during the growth phase using chemical vapor deposition (CVD) to form homoepitaxial layers, or alternatively, through the method of ion implantation. However, strong bonding and the compact lattice of SiC hinder the diffusion of dopants, making traditional diffusion doping methods ineffective. Among impurities in 4H-SiC, nitrogen (N) and aluminum (Al) are the choices for donor and acceptor. N impurity is also an unintentional background dopant impurity in SiC. High-purity 4H-SiC grown by the epitaxial technique is presently available commercially, with reported background doping concentrations as low as 1E14 cm^{-3} with wafer sizes of up to 8 inches [21, 22]. Another type of acceptor impurity explored in the past for SiC was boron (B) [23]. However, due to relatively high activation energy (330 meV), it did not find widespread application in SiC electronics.

Table 2.2 Parameters used for the calculation of the bandgap narrowing effect in n- and p-type 4H-SiC

Doping	A_C	B_C	A_V	B_V
n-type	-1.50×10^{-2}	-2.93×10^{-3}	-1.9×10^{-2}	$-8.74E \times 10^{-3}$
p-type	-1.57×10^{-2}	-3.87×10^{-4}	-1.3×10^{-2}	$-1.15E \times 10^{-3}$

Introduction of high concentration of impurities in a semiconductor impacts material bandgap "bandgap narrowing effect" and can be described by Eq. (2.3) where ΔE_C *and* ΔE_V are the downward and upward shifts in conduction and valence bands with dependence on $N^*_{A,D}$ -ionized acceptor concentrations in p-doped or ionized donor concentrations in n-doped materials; the remaining of parameters are given in Table 2.2 [24]:

$$E_{gap}\left(T, N^*_{A,D}\right) = E_{gap}(T) + \Delta E_C(N^*_{a,d}) + \Delta E_V(N^*_{a,d}) \tag{2.3a}$$

$$\Delta E_C\left(N^*_{A,D}\right) = A_C N^{*\,\frac{1}{3}}_{A,D} 10^{-6} + B_c N^{*\,\frac{1}{2}}_{A,D} 10^{-9} \tag{2.3b}$$

$$\Delta E_V\left(N^*_{A,D}\right) = A_V N^{*\,\frac{1}{3}}_{A,D} 10^{-6} + B_V N^{*\,\frac{1}{2}}_{A,D} 10^{-9} \tag{2.3c}$$

Doping impurities in 4H-SiC such as Al and N have activation energies that exceed the thermal energy at room temperature ($kT = 0.025$ eV), resulting in incomplete ionization of such impurities. The concentration of activated donor (N^*_D) and acceptor (N^*_A) incomplete ionization is given by

$$N^*_D = \frac{N_D}{1 + g_D N_V \exp\left(\frac{E_{FN} - E_D}{kT}\right)} \tag{2.4a}$$

$$N^*_A = \frac{N_A}{1 + g_A N_C \exp\left(\frac{E_A - E_{FD}}{kT}\right)} \tag{2.4b}$$

where N_A and N_D are the net compensated doping in the p-type and n-type materials, E_A and E_D are the acceptor and donor energy levels, k is the Boltzmann constant, g_A and g_D are the degeneracy factors, weakly temperature dependent and typically taken as constants, with values 4 and 2 accordingly. The activation energies of acceptor (E_A) and donor (E_D) impurities in 4H-SiC are dependent on their doping (N_A, N_D) and can be found using [25].

$$E_{A,D} = E_{A,D0} - \gamma_{A,D} \times N^{\frac{1}{3}}_{A,D} \tag{2.5}$$

Table 2.3 Parameters used to determine activation energies of impurities

Doping	$E_{A,D0}$ (eV)	$\gamma_{A,D}$ (eV × cm)
n-type, Acceptors (Al)	0.205	1.7×10^{-8}
p-type, Donors (N)	0.105	4.26×10^{-8}

with parameters presented in Table 2.3. In the case of the hexagonal structure of 4H-SiC, impurities can incorporate in cubic or hexagonal surrounding sites, which give rise to varying ionization energies for the same impurity. However, in many cases, it is sufficient to use average ionization energy for acceptors and donors, $E_{A,D0}$. Knowing the concentration of active dopants, the equilibrium carrier concentrations for holes (p) and electrons (n) in p- and n-type conductivity 4H-SiC can be found solving general charge neutrality equations [26].

The subsequent critical electrical characteristic to consider is the breakdown electric field. As shown in Table 2.1, SiC exhibits the capability to endure electric fields that are over ten times greater than those of Si and GaAs. This has several significant implications: First, it allows for a higher device packing density, meaning that more individual components can be placed closely together on a single wafer. This leads to the production of smaller and faster chips, which benefit from lower parasitic capacitance and inductance. Second, and of paramount importance in today's context, is that the layers designed for voltage blocking can be significantly thinner. This enables the creation of high-voltage devices with lower on-resistance, thereby minimizing power losses of the I^2R type, reducing forward-voltage drops, and enhancing operational efficiency. Third, the need to use multiple devices in series to achieve the required voltage-blocking capability is eliminated. Consequently, the circuit's component count is reduced, which in turn enhances the efficiency, power density, and reliability of the system.

The mobility of carriers is another important parameter. Low-field mobility as a function of doping concentration and temperature can be expressed through commonly used Caughey and Thomas model [27]:

$$\mu_{p,n} = \mu^{\min} + \frac{\mu^{\max} - \mu^{\min}}{1 + \left(\frac{N_{imp}}{N^{crit}}\right)^{\delta}} \tag{2.6a}$$

where

$$\mu^{\min} = \mu_0^{\min}\left(\frac{T}{300}\right)^{\alpha} \tag{2.6b}$$

$$\mu^{\max} = \mu_0^{\max}\left(\frac{T}{300}\right)^{\beta} \tag{2.6c}$$

Table 2.4 Values used for determining mobility dependence in 4H-SiC

Parameter	Holes	Electrons
μ_0^{min}, [cm^{2}V^{-1} s^{-1}]	15.9	40
μ_0^{max}, [cm^{2}V^{-1} s^{-1}]	125	950
N^{crit}, [cm^{-3} V^{-1} s^{-1}]	1.76E19	1.94E17
α	−0.5	−0.5
β	−2.15	−2.40
δ	0.34	0.61

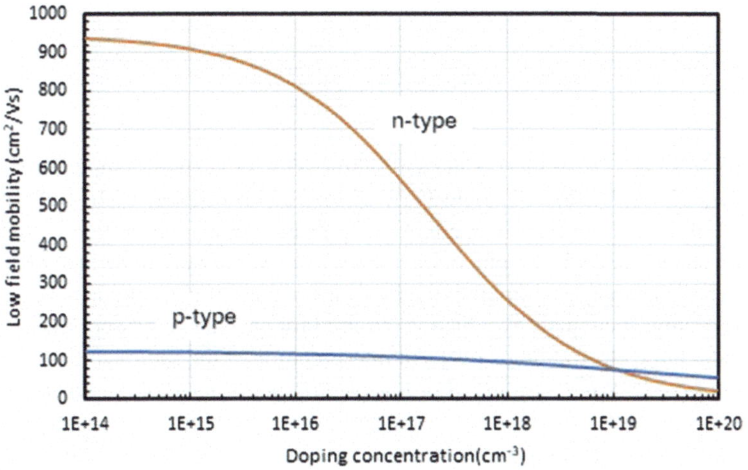

Fig. 2.14 Relationship between doping concentration and low-field mobility in the (0001) plane in n- (nitrogen—red line) and p- (aluminum—blue line)-doped 4H-SiC at room temperature

with N_{imp} as the total number of impurities (dopants and compensation impurities) in materials with specific parameters for p- and n-type conductivity materials [25] presented in Table 2.4 for the case ⊥ c-axis. The impact of doping concentration on mobility in n- and p-type-doped 4H-SiC materials is presented in Fig. 2.14. The crystallographic orientation in 4H-SiC influences mobility values; thus, mobility // c-axis is around ~20% more ($\mu\perp/\mu_{//} = 0.8$) [25].

The movement of carriers is affected by the presence of an electric field, which accelerates them, increasing their speed. However, there comes a point where their velocity reaches a saturation level due to the increasing strength of the electric field. To address this, it is essential to decrease the effective mobility, as the magnitude of drift velocity is determined by multiplying mobility by the component of the electric field aligned with the current direction (Eq. 2.7). The field-dependent mobility can be expressed as [28].

$$\mu_{p,n}^{high} = \frac{\mu_{p,n}}{\left[1+\left(\frac{\mu_{p,n}E}{\vartheta_{sat}(T)}\right)^{\alpha_v(T)}\right]^{\frac{1}{\alpha_v(T)}}} \tag{2.7a}$$

where

$$\vartheta_{sat}(T) = \vartheta_{sat}(300\,\mathrm{K})\left(\frac{T}{300}\right)^{\partial_{sat}} \tag{2.7b}$$

$$\alpha_v(T) = \alpha_v(300\,\mathrm{K})\left(\frac{T}{300}\right)^{\beta_{sat}} \tag{2.7c}$$

and ϑ_v(300 K) = 2.2*e*7 (cm/s) is the saturation velocity at room temperature, ∂_{sat} = −0.44, β_{sat} =1, α_v(300 K) =1.2, and *E* the electric field component aligned with the direction of the current flow. The effect of the electric field on mobility degradation is presented in Fig. 2.15. Despite being commonly overlooked in the face of parameters such as bandgap and breakdown field, the saturated electron drift velocity is also a very important quantity which is required for analysis of power devices operating under large electric fields [29]. An interpretation of the saturated drift velocity treatment from [30] is given here. For low electric fields, most of the kinetic energy gained during drift is transferred to the lattice via phonons or impurities, which serves to keep the electron distribution and the lattice at the same temperature; as

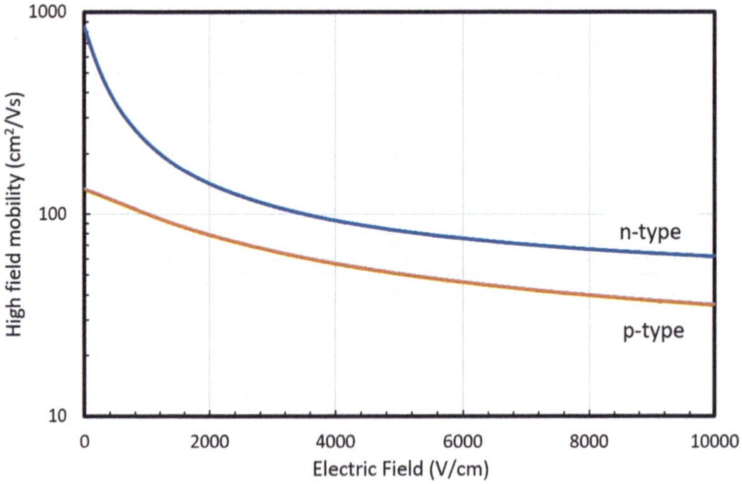

Fig. 2.15 Relationship between electric field and mobility in the (0001) plane in n- (nitrogen—red line) and p- (aluminum—blue line)-doped 4H-SiC at room temperature

such, the average energy of electrons remains close to the equilibrium value $(3kT/2)$, and the drift velocity bears a linear relationship to the field as seen in Eq. (2.8):

$$\upsilon_D = \mu E \tag{2.8}$$

For moderate fields, the energy transfer via phonons and impurities becomes less effective, and drifting electrons gain energy more quickly than it can be transferred to the lattice; the drift velocity and electric field are no longer linearly related. Finally, at high fields, the electrons have gained enough energy such that it can now be transferred through higher-frequency optical phonons. Because this is an efficient energy transfer mechanism, the drift velocity reaches a maximum value that is not affected by increasing the field further; this value of drift velocity is known as the *saturated drift velocity*. The importance of υ_{sat} is that it is the maximum velocity that may be attained under high electric fields—in other words, at high applied voltages common for power microwave electronics applications. The fact that υ_{sat} for SiC is twice that of conventional Si and GaAs means that carriers are swept into and out of devices twice as quickly, resulting in higher-frequency switching—all at higher voltage and power levels.

Another important parameter is the minority carrier lifetime that affects the performance of devices. In 4H-SiC, which is an indirect bandgap semiconductor, the dominant recombination-generation process is the Shockley-Read-Hall (SRH) recombination, which is determined by the type and density of recombination defects as well as the dopant concentration. While dependence of minority carrier lifetime in SiC on impurity concentration is not well established, the Scharfetter empirical model proposed for a wide range of semiconductors can be utilized to estimate the dependence of carrier lifetime on doping concentration and temperature for 4H-SiC [25]:

$$\tau_{e,h} = \frac{\tau_{e,h}^{\max}}{1+\left(\frac{N_{imp}}{N_{ref}}\right)^{\alpha_{SRH}}}\left(\frac{T}{300}\right)^{\beta_{SRH}} \tag{2.9}$$

where N_{ref}, α_{SRH}, β_{SRH} are the fitting parameters and $\tau_{e,h}^{\max}$ is the maximum carrier lifetime values of electrons or holes reported in low-doped p- or n-type-doped material. Thus, commercially as-grown 4H-SiC epitaxial layers have lifetime values

Table 2.5 Fitting parameters used in Eq. 2.9 for evaluating SRH lifetime dependence on doping and temperature in 4H-SiC

Parameter	4H – SiC	
	n-type	p-type
$\tau_{e,h}^{max}$	6E-6	2E-6
N_{ref}, [cm^{-3}]	5E15	5E15
α_{SRH}	0.67	0.55
β_{SRH}	1.72	1.72

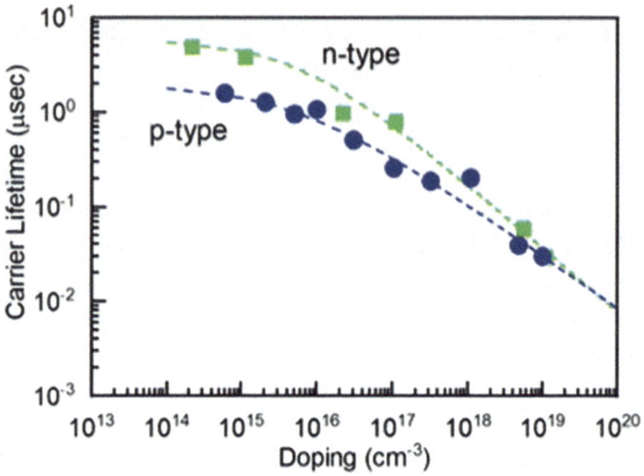

Fig. 2.16 Relationship between minority carrier lifetime (te,h) and total doping concentration (N_{imp}) at room temperature in n-type and p-type materials. Experimental data points adopted from [31]

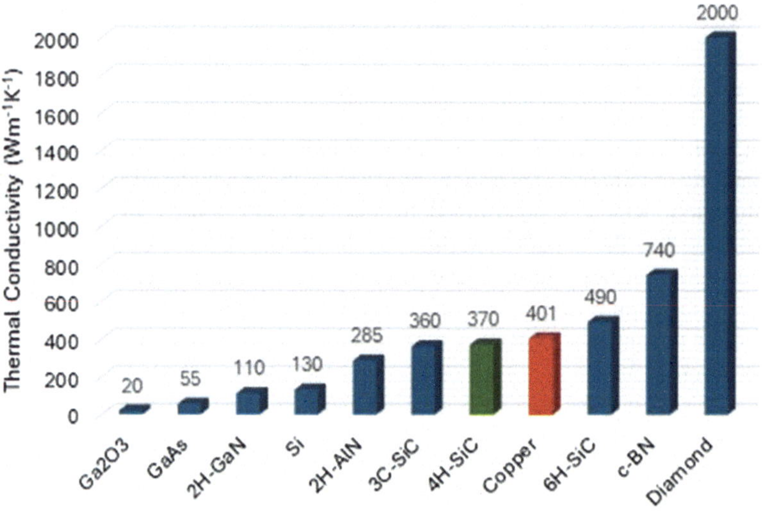

Fig. 2.17 Thermal conductivity of several semiconductor materials, with copper metal shown for comparison [6, 12, 21]

around 1-2 µs, though significant improvements in the lifetime of n-type-doped epitaxy were reported after annealing via the "carrier lifetime enhancement process" [31]. Experimental data reported in [31] for n- and p-type 4H-SiC subjected to the "carrier lifetime enhancement process" were adopted to generate fitting parameters (Table 2.5.) in Eq. 2.9. Figure 2.16 displays the fitting curves in conjunction with the experimental data reported in [31].

Silicon carbide sets itself apart from traditional semiconductors such as Si, GaAs and wide-bandgap competitors GaN and Ga_2O_3 through its superior thermal conductivity, comparable with metal, and only surpassed by diamond and c-BN (Fig. 2.17) [32]. Such high thermal conductivity enables SiC to effectively channel heat away from areas of intense heat generation. This capability is essential for electronic devices that produce substantial heat, such as those found in power electronics. The adept handling of heat by SiC components not only boosts their reliability but also extends their operational lifespan. The thermal conductivity of 4H-SiC exhibits slight variation between the a-axis and c-axis directions and vary with conductivity type. Lower thermal conductivity was reported for N-doped material (280 Wm^{-1} K^{-1}) in comparison with intrinsic SiC [34]. In general, the thermal conductivity of α-SiC (hexagonal type) as the function of temperature can be fitted through [28].

$$k(T) = \left(A_k + B_k + C_k + T + T^2\right)^{-1} \tag{2.10}$$

where $A_k = 2.5E\text{-}5$, $B_k = 2.75E\text{-}6$, $C_k = 1.3E\text{-}8$ [mKW^{-1}].

Another important parameter is the heat capacity of a semiconductor which determines how much heat energy the semiconductor can absorb or release while maintaining its temperature. This property is significant because it provides information on how quickly semiconductors heat up or cool down in response to changes in operating conditions, such as increased power consumption or ambient temperature. In case of α-SiC, heat capacity as a function of temperature could be approximated as [28].

$$c(T) = A_c + B_c \bullet T + D_c \bullet T^{-2} \tag{2.11}$$

where $A_c = 1026$, $B_c = 0.201$, and $D_c = -3.66E7$ [Jkg^{-1} W^{-1}]. The thermal conductivity decreases monotonically with the increase of temperature at temperatures above 300 K.

Resilience of SiC at elevated temperatures allows devices crafted from this material to perform reliably in harsher environments without a drop in efficiency (Fig. 2.18). The lower thermal resistance of SiC-based devices translates to less stringent cooling needs, which is advantageous for energy efficiency. This aspect is especially critical in power electronics, where reducing energy wastage is a key objective. Another important aspect of SiC is its excellent thermal shock resistance. Due to its high thermal conductivity and low thermal expansion coefficient (~5.1 ppm/K at room temperature [34]), 4H-SiC has excellent resistance to thermal shock, meaning it can withstand rapid changes in temperature without cracking or failing. Also, SiC has thermomechanical match to the most package materials as shown in Fig. 2.19, better than Si. Like its other properties, it has been shown that the coefficient of thermal expansion (CTE) and thermal conductivity for 4H-SiC exhibits slight variation between the a-axis and c-axis directions [34, 35] (Fig. 2.18).

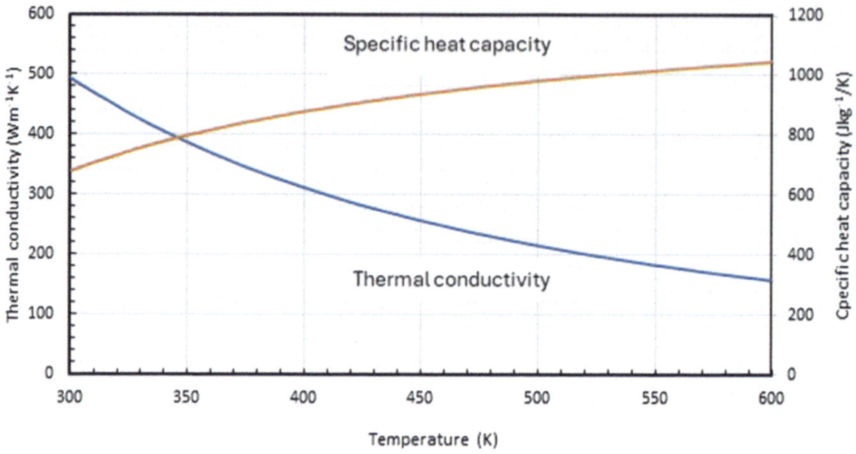

Fig. 2.18 Thermal conductivity and heat capacity of α-SiC as a function of temperature

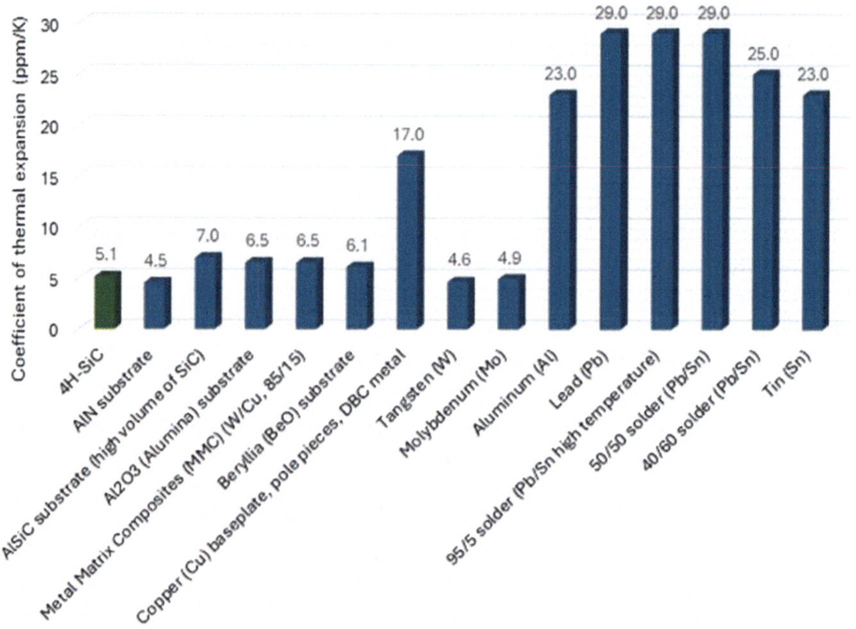

Fig. 2.19 Coefficient of thermal expansion for 4H-SiC and electronics packaging materials [28]

2.3 Substrates and Epitaxial Growth

Filippo Di Giovanni

In general, semiconductor boules are grown by crystal pulling or seeded solidification from melts consisting of elemental (Si and Ge) or compound semiconductors such as III–V (InP, InAs, GaAs, GaN, and InSb) and some II–VI formed by group IIB metallic elements Cd, Zn, and Hg with group VI nonmetallic elements (O, S, Se, and Te), the latter being extensively used in thin-film photovoltaics, nano-photodetectors, and lasers. Unfortunately, thermodynamic considerations cause these approaches to be impractical at present for the production of industrial quantities of monocrystalline SiC.

Bulk crystal growth is the principal process for producing monocrystalline wafers, the base material for SiC component fabrication. The latest advance in SiC device development is based on the availability of relatively large SiC wafers in 4″ and 6″, (8″ in pre-production stage) with adequate quality. As of today, the standard and mature technique for SiC bulk growth is PVT (physical vapor transport) *seeded sublimation*, referred to as the modified Lely method, introduced by J.A. Lely for the first time in 1955. Sublimation is the direct phase transition from solid to gas.

Figure 2.20 shows the phase diagram of the SiC binary system. There is no SiC liquid phase, so it is impossible to employ congruent melt growth for SiC bulk growth at practical values of system pressures. SiC sublimes at very high temperatures, above 1800–2000 °C and this represents the key step [21–24].

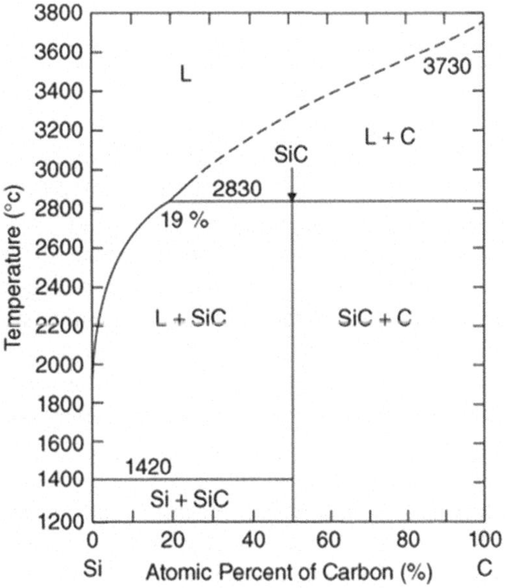

Fig. 2.20 Phase diagram of SiC [38]

The phase diagram indicates that up to 15% of carbon can be dissolved in a Si melt at around 2700 °C. Sublimation growth of SiC consists of three steps: (a) sublimation of the SiC source, (b) mass transport of sublimed species, and (c) surface reaction and crystallization. In the gas phase, the dominant species are not stoichiometric SiC molecules but Si_2C and SiC_2 molecules and atomic Si.

A brief description of the apparatus for the Lely method is described here. In essence, the SiC source is placed along the inner walls of a cylindrical graphite crucible. The source material is normally SiC powder produced via the Acheson process (see chapter on Crystal Structure of Silicon Carbide). By heating the crucible to the temperature of about 2500 °C, the SiC source sublimes and is transported to the inner part of the crucible. Under this nearly isothermal condition, many SiC platelets nucleate randomly along the vapor transport paths in the cavity. Nucleation refers to the initial process that occurs in the formation of a crystal from a solution, a liquid, or a vapor, in which a small number of ions, atoms, or molecules become arranged in a pattern characteristic of a crystalline solid, forming a site upon which additional particles are deposited as the crystal grows. Nucleation is, therefore, a self-arranging process that leads to a new thermodynamic phase or a self-assembled structure. Nucleation processes can be either heterogeneous or homogeneous, the latter being rare. In the former, the surface of some different substance, such as an impurity particle, a grain boundary, or the wall of the container, acts as the center upon which the first atoms, ions, or molecules of the crystal become properly oriented; in homogeneous nucleation, a few particles come into correct proximity in the course of their movement fluctuations through the bulk of the medium. Homogeneous nucleation thus occurs spontaneously and randomly, but it necessitates superheating or supercooling of the medium.

The grown SiC platelets are of very good quality and the typical dislocation density of good platelets is only several 100 cm^{-2}. However, the platelets are very small and irregular in shape, with typical areas of 1–2 cm^2 and thickness of 0.3–0.5 mm. The platelets' polytype is mainly 6H-SiC, but occasionally 4H- or 15R-SiC polytypes are also mixed in. Although these SiC platelets are of good quality, they cannot be used for device development. They can instead be reutilized as seed crystals for the early stage of the seeded sublimation growth, as explained below [39].

The modified Lely method illustrated in Fig. 2.21 consists of the placement of a seed crystal at a slightly cooler place inside the crucible. The SiC source (in the form of powder or sintered polycrystalline SiC) is placed at the bottom of a cylindrical dense graphite crucible, and a SiC seed crystal is placed near the lid of the crucible. The seed crystal is a monocrystalline silicon carbide slice that initiates the growth of a larger monocrystal, hence the term "seed." Like the seed of a plant, this crystal passes on important "genetic information" to the growing crystal. The distance between the top of the SiC source and the seed crystal is around 30 mm. The crucible is heated, mainly by a radio frequency (RF) induction coil, to 2300–2400 °C. The crucible is thermally insulated by graphite *felt* (an ideal material for vacuum furnaces or process temperatures above 2000 °C with ash value <20 ppm) or with porous graphite. By choosing an appropriate frequency, direct heating of this insulation can be avoided. The seed temperature is set at about 100 °C

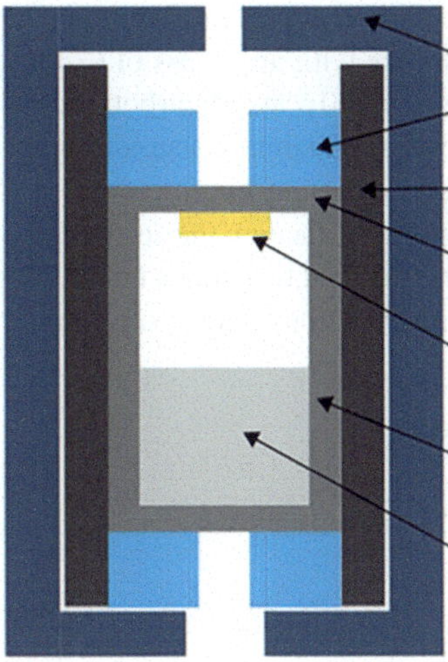

Fig. 2.21 Modified Lely method crucible for bulk SiC growth

lower than that of the source, so that sublimed SiC species condense and crystallize on the seed. Growth is usually performed at low pressure to enhance the mass transport from the source to the seed. A high-purity inert gas such as argon (or helium) is employed for growth [40].

The growth rate in sublimation growth is mainly determined by the flux of the source supply (sublimation rate) and the transport efficiency from the source to the seed. The sublimation rate is a function of the source temperature, while the transport efficiency depends heavily on the growth pressure, the thermal gradient, and the distance between the source and the seed.

Because the mass transport is diffusion-limited in sublimation growth, the growth rate is almost inversely proportional to the growth pressure. In other words, as the pressure is reduced, the vapor diffusion rate increases, and constituents move faster along the concentration gradient from the source toward the seed. Here, the concentration gradient is basically determined by the source and seed temperatures (temperature gradient). Experimentally, a growth rate of 150 mmh^{-1} is possible with a pressure of 1 Torr (or 1/760 atm) and 2000 °C of source temperature; at the same pressure, but with a source temperature of 2125 °C, the rate can be increased to 1000 mmh^{-1}.

Some precautions must be taken, however. The seed temperature must be controlled to minimize the loss of Si from the surface by maintaining the overpressure of the Si vapor near the seed. The source temperature and pressure must be controlled to develop the appropriate temperature gradient and to ensure that the proper amounts of Si and SiC compounds are transported from the source to the seed. Any fluctuations in the temperature profile and pressure can result in the buildup of Si droplets, surface graphitization, and C inclusions. All these phenomena lead to the formation of macro- and micro-defects in SiC boules.

The main species moved from the SiC source to the seed are Si, Si_2C, and SiC_2 at a growth temperature of 2300–2400 °C. Thus, the gas phase in sublimation growth is usually rich in Si because of preferential evaporation of Si from the SiC source. This leaves the source with an excess of C, and graphitization of the source is the result during the growth. To avoid carbon inclusions in the growing crystal, silicon is added to the source to maintain a stoichiometric or Si-rich source surface. This is important because sufficient overpressure of Si is also required to avoid graphitization of the growing surface on the seed. Carbon evaporation and transportation in the growth can be reduced by using a tantalum-carbon coating. Considering these phenomena, the main reactions during sublimation growth are summarized as follows:

1. $Si_2(g) + SiC_2(g) \leftrightarrow 3SiC(s)$.
2. $SiC_2(g) + 3Si \leftrightarrow 2Si_2C(g)$.
3. $Si_2C(g) \leftrightarrow 2Si_2(g) + C(s)$.
4. $Si(g) \leftrightarrow Si(l)$.

For the fabrication of vertical devices such as MOSFETs, low-resistivity wafers are required to minimize the series resistance, while high-resistivity wafers are desired for fabrication of lateral high-frequency devices, to reduce the parasitic impedance. In the sublimation growth of SiC, nitrogen and aluminum are the dopants of preference for growing n- and p-type boules, respectively. Semi-insulating SiC boules can also be obtained. However, the purity of SiC boules grown by the sublimation method should be carefully considered. The purity of SiC boules grown by sublimation is strongly dependent on the purity of the SiC source and graphite parts. The nature and amount of impurity atoms (i.e., undesired dopants) vary for different manufacturers; typical impurities include Ti, V, Cr, Fe, Co, Ni, and S. The density of these metallic impurities is in the range of 10^{13}–10^{15} cm^{-3}. The density of these impurities in the grown boules is usually lower than that of the source by a factor of 2–100.

2.3.1 High-Temperature Chemical Vapor Deposition (HTCVD) Growth of Bulk SiC

Currently, a PVT method is the most common technique for producing 4H-SiC wafers. One of the issues in this method is the limited supply of source materials due to the difficulty of adding source materials while the growth process is taking place. Moreover, the source materials tend to become carbon rich as growth proceeds since saturated vapor pressures of Si and SiC compounds are different, while varying process conditions during growth can cause a worsening of the crystal quality. Because of these reasons, 4H-SiC bulk crystals grown by the PVT method are limited to a certain crystal length. To overcome the limitations of the sublimation method of SiC, high-temperature chemical vapor deposition (HTCVD, also used in epitaxy) was proposed and developed.

In the HTCVD method, the SiC boule growth is performed in a vertical crucible made of graphite, where the source materials are in the form of precursor gases that are fed upward through a heating zone to the seed crystal stand placed at the top. The precursor gases are SiH_4 and a hydrocarbon, such as C_2H_4 and C_3H_8, diluted in a carrier gas. The geometry is similar to that of a vertical CVD reactor used for epitaxial growth, but the typical growth temperature is extremely high, 2100–2300 °C. Inside the hot zone, the precursors are completely decomposed, with several reactions taking place. As a result, Si and SiC clusters are formed via homogeneous nucleation because of the high supersaturation in the gas phase. These clusters act as a virtual source for SiC boule growth on the seed. Therefore, it is important to establish an appropriate temperature gradient from the gas inlet to the seed. Other advantages include (i) a higher source purity with the residual impurity density reduced by two orders of magnitude and (ii) a control of the C/Si ratio as an independent process parameter which is beneficial for enhancing micropipe closing and reducing intrinsic point defects.

2.3.2 *Epitaxial Growth and Doping of SiC*

Epitaxial growth is used to produce active layers of SiC-based device structures with desired doping density and thickness, because control of doping and thickness in bulk growth is difficult. In general, epitaxial growth (or epitaxy) refers to the growth of thin solid layers on a surface from a chemical reaction with well-controlled quality and thicknesses. The template used for the epitaxial growth of semiconductor materials consists of a substrate, and both homo-epitaxy (substrate and epilayer are of same materials) and hetero-epitaxy (substrate and epilayer are of different materials) are currently used in the crystal growth technology.

As mentioned earlier, unlike other semiconductors, silicon carbide has no liquid phase, so the only method to realize SiC wafers implies the use of gaseous phases. Homoepitaxial growth of SiC is one of the most important processes in the fabrication of SiC devices. Although various methods such as molecular-beam epitaxy, liquid phase epitaxy, sublimation epitaxy, and vapor-liquid-solid method have been attempted in SiC epitaxial growth, chemical vapor deposition (CVD) is still the leading and the most effective technique to grow thick and high-quality epitaxial SiC layers. In CVD, a layer is deposited on a solid monocrystalline surface starting from a gaseous component that decomposes on the surface of the substrate. A proper epitaxy should yield precise thickness of the crystal layers, with the right doping level and homogeneity. During CVD, monosilane (SiH_4) and propane (C_3H_8) or ethylene (C_2H_4) are usually used as the precursors and can be diluted into a carrier gas such as hydrogen (H_2) and argon (Ar).

Researchers have proposed several models of reactors with horizontal and vertical flow of the gases and different designs of the susceptors including nonrotating and rotating substrates and multi-wafer with planetary rotation with hot-wall or cold-wall configurations. The hot-wall reactor looks superior because of its accrued

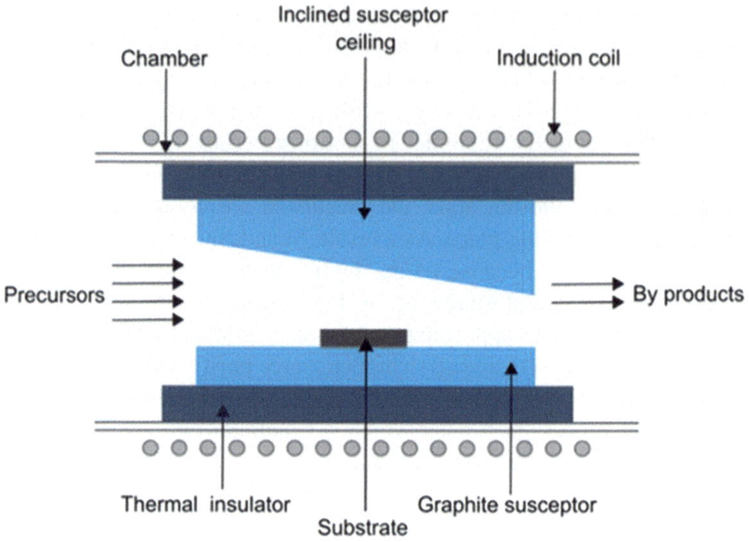

Fig. 2.22 Schematic of a CVD reactor

efficiency, uniform heating due to reduced temperature gradients, and more effective transformation of precursors into lighter products. CVD science and technology include specific knowledge in various fields of science and engineering, namely, precursor and growing material chemistry, thermodynamics, technological plasmas, fluid dynamics, radio frequency (RF) power supply concepts, vacuum technology, and mechanical design.

The typical growth temperature and growth rate are 1500–1650 °C and 5–15 mmh^{-1}. The CVD growth process for SiC usually consists of (a) in situ etching and (b) main epitaxial growth step. The in situ etching is performed with pure H_2, HCl/H_2, hydrocarbon/H_2, or SiH_4/H_2 at very high temperature, typically the same temperature as that of the main growth. The purpose of in situ etching is to remove the below-surface damage and to obtain regular step structures. Immediately after the etching, the main growth of n-type or p-type SiC (or their multilayers) is performed.

We have seen that an important aspect of the quality of a crystal is related to the presence of lattice defects that adversely impact the operation of devices manufactured from that crystal. This is due to the introduction of new electron levels into the bandgap which, in turn, interfere with carriers in both the valence and conduction bands. To obtain high-quality materials, it is imperative to use a substrate with the lowest defect density and an epitaxial process capable of reducing the concentration of defects in the substrates. Figure 2.22 shows a schematic of a hot-wall CVD reactor [26, 27, 29].

A common method for high-temperature heating is to use an RF (radio frequency) induction coil that contains a slab of graphite that is heated by the induced alternating current generated by an alternating magnetic field. The substrate is laid on the heated piece called the susceptor. Many steps, each one of them well controlled by optimized reactor geometry, are involved: mass transport of precursors and reactants, adsorption/desorption on the surface with consequent diffusion to growth sites, and finally, mass transport for removing by-products. Instead of using separate precursors whereby one species provides silicon Si, and the others supplies carbon C, single-source precursors such as tetramethylsilane [$Si(CH_3)_4$] have been successful.

In silicon semiconductors, doping can be achieved through high-temperature diffusion steps. In the case of SiC, however, only ion implantation followed by annealing is possible. Because the high energy associated with ion implantation causes damage in the crystal (point defects), another doping method is to add a third precursor during epitaxial growth (in situ doping) to get a large range of concentrations for both n-type and p-type (10^{14}–10^{19} cm^{-3}). This is rather rare for other wide-bandgap semiconductors such as GaN for which p-type doping is difficult. An important factor in SiC is the deposition across the reactor walls which depletes the gas from precursors. This effect is important to accurately estimate growth rates [31, 32, 34, 35].

2.4 Substrates: Understanding the Foundation of Semiconductor Device

Ravi Bollina, William Gemmill, Terry Knight, Christian Jentgens, and Helge Willers

The manufacture of semiconductor devices is a modern engineering and chemical engineering breakthrough that combines the precision of atomic-level structures with the robustness required for everyday electronic devices. At the heart of this process is the preparation of the substrate, the base material on which the devices are built. This chapter looks at the complex process of silicon carbide wafer substrate preparation, focusing on silicon carbide (SiC)—a material that has excellent properties for high-power and high-frequency applications. We will examine the step-by-step process of SiC wafer substrate preparation, including ingot growth and wafer processing and epitaxial growth, to illustrate the technical refinement involved in producing the base layer for semiconductor devices.

2.4.1 Why SiC Wafer Costs Matter

As opposed to silicon, the sublimation process to produce SiC requires significant energy to reach 2200 °C, while the final usable boule is no more than 25 mm in length, and growth times are very long. The consequence of this is a factor of

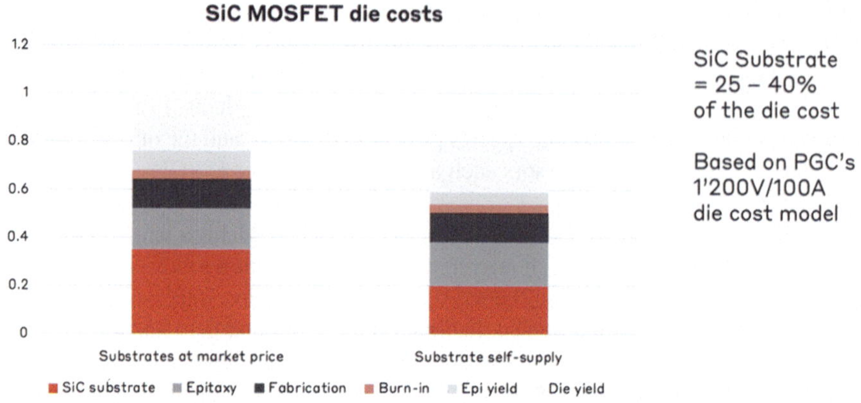

Fig. 2.23 The cost of the SiC substrate, which comprises the bulk part of the die cost, is a critical factor in reducing overall die expenses. [Source: PGC consultancy. Gammon, P. (October 19, 2021). Taking stock of SiC Part 1: A review of SiC cost competitiveness and a roadmap to lower costs. PGC Consultancy. https://www.pgcconsultancy.com/post/taking-stock-of-sic-part-1-a-review-of-sic-cost-competitiveness-and-a-roadmap-to-lower-costs]

30–50× increase in the cost of a SiC wafer compared with silicon (Si). Other costs such as epitaxy (growth of a high-quality layer of the SiC device on the surface of the substrate) and the fabrication costs are low compared to the SiC substrate cost. However, they still require higher temperatures and more expensive consumables than processing Si power devices.

Another fact to consider, as PGC Consultancy points out, is efficiency at all stages of manufacture. Yield is related to the number of unusable wafers taken from the boules and the number of unusable die after epitaxy and fabrication. A major factor affecting die yield after fabrication is the quality of the material, which includes a whole range of defects that lead to non-fabrication of the device as well as other less obvious defects that affect reliability. Another problem in fabrication, as PGC Consultancy indicates, is the reliability of the gate oxide of the MOSFET.

Shown in Fig. 2.23 is the cost breakdown for a SiC MOSFET die, with the yield derived from a best-in-class 2021 commercial 1200 V/100A device produced on a 150 mm substrate. The two stacks approximate two different substrate supply models, comparing the majority that buy their substrates at market prices to those companies that are vertically integrated and hence can self-supply the substrates at cost.

2.4.2 *SiC Ingot Growth: The Starting Point*

The journey of substrate preparation begins with the growth of SiC ingots. SiC is favored for its exceptional thermal conductivity, wide bandgap, and high electric field breakdown strength. The ingot growth process typically involves a method

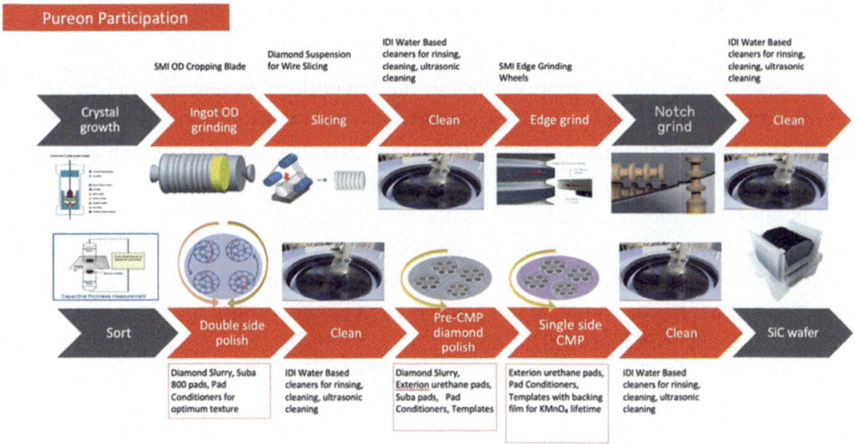

Fig. 2.24 Illustration of the legacy SiC wafer process. (Pureon AG, 2024)

known as physical vapor transport (PVT) where SiC powder is sublimated at high temperatures and then condensed on a cooler seed crystal in a controlled environment. The quality of the seed crystal is paramount, as imperfections can propagate through the ingot. The controlled temperature gradient, pressure conditions, and growth atmosphere are meticulously maintained to ensure uniform crystal growth and to minimize defects such as micropipes and dislocations, which can severely impact device performance. This process gradually builds up a large, single-crystal SiC ingot, which is the precursor to individual wafers. The quality of the boule determines the quality of the final wafer [21, 46–48].

2.4.3 *Shaping the Ingot: Grinding and Sawing*

Once the SiC ingot is grown, it undergoes outside grinding to achieve the desired cylindrical shape and remove any surface irregularities. It is about ensuring a uniform, defect-free surface. This step is crucial for the subsequent precision-cutting processes. Following the grinding, the ingot is sliced into thin wafers using a wire saw. Loose abrasive sawing, while potentially offering smoother surfaces, requires careful management of the slurry to avoid contamination and surface damage. The choice between the two depends on the desired surface quality and efficiency of the process.

A legacy SiC wafering process looks like the one shown in Fig. 2.24. The process begins with the wire saw slicing using a multi-wire saw slicing machine and then continues to lapping and polishing using diamond slurries known as diamond mechanical polish (DMP) and finally the CMP process [49–51].

2.4.4 Multi-wire Sawing Process for Silicon Carbide

A single wire is fed from a supply spool through a pulley and tension control unit to the four wire guides. These guides are grooved with a constant pitch, and by winding the wire around them, a net of multiple wire strands called a wire web is created within the 500–700 parallel grooves. A take-up spool collects the used wire. The wire is pulled by the torque exerted by the main drive and slave drive. The tension on the wire is maintained by the feedback control unit. To get a prescribed value, the silicon carbide boule on the holder is pushed against the moving wire web and sliced into hundreds of wafers at the same time. The wire either moves in one direction or oscillates back and forth.

A material is continuously removed through the interaction of the diamond particles below the moving wire and the silicon carbide surface. The abrasive action of the diamond particles depends on many factors such as wire speed, the force between the wire and crystal, the solid fraction of diamond in the suspension, the viscosity of the suspension, the size distribution, and the shape of diamond particles. The viscosity of the slurry depends on the temperature and solid fractions of the particles. Since the temperature rises due to the cutting process, the suspension has to be cooled and the temperature controlled during the soil. The viscosity of the slurry also changes because of the continuous abrasion of silicon carbide and iron from the wire. This gradually deteriorates the abrasive action. The slurry has to be replaced or mixed with a new slurry after some time.

The kerf loss is determined by the diameter of the wire, the size distribution of the diamond particles, and the vibrations of the wire. The objective of efficient sawing is to slice with a high throughput with a minimum kerf loss and with the high quality of the resulting wafers.

A typical wire slicing process looks like the one shown in Fig. 2.25. A diamond wire sawing slurry such as WSG-56 from Pureon (https://pureon.com/products/

Process step	Slicing
Customer specifications	• Bow: < 10μm • Warp: < 20μm • TTV: < 15μm
OEM tools	PW WaferTec DW 288
General process parameters	• Wire rate: 25-30m/sec • Tension: 40-50N • Flow rate: 18-220ml/min
Pureon solution	Slurry: WSG-56/500 Mono 3-6
Benefits vs POR	• Eco friendly (no oil) • Improved geometry • Faster cutting times

Fig. 2.25 Wire saw slicing process conditions. (Pureon AG, 2024)

Range for Inputs

Parameter	Value
Wire Feed Rate	25-30m/sec
Slurry Flow	18-22L/min
Tension	40-50N
Rocking Angle	+/- 10-15°
Stack Length	150-300mm
Diamond Concentration	500cts.
Wire Diameter	120-160μm

Expected Outputs

Parameter	Value
Cycle Time	80-120hrs. for 150mm wafers
Average Bow	10-20μm
Average Warp	20-40μm
Average TTV	< 15μm
Lifetime of Slurry	5-9cycles
Kerf Loss	130-170μm

Fig. 2.26 Description of typical input parameters and outputs, commonly used in the semiconductor industry. (Pureon AG, 2024)

Fig. 2.27 Picture showing the inside of a double-sided polishing machine. Carrier rings on a profiled lapping pad prior to loading the silicon carbide wafers. (Pureon AG, 2024)

wafering-slicing-and-wire-sawing/wsg/) used to slice a SiC ingot provides optimum slicing with the wafer parameters suitable for the further processing steps such as diamond mechanical polish (https://pureon.com/products/lapping-and-polishing-slurries/) (Figs. 2.26, 2.27, 2.28, 2.29, 2.30, 2.31, and 2.32).

2.4.5 *Wafer Lapping: Surface Uniformity and Smoothness—Diamond Mechanical Polish*

The process of lapping is more than just an intermediate polishing step; it is a critical process for ensuring wafer flatness and uniform thickness, which are vital for the subsequent epitaxial growth. The lapping process, which uses a rotating plate and

Fig. 2.28 Wafers are loaded into a double-sided polishing machine. (Pureon AG, 2024)

Process step	**DMP1**
Customer specifications	• Bow: < 10 μm • Warp: < 20 μm • TTV: < 3-5μm
OEM tools	Wolters AC1200
General process parameters	• Plates: 0.9-1.1 m/s • Carrier: 3-6 orbits • Pressure: 1.8-2.5 PSI • Flow rate: 40-80 ml/min
Pureon solution	Slurry: OPW-21 or New generation slurry Pad: MHS15S
Benefits vs POR	• Reduced thinning time through optimized pad & slurry with lower cost of ownership

Fig. 2.29 Expectations of the DMP step for material thinning of SiC wafers. (Pureon AG, 2024)

Range for Inputs

Parameter	Value
Lower plate speed	0.9-1.1m/sec CCW
Upper plate speed	0.9-1.1m/sec CW
Carrier	3-6orbits/min
Flow rate	40-80ml/min
Down force	1.5-2.5PSI
Pad	MHS15S
Slurry	OPW-21/FG

Expected Outputs

Parameter	Value
Removal rate	0.4-0.8μm/min
TTV	< 3μm
Bow	10-20μm
Warp	20-40μm
Surface quality	< 3nm Ra
Typical cycle time	60-180mins
Pad lifetime	40-80 cycles

Fig. 2.30 Process inputs and the expected outputs from a DMP step using a Pureon diamond slurry on a Pureon lapping pad. (Pureon AG, 2024)

abrasive slurry, is carefully controlled to achieve a balance between material removal and surface finish. The uniformity achieved through lapping ensures that the epitaxial layers grown on the wafer will be uniform in thickness and free from defects, directly impacting the electrical properties and performance of the final devices.

2.4.6 CMP Polishing: The Final Touch for EPI-Ready Wafers

The transition to an "EPI-ready" wafer is marked by the final polishing step, employing chemical mechanical polishing (CMP). This process combines mechanical polishing with a chemical slurry that selectively removes a material, resulting in an atomically, ultrasmooth finish. This in turn is crucial for epitaxial growth, as any surface irregularities can lead to defects in the epitaxial layers, affecting device characteristics such as carrier mobility and breakdown voltage. The result is a wafer with minimal surface defects, ready for the deposition of epitaxial layers.

Chemical mechanical polishing (CMP) is a well-known technology that can produce surfaces with outstanding global planarization without subsurface damage. A good CMP process for silicon carbide (SiC) requires a balanced interaction between SiC surface oxidation and the oxide layer removal. The oxidants in the CMP slurry control the surface oxidation efficiency, while the polishing mechanical force comes from the abrasive particles in the CMP slurry and the pad asperity, which is attributed to the unique pad structure and diamond conditioning. To date, to obtain a high-quality as-CMP SiC wafer, the material removal rate (MRR) of SiC should only be a few micrometers per hour, which leads to significantly high operation costs. In comparison, conventional Si CMP has the MRR of a few micrometers per minute. To increase the MRR, improving the oxidation efficiency of SiC is essential. The higher oxidation efficiency enables the higher mechanical forces, leading to a higher MRR with better surface quality. However, the disparity on the Si-face and C-face surfaces of 4H- or 6H-SiC wafers greatly increases the CMP design complexity.

2.4.7 A Novel Approach to Reinvent the Wafering Process: Rapid Thinning Approach

In competition with traditional wire sawing and diamond mechanical polishing processes or grinding processes, a new rapid thinning process enters the market (see Fig. 2.31).

The rapid thinning process combines a novel pad technology with a diamond slurry for rapid thinning of the wafer right of the saw. The material removal can be as high as 1.5 μm/min and the surface roughness is less than 6 nm. The wafer now can go directly into the CMP process, therefore saving processing time (increasing the throughput) and lowering the cost of the final SiC wafer.

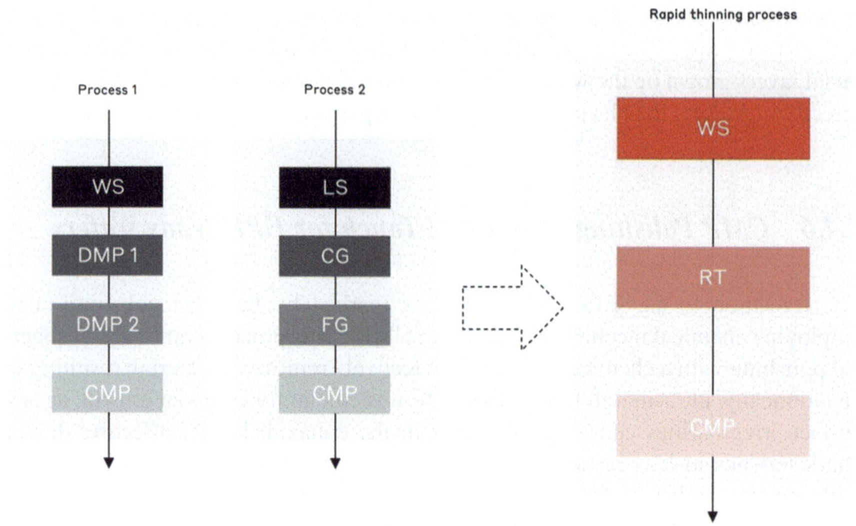

Fig. 2.31 WS (wire sawing), DMP1/DMP2 (diamond mechanical polishing), CMP (chemical mechanical polishing), laser slicing (LS), CG (coarse grinding), FG (fine grinding), and RT (rapid thinning). The current process of record for many SiC wafer manufacturers involves either process one or two Process 1 and 2). The rapid thinning approach is shown on the right side. (Pureon AG, 2024)

The cost of ownership for the 6-inch SiC wafers by rapid thinning lowers the cost of SiC wafers by almost 40% when compared to a grinding approach [52].

2.4.8 *Epitaxial Growth: Building the Device Layer*

Epitaxial growth on the prepared SiC substrate is the next critical phase, where thin layers of crystalline material are deposited to form the device's active regions. The quality of epitaxy directly influences device performance, making the purity of the epitaxial layer, the control of dopant levels, and the minimization of defects' essential goals. Chemical vapor deposition (CVD) is a common technique for SiC epitaxy, allowing for precise control over the growth parameters and incorporating dopants to modulate the semiconductor's electrical properties. The epitaxial layer's quality is paramount, as it forms the active region of semiconductor devices.

2.4.8.1 The Role of Dopants in Epitaxial Layers

Doping during epitaxial growth introduces specific impurities into the SiC to modify its electrical properties, a process that must be meticulously controlled. The choice of dopant, its concentration, and its distribution within the epitaxial layer are all critical factors that influence the semiconductor's behavior. For example, nitrogen and phosphorus can donate free electrons to the SiC, making it n-type, while boron and aluminum accept electrons, creating p-type SiC. The ability to control these properties allows for the engineering of complex device structures such as p-n junctions, MOSFETs, and Schottky diodes.

2.4.8.2 Challenges in Epitaxial Growth

Despite advances in technology, epitaxial growth on SiC substrates faces challenges. This includes defect propagation from the substrate, auto-doping from impurities in the SiC material, and the formation of epitaxial defects such as stacking faults and dislocations. Advanced growth techniques and post-growth treatments are continually being developed to address these issues, aiming to improve the crystalline quality of the epitaxial layers and, by extension, the performance of the SiC-based devices.

2.4.8.3 Schematic Overview of SiC and GaN Power Transistor Architectures

In the context of advanced semiconductor devices, Fig. 2.32 presents a schematic representation of a semiconductor structure utilized in commercial SiC and gallium nitride (GaN) power transistors. The SiC-based device includes the SiC vertical planar-gate metal-oxide-semiconductor field-effect transistor (MOSFET), showcasing the implementation of SiC in high-power settings due to its high breakdown voltage and thermal conductivity. The diagram encapsulates the structural intricacies that contribute to the performance characteristics of a device type, such as the gate design, the channel formation, and the source-drain configuration, which are

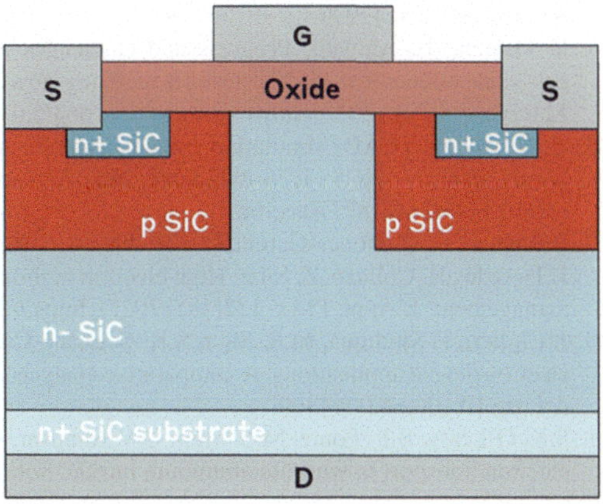

Fig. 2.32 Schematic representation of semiconductor structures used in commercial SiC and GaN power transistors. Here: SiC vertical planar-gate depletion MOSFET

critically dependent on the quality and preparation of the underlying wafer substrates and epitaxial layers discussed previously.

The preparation of SiC wafers for semiconductor device manufacture is a complex, multistep process that requires precision at every stage. From the growth of the ingot to the final polishing of the wafer, each step is crucial in ensuring the substrate's quality. In turn, this influences the performance of the final semiconductor devices. Each step is also associated with technological challenges that require precision, control, and a deep understanding of material properties. As the demand for high-power, high-frequency, and high-temperature semiconductor devices grows, the importance of these foundational processes becomes ever more critical. The ongoing research and development in this field promise not only to enhance the capabilities of current semiconductor technologies but also to pave the way for new innovations. The advancements in substrate preparation and epitaxial growth techniques continue to push the boundaries of what is possible in electronics, enabling devices that are more efficient, powerful, and capable of operating under extreme conditions (Figs. 2.26, 2.27, 2.28, 2.29, 2.30, 2.31, and 2.32).

References

1. K. M. Speer, The Silicon Carbide Vacuum Field-Effect Transistor (VacFET), Doctoral dissertation, Case Western Reserve University. OhioLINK Electronic Theses and Dissertations Center. http://rave.ohiolink.edu/etdc/view?acc_num=case1301445437 (2011)
2. J.M.A. Beattie, J.P. Goss, M.J. Rayson, P.R. Briddon, Structure and electron affinity of the 4H-SiC (0001) surfaces: A methodological approach for polar systems. J. Phys. Condens. Matter **33**(16), 165003 (2021). https://doi.org/10.1099/1361-648X/abf0be
3. O. Madelung, *Semiconductors: Data Handbook*, 3rd edn. (New York/Berlin/Heidelberg, Springer, 2003)
4. T. Kimoto, J.A. Cooper, *Fundamentals of Silicon Carbide Technology* (Wiley, Singapore, 2014)
5. M.E. Levinshtein, S.L. Rumyantsev, M.S. Shur, *Properties of Advanced Semiconductor Materials* (Wiley, New York, 2001)
6. J. Zhang, J. Shi, D.-C. Qi, L. Chen, K.H.L. Zhang, Recent progress on the electronic structure, defect, and doping properties of Ga2O3. APL Mater. **8**(2), 020906 (2020). https://doi.org/10.1063/1.5142999
7. Y. Oshima, E. Ahmadi, Progress and challenges in the development of ultra-wide bandgap semiconductor α-Ga2O3 toward realizing power device applications. Appl. Phys. Lett. **121**(26), 260501 (2022). https://doi.org/10.1063/5.0126698
8. H.Y. Wong, TCAD simulation models, parameters, and methodologies for β-Ga2O3 power devices. ECS J. Solid State Sci. Technol. **12**(5), 055002 (2023). https://doi.org/10.1149/2162-8777/accfbe
9. P. Bagheri, C. Quiñones-Garcia, D. Khachariya, S. Rathkanthiwar, P. Reddy, R. Kirste, S. Mita, J. Tweedie, R. Collazo, Z. Sitar, High electron mobility in AlN:Si by point and extended defect management. J. Appl. Phys. **132**(18) (2022). https://doi.org/10.1063/5.0124589
10. J. Chilleri, P. Siddiqua, M.S. Shur, S.K. O'Leary, Cubic boron nitride as a material for future electron device applications: A comparative analysis. Appl. Phys. Lett. **120**(12) (2022). https://doi.org/10.1063/5.0084360
11. S.K. O'Leary, B.E. Foutz, M.S. Shur, U.V. Bhapkar, L.F. Eastman, Monte Carlo simulation of electron transport in wurtzite aluminum nitride. Solid-State Comm. **105**(10), 621–626 (1998)

12. http://www.ioffe.ru/SVA/NSM/Semicond
13. N. Donato, N. Rouger, J. Pernot, G. Longobardi, F. Udrea, Diamond power devices: State of the art, modelling, figures of merit and future perspective. J. Phys. D. Appl. Phys. **53**(9), 093001 (2020). https://doi.org/10.1088/1361-6463/ab4eab
14. N. Izyumskaya, D.O. Demchenko, S. Das, Ü. Özgür, V. Avrutin, H. Morkoç, Recent development of boron nitride towards electronic applications. Adv. Electron. Mater. **3**(5), 1600485 (2017). https://doi.org/10.1002/aelm.201600485
15. K. Hirama, Y. Taniyasu, S. Karimoto, H. Yamamoto, K. Kumakura, Heteroepitaxial growth of single-domain cubic boron nitride films by ion-beam-assisted MBE. Appl. Phys. Express **10**(3), 035501 (2017). https://doi.org/10.7567/APEX.10.035501
16. D.F. Storm, S.I. Maximenko, A.C. Lang, N. Nepal, T.I. Feygelson, B.B. Pate, C.A. Affouda, D.J. Meyer, Mg-facilitated growth of cubic boron nitride by ion beam-assisted molecular beam epitaxy. Phys. Status Solidi (RRL) Rapid Res. Lett. **2200036**, 2200036 (2022). https://doi.org/10.1002/pssr.202200036
17. S.I. Maximenko, J.E. Moore, C.A. Affouda, P.P. Jenkins, Optimal semiconductors for 3H and 63Ni betavoltaics. Sci. Rep. **9**(1), 10892 (2019). https://doi.org/10.1038/s41598-019-47371-6
18. C. Persson, U. Lindefelt, Relativistic band structure calculation of cubic and hexagonal SiC polytypes. J. Appl. Phys. **82**, 5496–5508 (1997). https://doi.org/10.1063/1.365578
19. National Research Council, *Materials for High-Temperature Semiconductor Devices* (National Academy Press, Washington, D. C, 1995)
20. A. Agarwal, M. Das, S. Krishnaswami, J. Palmour, J. Richmond, S.-H. Ryu, SiC power devices—An overview. Mater. Res. Soc. Symp. Proc. **815**, J1.1.1 (2004)
21. T. Kimoto, Y. Yonezawa, Current status and perspectives of ultrahigh-voltage SiC power devices. Mater. Sci. Semicond. Process. **78**(August), 43–56 (2017)
22. https://www.globenewswire.com/news-release/2023/04/04/2640950/0/en/2023-Analysis-Report-8-inch-SiC-Wafer-Mass-Production-Plans-and-Production-Progress-of-Leading-Manufacturers-Featuring-Wolfspeed-and-Coherent.html
23. V. Bolotnikov, P.G. Muzykov, T.S. Sudarshan, Investigation of two-branch boron diffusion from vapor phase in n-type 4H-SiC. Appl. Phys. Lett. **93**(5), 052101 (2008)
24. U. Lindefelt, Doping-induced band edge displacements and band gap narrowing in 3C–, 4H–, 6H–SiC, and Si. J. Appl. Phys. **84**, 2628–2637 (1998). https://doi.org/10.1063/1.368374
25. M. Lades, Modeling and simulation of wide bandgap semiconductor devices: 4H/6HSiC, Ph.D Dissertation (T.U. Munich, 2000)
26. S.M. Sze, K.K. Ng, *Physics of Semiconductor Devices*, 3rd edn. (2006)
27. D.M. Caughey, R.E. Thomas, Carrier mobilities in silicon empirically related to doping and field. Proc. IEEE **55**, 2192–2193 (1967). https://doi.org/10.1109/PROC.1967.6123
28. T. Ayalew, SiC semiconductor devices technology, modeling, and simulation, Ph.D. dissertation (T.U. Wien, 2004)
29. B. Baliga, *Jayant, Modern Power Devices* (Wiley, 1987)
30. C.M. Wolfe, N. Holonyak, G.E. Stillman, *Physical Properties of Semiconductors* (Prentice-Hall, Englewood Cliffs, 1989)
31. T. Kimoto, H. Niwa, T. Okuda, E. Saito, Y. Zhao, S. Asada, J. Suda, Carrier lifetime and breakdown phenomena in SiC power device material. J. Phys. D. Appl. Phys. **51** (2018). https://doi.org/10.1088/1361-6463/aad26a
32. J. L. Hudgins, G. S. Simin, E. Santi, M. A. Khan, "An assessment of wide bandgap semiconductors for power devices," IEEE Trans. Power Electron., vol. 18, no. 3, pp. 907–914, May 2003, https://doi.org/10.1109/TPEL.2003.810840
33. https://doi.org/10.1063/1.4790134
34. J.J. Neumeier, Y.V. Shvyd'ko, D. Haskel, Thermal expansion of 4H and 6H SiC from 5 K to 340 K. J. Phys. Chem. Solids **187**, 111860 (2024). https://doi.org/10.1016/j.jpcs.2023.111860
35. X. Qian, P. Jiang, R. Yang, Anisotropic thermal conductivity of 4H and 6H silicon carbide measured using time-domain thermoreflectance. Mater. Today Phys. **3**, 70 (2017)

36. T. Kimoto, J.A. Cooper, *Fundamentals of Silicon Carbide: Technology, Growth, Characterization, Devices, and Applications* (Wiley, Singapore, 2014)
37. C.J.H. Wort, R.S. Balmer, Diamond as an electronic material (Element Six, Ascot)
38. T. Kimoto, Material science and device physics for high-voltage power devices. Jpn. J. Appl. Phys. **54**, 040103 (2015)
39. V.D. Heydemann, N. Schulze, D.L. Barrett, G. Pensl, *Growth of 6H and 4H Silicon Carbide Single Crystals by the Modified Lely Process Utilizing a Dual-Seed Crystal Method* (Institut für Angewandte Physik, Universität Erlangen-Nürnberg, Erlangen, 1996)
40. F. La Via, M. Camarda, A. La Magna, Mechanisms of growth and defect properties of epitaxial SiC. Appl. Phys. Rev. **1**, 031301 (2014)
41. R. Singh, Reliability and performance limitations in SiC power devices. Microelectron. Reliab. **46**, 713–730 (2006)
42. Y. Tokuda, E. Makino, N. Sugiyama, I. Kamata, N. Hoshino, J. Kojima, K. Hara, H. Tsuchida, Stable and high-speed SiC bulk growth without dendrites by the HTCVD method
43. M. Musolino, E. Carria, D. Crippa, S. Preti, M. Azadmand, M. Mauceri, M. Isacson, M. Calabretta, A. Messina, Development of n-type epitaxial growth on 200 mm 4H-SiC wafers for the next generation of power devices
44. T. Rana, M.V.S. Chandrashekhar, H. Song, T.S. Sudarshan, A comparative study of SiC epitaxial growth in vertical hotwall CVD reactor using silane and dichlorosilane precursor gases
45. J. Ul Hassan, Epitaxial growth and characterization of SiC for high power devices, Linköping Studies in Science and Technology Dissertation No. 1243
46. Y. Cui, Research on the growth and application of silicon carbide in China. Proc. SPIE (2022). https://doi.org/10.1117/12.2641064
47. C. Basceri, I. Khlebnikov, Y. Khlebnikov, P. Muzykov, M. Sharma, G. Stratiy, M. Silan, C. Balkas, Growth of micropipe-free single crystal silicon carbide (SiC) ingots via physical vapor transport (PVT). Mater. Sci. Forum **527–529**, 39–44 (2006). https://doi.org/10.4028/www.scientific.net/MSF.527-529.39
48. T. Kimoto, J.A. Cooper, *Fundamentals of Silicon Carbide Technology: Growth, Characterization, Devices, and Applications* (Wiley, 2014). https://doi.org/10.1002/9781118313534
49. M. Gola, Polishing of hard machining semiconductor materials made of silicon carbide. ResearchGate (n.d.). Retrieved from https://www.researchgate.net/publication/281499874_Polishing_of_hard_machining_semiconductor_materials_made_of_silicon_carbide
50. H. Jamil Noorina, W. Xian, W. Arif, C.P. Faizul, R.M. Zaki, Production of silicon carbide via grinding and heat treatment process. Key Eng. Mater. **594–595**, 740 (2013). https://doi.org/10.4028/www.scientific.net/KEM.594-595.740
51. I.S. Kainarskii, É.V. Degtyareva, I.P. Safronova, The method and degree of grinding as factors in the grinding and washing kinetics of silicon carbide. Powder Metall. Met. Ceram. **13**, 188–192 (1974). https://doi.org/10.1007/BF01286259
52. R. Bollina, PowerAmerica Annual Meeting, Raleigh, 16–19 2024 Mar

Kevin Speer is the technology director of Microchip Technology's silicon carbide (SiC) business unit. He leads the technology development for Microchip's SiC power solutions including device design, applications, advanced packaging, and materials. Since 1999, Dr. Speer has published a range of SiC research spanning materials science to applications, including crystal growth, defect-related device degradation, SiC MOSFET and diode device design and processing, and advanced power packaging. Dr. Speer has held strategic technical marketing roles with SemiSouth Laboratories, Infineon, and Littelfuse. Prior to Microchip, Dr. Speer founded Speer Semiconductor, a business aimed at empowering stakeholders across the value chain through product development and road-mapping, customer outreach, strategy and investment diligence, and market analysis. Dr. Speer holds a Bachelor of Science in Electrical Engineering (BSEE) from the University of Arkansas, a Master of Science in engineering from Case Western Reserve University (CWRU), and a Ph.D. in electrical engineering from CWRU as a NASA Research Fellow.

Yogesh Sharma works as a senior manager of strategic marketing in SiC with Microchip. He has been working in the field of wide-bandgap (WBG) semiconductors for more than 15 years. In the past, he worked with companies such as Dynex, Bosch, and Nexperia. Dr. Sharma held various positions in R&D, applications, technical marketing, and business development. Dr. Sharma started his integrated Ph.D. program at Auburn University in wide-bandgap semiconductors, SiC, and GaN. After completing his Ph.D. degree, he began working in wide-band semiconductor technology and other high-power semiconductor devices. He has authored and co-authored over 70 journal or conference papers with more than 1000 citations. He is the holder of multiple patents and acted as a reviewer for various international journals.

Sergey Maximenko is the senior technical staff engineer of Microchip Technology's silicon carbide (SiC) business unit. He leads the development of new SiC device processes and their integration into both new and established product platforms while also shaping Microchip's strategic direction in SiC materials. Previously, he held a position at the US Naval Research Laboratory where he developed cutting-edge technologies tailored for energy and space applications. He has co-authored more than 80 publications in the field of modeling, crystal growth, and characterization of semiconductor materials and devices. Dr. Maximenko earned his Ph.D. in electrical engineering from the University of South Carolina, specializing in high-power SiC electronics with a focus on material characterization and growth. He further enhanced his expertise through a postgraduate tenure at the US Naval Research Laboratory in Washington, D.C., as a recipient of the National Academy of Sciences Associateship award.

Filippo Di Giovanni has previously been with STMicroelectronics as marketing manager for power transistors and, more recently, also for wide-bandgap (WBG) products such as silicon carbide and gallium nitride, plus power modules. Doctor Di Giovanni's current interests span from power technologies to their automotive and Industrial applications. His working experience also includes attendance to conferences and workshops dedicated to power conversion and coordination activities for European-funded projects. At the end of 2023, after retirement, Filippo is a full-time technical writer about semiconductor technology and market analysis for professional magazines.

Ravi Bollina is a chief sales officer at Pureon and an executive board member. He is an accomplished professional in material science technologies and holds a Ph.D. in engineering science and mechanics from Penn State University. Dr. Bollina specializes in silicon carbide, gallium nitride, and wide-bandgap semiconductors. Beyond that, his expertise extends to semiconductor fabrication, diamond powder processing, and solutions.

William Gemmill is the chief innovation officer at Pureon where he is responsible for the development of new technologies and commercialization of new products. As an executive board member, he is accountable for the global innovation pipeline supporting Pureon's strategy. His career spans over 15 years of industry experience, formulating cleaning and etching chemistries for the semiconductor industry, polishing slurry formulations, and processing solutions for both the semiconductor and precision optics industries.

Terry Knight is a technical services manager at Pureon. With over 35 years of experience in the silicon wafers, disk drive, CMP (chemical mechanical polishing), and optics industries, he brings a wealth of knowledge and expertise to the field. Holding a B.S. in Electrical Engineering from Arizona State University, his expertise includes process knowledge, process optimization, and process recommendations for several of the manufacturing steps from slicing to final polish for substrates and device and optics applications.

Christian Jentgens is the head of research and development at Pureon where he is responsible for the development of new technologies and commercialization of new products. Among others, he is responsible for the invention and development of Pureon's innovative diamond-bonded SQUADRO-G grinding pads. His career spans over 22 years of industry experience, formulating consumables for the semiconductor industry, polishing slurry formulations, and processing solutions for both the semiconductor and precision optics industries.

Helge Willers is a sales director and product manager for grinding pads at Pureon. In his role, he is responsible for innovations as well as further development of grinding and composite polishing pads. He played an important role in the building of the Pureon product family of grinding pads and is a well-known expert for the precision, electronics, and semiconductor industry. His career spans over 27 years of industry experience.

Chapter 3
Manufacturing Processes

John R. Silk, Victor Veliadis, Dario Tenaglia, Paolo Badalà, and Simone Rascuna

In the rapidly evolving landscape of semiconductor technology, the fabrication of silicon carbide (SiC) devices stands at the forefront of innovation. Authored by distinguished experts in the field, This chapter of this book offers a comprehensive exploration into the intricate processes essential for manufacturing high-performance SiC semiconductor devices.

John R. Silk, Director of Quality and Epi Operations at SemiQ, Inc., opens the chapter with a detailed discussion on bulk growth and epitaxial growth techniques in SiC semiconductor device manufacturing. His expertise delves into the complexities of SiC substrate preparation and epitaxial growth processes crucial for achieving superior device performance.

Continuing the exploration, Dr. Victor Veliadis, Executive Director and CTO of PowerAmerica and ECE Professor at NC State, provides insights into ion implantation in SiC devices. His contribution highlights the challenges and innovations involved in optimizing SiC fabrication for power electronics applications.

Dario Tenaglia and Paolo Badalà further delve into the crucial processes of etching and metallization in SiC device fabrication, offering a comprehensive understanding of techniques, challenges, and advancements necessary for enhancing device performance and reliability.

Concluding the chapter, Simone Rascunà, Advanced Research Senior Manager at STMicroelectronics, provides an insightful overview of the entire SiC fabrication

J. R. Silk
SemiQ, Lake Foest, CA, USA

V. Veliadis (✉)
PowerAmerica, Raleigh, NC, USA
e-mail: jvveliad@ncsu.edu

D. Tenaglia · P. Badalà · S. Rascuna
STMicroelectronics, Catania, Italy

M. Di Paolo Emilio (ed.), *SiC Technology*, https://doi.org/10.1007/978-3-031-63418-5_3

process, from ion implantation to metallization. Rascunà's expertise sheds light on the intricate complexities of SiC technology, further enriching the reader's understanding of this cutting-edge field.

3.1 Bulk Growth and Epitaxial Growth Techniques

John R. Silk

Epitaxy creates the crystalline structure on the SiC substrate that establishes the desired material properties. Silicon carbide has many stacking configurations, which complicate epitaxy growth compared to the processes for making ordinary silicon semiconductor devices. The 4H polytype is most commonly used for power devices, while 6H is generally preferred for RF applications.

A high-quality substrate preparation is essential for epitaxial growth, which must be meticulously clean to ensure no contaminants are present before subsequent deposition. Complicating this challenge is the fact that silicon carbide ingots and wafers are difficult to work with. The material is extremely hard and must usually be cut with a diamond saw, although laser dicing could be adopted in the future. In addition, unlike silicon ingots that can be grown with close-to-perfect purity and then sliced into very uniformly flat wafers, SiC wafers incorporate many crystal defects and display great variation in flatness at local sites and across the wafer. The crystal defects reside in both the bulk and on the surface. They cannot all be removed, which makes some sites unusable for device fabrication.

3.1.1 SiC Substrate Polytypes and Purity

SiC semiconductor device production begins with growing the high-purity silicon carbide crystal to create the ingot (boule in SiC parlance) from which wafers are sawn (referred to as wafering) and subsequently prepared for epitaxy. Briefly, this boule is grown by a sublimation process using polycrystalline silicon carbide or a silicon/carbon powder mix as a feedstock. Heating to a high temperature in a vacuum or inert gas atmosphere causes sublimation. A seed crystal is introduced and the vapor is cooled causing SiC molecules to deposit on the seed crystal, forming a larger crystalline structure.

Unlike growing a silicon ingot, which is relatively straightforward, SiC crystals have many different stacking configurations. The crystal can form with hexagonal or cubic structures, giving rise to many different polytypes. The specific conditions under which the crystal is formed, including temperature, pressure, and any impurities present, determine which polytype will predominate. Some common configurations are 3C, 4H, and 6H, although over 200 are possible, having distinct electrical, thermal, and mechanical properties. Power semiconductors are typically fabricated using 4H polytype wafers. 6H polytype properties are suited to RF devices. The bandgap energy of 4H SiC is 3.26 eV, which compares with 3.03 eV for 6H SiC and 1.12 eV for silicon. This, combined with a relatively high breakdown field strength of 3.5MV/cm, compared to 3.0MV/cm for 6H SiC, allows the 4H polytype superior properties for handling high applied voltages. While 4H and 6H SiC have the same thermal conductivity of 4.9 W/mK, far superior to silicon's 1.31 W/mK, 3C SiC's conductivity is lower at 3.2 W/mK. SiC electron saturation velocities are at least double that of silicon, supporting superior high-frequency performance.

The SiC boule produced from the crystal formation process is sliced into wafers using diamond-encrusted wire saws, necessary due to the extremely hard nature of silicon carbide. The wafers then undergo grinding and extensive polishing to achieve a flat and smooth, ideally defect-free, surface. Typical wafer diameters are currently 150 mm and 200 mm.

Manufacturers of semiconductor devices source high-quality SiC boules or wafers from their preferred suppliers, specifying the required quality in terms of the tolerable limits for imperfections. No substrate is a perfect crystal; some imperfections resulting in areas that are unsuitable for device fabrication must be expected. Typically, the more exacting the specification, closer to perfection, the more expensive the supplied material. The following chapter will discuss this in more detail.

3.1.2 SiC Device Epitaxy

Epitaxy is a fundamental process in semiconductor device manufacturing, which is crucial for building layered structures. Its purpose is to grow a layer of uniform thickness and accurately controlled electrical properties as a suitable substrate for the subsequent device processing. In particular, this process grows crystalline layers where the deposited atoms maintain the same structural orientation as the underlying substrate. In this way, thin, single-crystal layers of a high quality are formed.

Epitaxial layers are typically grown using a chemical vapor deposition (CVD) process. Alternative techniques include sublimation, which requires higher process temperature than CVD and so can be more expensive, and molecular beam epitaxy (MBE), which is relatively slow.

In CVD, the epitaxy is grown on the silicon carbide wafer using a stepped flow process. The SiC substrate surface is treated to create steps and terraces. For optimal growth, the substrate, as taken from the boule, is cut off-axis by approximately 4° for the 4H polytype. This is a relatively low angle compared to early epitaxy processes and has been found to result in longer terrace lengths. While cutting the substrate off-axis tends to drive up costs, the shallower angle requires a higher processing temperature to ensure proper diffusion.

To perform the epitaxy, a precursor gas is introduced into the epitaxial growth chamber. These gases are silicon-based, such as silane +HCl or chlorosilanes (TCS), as well as carbon sources such as propane or halogenated hydrocarbons (CH, Cl).

Maintaining a controlled gas phase is essential for precise deposition. The carbon-to-silicon ratio is carefully managed for doping control.

The precursor gas atoms adsorb onto the SiC substrate and ideally diffuse along the surface until reaching a step. Under these conditions, chemical reactions occur that lock the atom in place. In this way, the crystal grows along the step edges. The desired layer thickness and crystal quality are achieved through careful control of the growth parameters such as the flow rates of the precursor gases and the temperature and pressure inside the chamber.

Doping of SiC layers is often necessary to tailor the electrical properties. Doping introduces specific elements to the silicon carbide crystal lattice. The dopant gases are typically nitrogen for *N*-type devices and aluminum for *P*-type devices. Precision in doping levels is crucial for semiconductor performance.

The precursor gas flow rate and doping composition are critical process parameters that govern the epitaxial growth. The growth rate is directly proportional to the flow of silicon precursor and can be increased by introducing chlorine through the addition of chlorosilanes, halocarbons, or HCl. On the other hand, the relative quantities of carbon and silicon precursors affect dopant incorporation. A ratio of about 1:3, carbon to silicon, is usually optimal, depending on the reactor.

After epitaxial growth, annealing processes may be applied to improve the crystal structure and remove defects.

3.1.3 Process and Equipment Types

The CVD reactor equipment must handle heating of the wafer as well as temperature control, introducing the precursors, managing the gas flow, and maintaining the correct pressure for CVD to occur. SiC device fabrication began using small, single-wafer reactors. As the devices have become commercialized, multi-wafer reactors have become available, capable of handling wafer diameters of 6 inches and more recently 8 inches. Single- and multi-wafer processing each have their advantages: single-wafer processing allows the settings to be highly optimized, resulting in high yield, although the processing cost is relatively high; multi-wafer processing, on the other hand can deliver economies of scale, although uniformity is much more difficult to achieve.

The reactor contains a heating system that comprises a heat source, such as an induction coil, with a susceptor designed to transfer heat efficiently and uniformly to the wafers. In a simple reactor, the walls may be unheated. Alternatively, heating the reactor walls and/or ceiling enables growth rates as fast as 100 μm per hour. Some reactors contain a rotating mechanism to turn the wafers in a planetary motion to ensure uniform heating.

The heating system raises the wafer temperature to 1600 °C for an etching process to be applied, which takes several minutes, and then further raised to 1650 °C for epitaxial layer growth. Maintaining the appropriate pressure within the reactor chamber is crucial for the growth process. This is often done under low-pressure

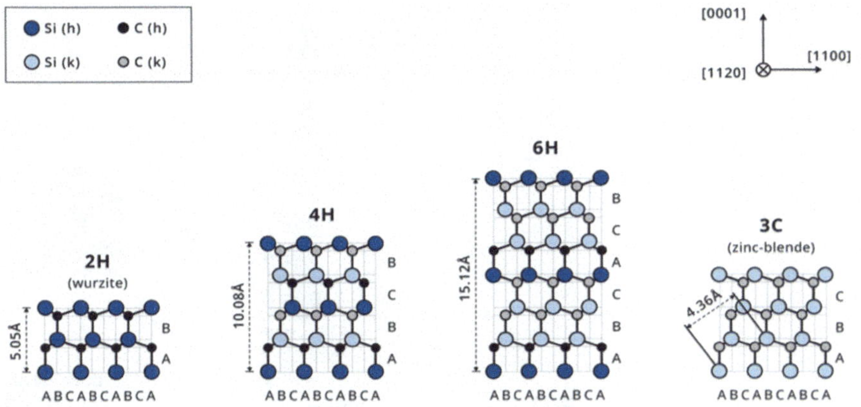

Fig. 3.1 Structures of SiC polytypes. 4H-SiC and 6H-SiC are best suited to power devices and RFICs, respectively

conditions, typically below 100 mbar, to facilitate controlled deposition. While CVD can be performed satisfactorily at atmospheric pressure, allowing simplified equipment design and lower operating costs, low-pressure CVD (LPCVD) can allow precise control of the deposition parameters delivering advantages such as greater thickness uniformity and purity, reduced defects, and superior step coverage.

After deposition, the wafer may go through a cooling phase, and additional annealing steps might be employed to optimize crystal quality and remove defects (Figs. 3.1, 3.2, and 3.3).

3.1.4 Monitoring and Measurement

Real-time monitoring tools, such as spectroscopy or ellipsometry, are employed for in situ monitoring of the epitaxial process. Feedback from these tools enables adjustments to parameters, ensuring the precise control of film thickness, composition, and crystal quality.

Typical process metrics include epitaxial thickness and doping level, checking that the desired level has been achieved for each and assessing uniformity. In addition, the wafer shape and any defects in the wafer or surface are assessed and measured.

3.1.5 Thickness Characterization

Epitaxial film thickness can be measured by spectral reflectance, which is a relatively simple and fast method that has the advantages of being noncontact and nondestructive. The technique can typically measure thickness ranges from 2 to 100 μm,

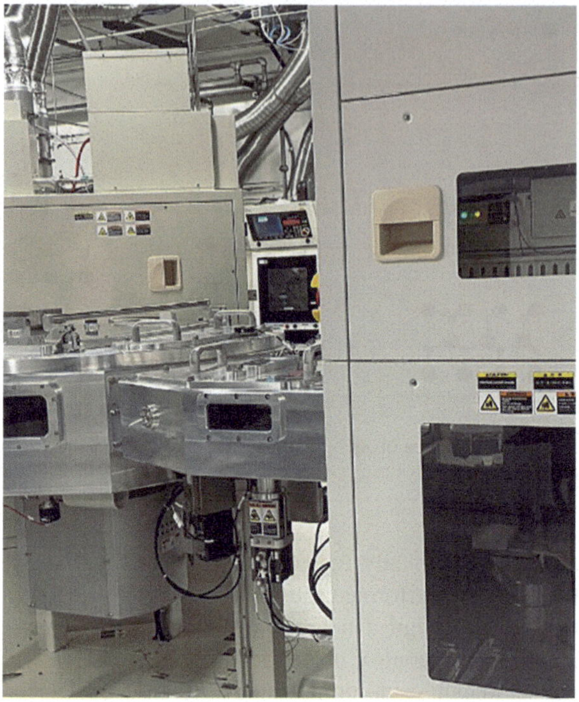

Fig. 3.2 An epireactor for commercial multi-wafer SiC fabrication

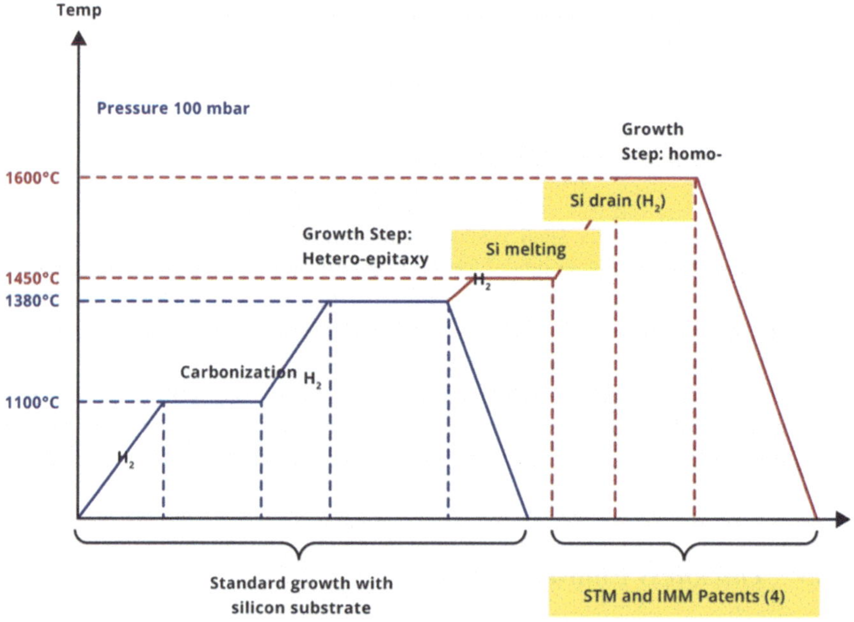

Fig. 3.3 Temperature profile of SiC epitaxy process

although this can be dependent on the doping level. Alternatively, techniques such as FTIR reflectance and ellipsometry can be applied for thickness characterization. These techniques are also noncontact and nondestructive.

3.1.6 Doping Characterization

Nondestructive doping characterization can map the wafer and quickly give insights into doping concentration and layer depth across the surface of the wafer. The concentration typically falls within the range of 5 × 10e14–5 × 10e17 carriers/cm^3.

3.1.7 Shape Characterization

The epitaxial wafer shape is often measured using capacitive probes. This technique is relatively simple, fast, noncontact, and nondestructive. Probing can generate contour maps for parameters that apply to the entire wafer, such as bow, warp, and total thickness variation as well as those that apply to smaller areas, such as local thickness variation.

3.1.8 Conclusion

Although SiC provides a foundation for high-performing semiconductors, needed to achieve future energy-conversion efficiency and system reliability targets in applications such as green power and e-mobility, the material is extremely difficult to work with. From producing crystal ingots and wafers of suitable quality to achieving accurate and consistent epitaxy, many technical barriers have impeded progress. Despite the challenges, material suppliers and semiconductor device manufacturers are now able to produce SiC power devices and RFICs that deliver their long-promised performance advantages over silicon devices at cost-effective price points.

3.2 Wafer Fabrication and Defects

John R. Silk

This paragraph discusses how SiC device structures are created and the issues that can cause performance and parameter variation, with the potential for device rejection after testing.

3.2.1 Creating the Foundations: Crystal Growth and Epitaxy

Up to the point device fabrication begins on SiC wafers, the processes consisted of crystal growth under controlled conditions, subsequently cutting wafers from the boule, grinding, polishing, and epitaxy.

Devices are then fabricated on the epiwafer, beginning with photolithography to define the features and geometry. This applies a photoresist material that is exposed using the device mask set and then developed. Chemical or plasma etching then selectively removes materials from the surface to create the device structure. Next, doping is selectively infused into specific areas to establish the desired electrical properties, and additional layers are deposited on top of the epilayer using chemical or physical vapor deposition. Annealing is then performed to activate the dopants, repair crystal defects, and optimize the electrical properties of the fabricated semiconductor device. After this comes metallization, which enables external electrical connections to the semiconductor device, before testing, dicing, back-end packaging, and more testing.

3.2.2 Wafer Fabrication and Defects

The two major issues that affect defects are imperfections in the SiC boule, or crystal, from which the wafers are sawn, and the resolution and repeatability of the fabrication equipment.

3.2.3 Substrate and Epitaxial Defects

The characteristics of silicon carbide—principally its more complex crystal structure and relative hardness—mean producing SiC wafers is more challenging than with silicon. While growing and machining a silicon ingot can expect to result in about 99% perfect sawn wafers, imperfections are expected in SiC wafers. These include variations in local site flatness that affect the focusing of the lithographic equipment, as well as defects in and below the substrate surface. Subsequently, the reactor causes variations in epitaxial doping and thickness that result in a spread of device parameters. Severe defects at some sites on any given wafer may result in a number of dies being produced that cannot function at all.

Clearly, a compromise must be found, between perfection and acceptability. When arranging the supply of SiC epiwafers, chip makers need to specify the extent of acceptable imperfections. Epiwafers are supplied with a certificate of analysis (CofA) that confirms conformance with the agreed specification. In general, the more imperfections, the fewer the number of good dies the wafer will yield, which could prevent the manufacturer from producing an economically viable number of components or may lead to a very high per-unit component price. On the other hand, buyers that set a high specification can expect to pay more for bare wafers.

Both epiwafer suppliers and SiC device manufacturers have needed to learn how to determine workable specifications that will ensure viability for both parties. A suitable specification should allow the epiwafer supplier to deliver the desired quality cost-effectively while allowing the device manufacturer to produce enough good dies from the wafer to be commercially viable. Figure 3.4 shows the number of defective sites found on the surface of an epiwafer considered to be a high-quality wafer suitable for high-yield fabrication.

The site analyses for two wafers of lower quality, although taken from the same boule as the wafer of Fig. 3.4, show how substrate defects can heavily influence the overall wafer yield. These are shown in Figs. 3.5 and 3.6.

As the SiC industry has matured, device manufacturers have requested epiwafer suppliers to provide more information in the CofA. Early in SiC's history, suppliers and device makers paid attention to only a few wafer parameters. Surface flatness, for example, was typically stated for the entire wafer. In practice, device manufacturers need assurances about flatness in more localized areas across the wafer surface. Accordingly, local site flatness has become a typically requested parameter.

Crystal defectivity measurements are taken from a very limited sample of wafers (substrates), usually from the top and bottom of a boule, which are etched to highlight the defects. The etching is a destructive process that provides an imperfect, but workable, indication of the substrate quality, prior to performing the expensive

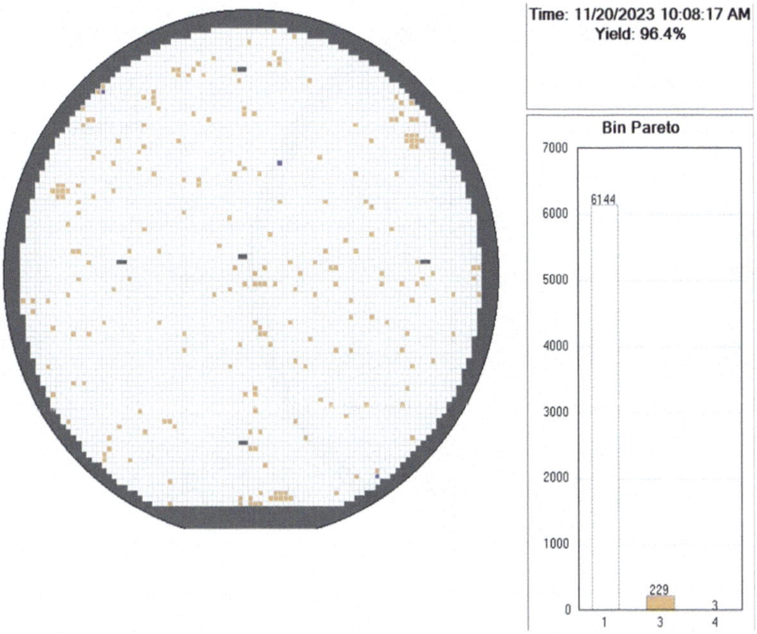

Fig. 3.4 Defective site analysis

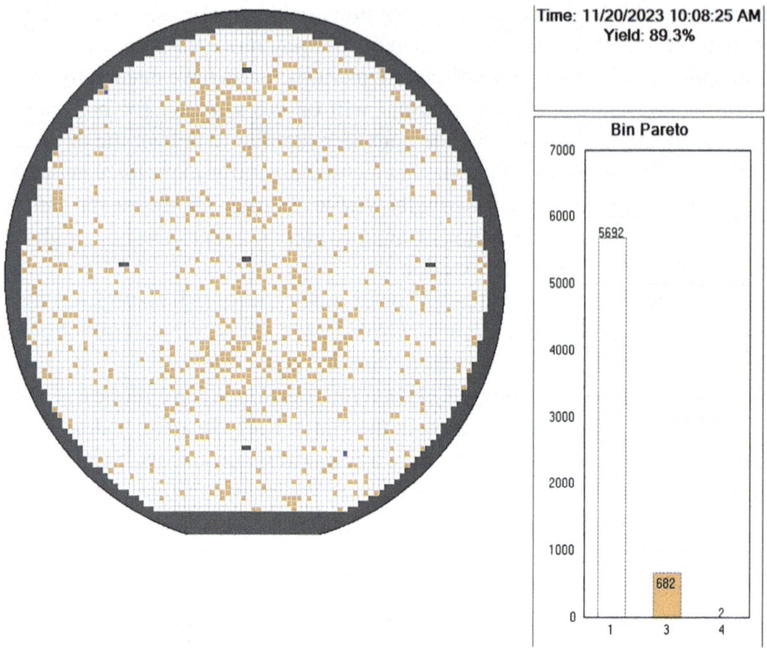

Fig. 3.5 Defective sites on the SiC wafer

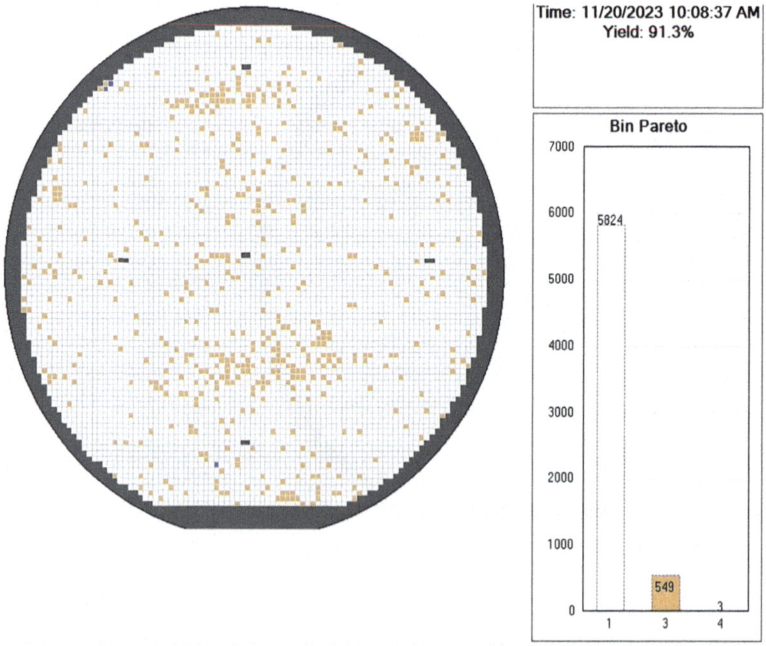

Fig. 3.6 SiC wafer from the adjacent slide. The close correlation between defect sites illustrates the effect of substrate quality on device yield

Fig. 3.7 Parameters as specified in a typical epiwafer CofA

Epi-Layer1 (buffer) doping concentration
Epi-Layer2 thickness
Epi-Layer2 Thickness Maximum Deviation (TMD)
Epi-Layer2 doping type
Epi-Layer2 dopant
Epi-Layer2 doping concentration
Epi-Total Thickness (all epi layers) (Layer1 + Layer2)
Post-epi Characteristics
Edge exclusion
Warp, epi-wafer
Bow, epi-wafer
Total thickness variation (TTV), epi-wafer (GBIR)
Epi-wafer Site Flatness; SBIR (LTV) MAX (10mm x 10mm site size)
Epi-wafer Site Flatness; SBIR (LTV) (10mm x 10mm site size)
Surface roughness
Front Surface visual inspection requirements
Edge exclusion
Scratches-cumulative
Edge chips, cracks Size>0.5mm width and depth
Surface Defects Density
Defect Free Area ($5mm^2$)
BPD Density
BPD Free Area ($5mm^2$)

epitaxial growth process. Substrate and post-epitaxial growth inspection techniques such as ultraviolet irradiation to detect luminescence are known, although the costs of inspection equipment and processes must be considered.

Figure 3.7 shows the parameters now typically requested in a SiC epiwafer CofA.

The parameters described in Fig. 3.7 are becoming widely accepted within the industry. They enable device makers to ensure that the supplied wafers meet the specified criteria, facilitating the production of components with satisfactory performance and quality.

It is worth noting, here, that SemiQ specifies a site size of 10 × 10 mm for local site flatness, as shown. The measured flatness is expressed as the SBIR, or (S)ite (B)ackside (I)deal Focal Plane (R)ange. Considering the surface topology as a set of peaks and troughs, above and below the center point of the site, SBIR is the sum of the highest peak height and the lowest trough depth. Although an SBIR value of 2 μm has been considered adequate, 1.2 μm is a more acceptable range to minimize flatness deviation as a cause of defects in fabricated devices.

Note also that the defect-free area listed among front surface visual inspection requirements specifies the percentage of 5 × 5 mm sites that are required to be defect-free. Historically, only the total count of defects on the wafer surface have been monitored. Similarly, the BPD-free area specifies the percentage of 5 × 5 mm sites that have zero BPD defects. This is most important in wafers intended for MOSFET production.

The CofA shown is the result of a general tightening of the parameters for acceptable SiC epiwafers. This has been established while ongoing improvements in processes and equipment have provided the means for wafer producers to deliver better wafers that display more uniform epitaxy, which, in turn, permits more good dies per wafer. Ultimately, this leads to lower per-unit prices for the resulting components and is helping to grow the market for SiC technology. More affordable device prices strengthen the economic argument for changing from silicon to realize the advantages of its superior energy efficiency and reliability in cost-sensitive applications, as well as those where ultimate performance is the dominant goal.

3.2.4 Defects in SiC Epiwafers

Raw wafers, as shown, can contain various types of defects resulting from imperfect crystal growth. Examples include basal plane dislocations (BPD) due to misalignment of atoms, which can introduce strains and affect the overall integrity of the crystal lattice. There can also be screw dislocations (SD), where defects form around an axis. After epitaxy, these dislocations can produce visible defects in the surface of the epilayers. Improving control over crystal-growth conditions, as well as applying post-growth treatments, can help to mitigate SD. In addition, epitaxy can introduce further defects.

Visible defects related to crystal dislocations include carrot defects, characterized by a conical or carrot-shaped inclusion in the SiC crystal. The defect forms typically as a result of screw dislocations intersecting the surface of the crystal during growth. These screw dislocations can act as conduits for impurities to be incorporated into the crystal causing carrot-shaped inclusions.

Other potential defects in the crystal structure include micropipes, which are hollow core dislocations within the SiC crystal, as well as irregular crystal stacking, missing atoms, extra atoms, and grain boundaries that can extend through the crystal lattice.

Epitaxy can produce additional defects, such as step bunching. This can occur where the growth of an epitaxial layer is impeded by a surface imperfection, effectively slowing down the formation of steps and causing problems such as interference and unwanted changes in the step height.

As far as the device characteristics are concerned, variability in the epitaxial doping concentrations and layer thicknesses translates into parametric inconsistencies. The breakdown voltage (BVDSS) of SiC MOSFETs is known to be particularly strongly affected.

Local site flatness, mentioned at the beginning of this paragraph, is important yet especially difficult to achieve with SiC wafers. A flat surface is critical to ensure proper focusing of the fabrication lithography equipment and thus the dimensions and parameters of the devices produced. The stepper tries to focus on the center of each site. Because the depth of field of the lithographic equipment is relatively shallow, wafer shape variation means the image can be less sharply focused in some places. Experience has shown that local site flatness is a more important parameter to specify in the CofA than the maximum deviation over the entire wafer.

In practice, as much as 40% of the wafer thickness can be ground away to establish a flat enough surface for acceptable lithography. The relative hardness of silicon carbide, compared to pure silicon, ensures that flatness concerns are difficult to completely eliminate.

Also, the industry is working to establish suitable specifications that will allow working closer to the wafer edge, to minimize so-called edge exclusion and thus produce more usable dies per wafer. Better data about the material characteristics within 4–6 mm of the wafer edge can help to increase utilization of this area, resulting in lower per-unit device cost.

Proper cleaning of the reactor after use is also extremely important, to remove particles that can otherwise accumulate on the reactor's internal components and mechanisms. Downfall of these particles can produce faults of various shapes and sizes in the wafer surface.

3.2.5 *Limitations in the Fabrication Equipment*

Processing multiple wafers at a time is the economical approach, and increasing the number that can be processed is desirable from a device cost standpoint. Processing a single wafer provides greater flexibility to perfect the setup but is more expensive and impacts the overall device cost.

Improvements in equipment, such as smaller step size in photolithographic motion control, effectively minimize and reduce instances of defects. In fact, the typical step size today has reduced to about 25% of the size in previous equipment generations.

Other improvements in equipment performance include the typically greater depth of field of the latest equipment generations. While the depth of field has not been an issue when fabricating in silicon, which has much greater inherent uniformity, SiC crystal growth is more difficult to control and the boule is more difficult to process, especially because of the material's extreme hardness.

Further potential hazards exist in wafer cutting. SiC's extreme hardness imposes a greater risk that the cutting disk can shatter. Laser cutting is an option being researched, which could allow an increased number of substrates to be wafered from the expensive boule. SemiQ views that laser cutting will be preferred in the future. Laser dicing is also being pursued, which would allow the wasted "street" between adjacent dies to be narrowed, increasing the die per wafer.

3.2.6 *Influence of Defects, Epitaxial Uniformity, and Local Shape on Device Parameters*

The long-term reliability of MOSFETs can be affected by the presence of surface and below-the-surface defects. Imperfections below the crystal surface are known to affect device parameters such as the body diode characteristics and the quality of the gate oxide. Diodes, by virtue of their simpler construction, are less affected by imperfections in the SiC boule.

Both epitaxial uniformity and local shape variation impact device parameters and can interact to create strange parametric distributions. Figure 3.8 illustrates the combined effects of wafer shape imperfections and reactor precision on VSD or a finished SiC MOSFET wafer. The "fingers" illustrate variability in the measured VSD at different locations on the wafer due to the effects of local site flatness on stepper focus. The gradual variation from the center of the wafer toward the edge is from doping and thickness variation in the epitaxial layer. Depending on the epireactor configuration, one can see variations in device parameters from the wafer top to flat, or a radial distribution of many critical parameters.

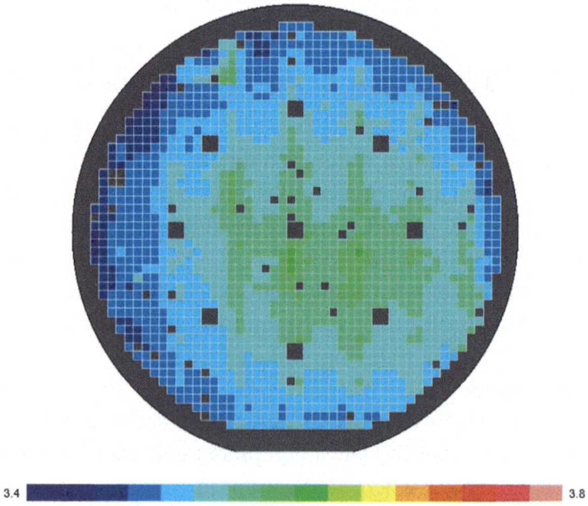

Fig. 3.8 Graphical illustration of SiC MOSFET VSD variation

The wafer shown in Fig. 3.8 contains 1200 V MOSFETs. Generally, building MOSFETs becomes more challenging at higher voltages, in SiC as with silicon. To produce devices that are stable and reliable, and to achieve high yield, demands extra care to ensure the substrate has high purity.

3.2.7 Conclusion

Growing and processing SiC crystals is considerably more difficult, being less predictable, and less controllable than in conventional silicon fabrication, due to the challenging material properties. Defects in the underlying substrate, in addition to normal industrial process-control challenges to ensure consistent epitaxy and device fabrication, can cause defects ranging from total device failure to parametric variation and device instability. MOSFET device yield and performance are more heavily impacted than diodes, due to their relative complexity.

Improvements to crystal-growth processes, wafering and grinding, the depth of field and stepper motion of the lithographic equipment, and the interaction between dopant vapor flow and wafer movement in the reactor are needed to continue increasing SiC device quality and economy.

3.3 Ion Implantation

Victor Veliadis

For power device applications, 4H-SiC is preferable over 6H-SiC due to its higher longitudinal and transversal electron mobilities. Off-axis substrate growth is desirable as it results in higher-quality material, with 4° off being the most common. In this section, an overview of SiC device implantation will be illustrated using scanning electron microscopy (SEM) images.

Multiple mature Si processes have been successfully transferred to SiC. However, SiC material properties necessitate development of new processes, whose parameters must be optimized and qualified. The primary SiC-specific processes are summarized below:

- *Etch*: SiC is inert against chemical solvents and only dry etching is practical. Masking materials, mask etch selectivity, gas mixtures, control of sidewall slope, etch rates, sidewall roughness, etc. need development.
- *Substrate thinning*: is performed to lower resistance of low-voltage devices (≤1.7 kV), and to facilitate handling of the thick epitaxy wafers used in high-voltage device fabrication (SiC material hardness requires special recipes).
- *Doping*: conventional thermal diffusion is not practical in SiC due to its high melting point and the low diffusion constant of dopants within SiC. Implantation

species, dose, energy, temperature, masking materials, etc. need to be evaluated. Post-implantation SiC recrystallization and implant ion-activation anneal method (furnace, RTA, etc.), temperature, ramp rate, duration, gas flow, etc. need to be developed. Protective cap layers, to minimize SiC wafer surface degradation during anneal, need to be deposited and removed.

- *Metallization*: evaluate coefficient of thermal expansion-matched metals and select resist type and develop undercut lift-off profile for metal evaporation and lift-off or for metal sputter deposition and dry etch.
- *Ohmic contact formation*: the high value of the SiC/metal barrier results in rectifying contacts. Post-metal deposition anneal is required for ohmic contact formation. Optimize anneal temperature, ramp rate, duration, and gas flow and maintain surface quality.
- *Gate oxides*: the poor SiC/SiO2 interface quality reduces MOS inversion layer mobility and causes threshold voltage instability. Develop passivation techniques to improve the SiC/SiO2 interface quality.
- *Transparent wafers*: complicate CD SEM measurements and metrology when the focal plane is determined by optical microscopy. Metrology tools need software/gain/hardware adjustments to move to wavelengths where SiC is opaque.
- *Relative lack of planarity*: in SiC wafers can complicate photolithography and other processing, particularly of high-voltage devices due to their thick drift epitaxial layers. High-temperature processes can further negatively affect wafer flatness.
- *Insulation dielectrics*: thick dielectrics are deposited in SiC as thick thermal oxides are not possible. Deposited dielectrics are prone to defect incorporation, and the impact on edge termination and device reliability must be assessed.

In this section, we will focus on the unique aspects of the SiC implantation process. Although SiC MOSFETs and junction barrier Schottky diodes are primarily used in power electronics, the implantation processes described below are applicable to all SiC devices including JFETs, BJTs, thyristors, IGBTs, and PiN diodes, all of which have been successfully demonstrated.

Due to the negligible diffusion coefficients of common doping species in SiC, and the high temperatures needed, exceeding 1800 °C, conventional semiconductor diffusion processes are not practical in SiC. Ion implantation is exclusively used in SiC fabrication, and selective doping of materials can be achieved by employing photolithography masking techniques. Depositing metals of high mass or using combinations of thermally grown and deposited oxides is typically done to mask the SiC wafer surface. Ion implantation is usually performed at elevated temperatures (600–900 °C) to minimize crystal lattice damage from ion bombardment, particularly for the high doping densities required for low-resistivity ohmic contacts. The elevated temperature imposes restrictions on the masking materials. Room-temperature implantation can work for low-implant doses ($<10^{15}$ cm^{-2}), while elevated temperature implantation is performed for the doping densities of 10^{19}–10^{20} cm^{-3}—needed for ohmic contact formation. Implant profiles can be predicted by Monte Carlo simulations such as SRIM (Stopping and Range of Ions in

Matter), and verified by SIMS (secondary ion mass spectroscopy). Post-implantation anneal at 1600–1800 °C is carried out for lattice damage recovery and to achieve high dopant electrical activation ratios (~95%). The as-implanted depth profiles are retained after the 1600–1800 °C anneal for Al, P, and N dopants as expected from their low diffusion constants. The lack of dopant diffusion makes it easy to form well-controlled shallow junctions and difficult to form deep junctions.

Aluminum and boron are commonly used as acceptor dopants in SiC, and high concentrations must be introduced for highly conductive *p*-type regions. Dose and energy levels determine *p*+ depth and its small lateral straggle and, therefore, impact capacitance, contact resistance, depletion widths and pinch-off voltages, and gate-integrity characteristics. Trade-offs between the values of these parameters determine the optimal ion implantation design. Boron has a higher lateral straggle [1] than that of aluminum and is typically not used in device fabrication due to its higher implantation-related defect generation, and the higher ionization energy of its acceptors (628 ± 1 and 285 ± 10 meV for boron vs. 190–250 meV for aluminum) that hinders the formation of low-resistance *p*-type SiC. Nitrogen and phosphorus are commonly used as donor dopants in SiC, with ionization energies of 53–61 meV and 99–125 meV, respectively. Similar considerations to the ones outlined above apply for *n*-type donor dopants in fabrication. The SiC implantation dopants and their ionization energy levels are schematically shown in Fig. 3.9.

As stated above, irradiating SiC with ions during implantation causes damage to the crystal structure. This is experimentally demonstrated in the photographs of Fig. 3.10. Prior to ion implantation, photographed in Fig. 3.2a, the quarter-piece wafer is transparent due to the ultraviolet wavelength of the 4H-SiC wide bandgap. The implantation's high-energy ion bombardment degrades the SiC wafer crystallinity, and the quarter-piece wafer is now opaque, photographed in Fig. 3.2b.

To recrystallize the material and activate the implanted atoms, thermal annealing must be performed. For improved surface quality and morphology, post-implantation anneal can be done under gas flow conditions. Furnace annealing or rapid thermal annealing utilizing high-intensity flash lamps or laser irradiation can be used. In the case of furnace annealing, temperatures above 1300 °C are common. A concern with annealing SiC at these high temperatures, or under high vacuum conditions, is surface degradation due to silicon's higher desorption rate. This causes a higher rate of silicon surface atom migration compared to that of carbon atoms. By using a protective cap layer over the wafer surface during the annealing process, surface

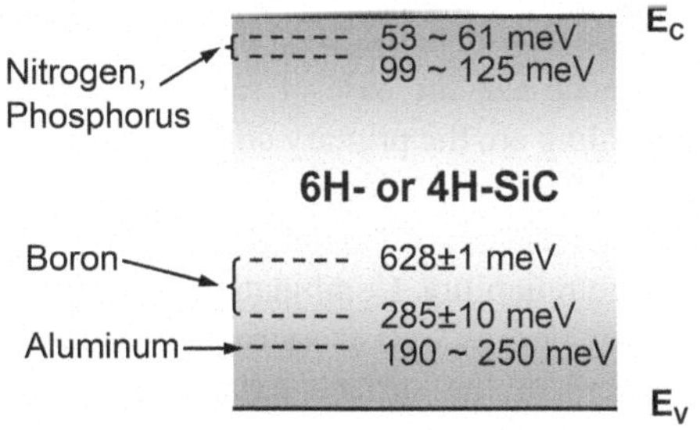

Fig. 3.9 Common SiC implantation acceptor and donor dopants and their respective ionization energies

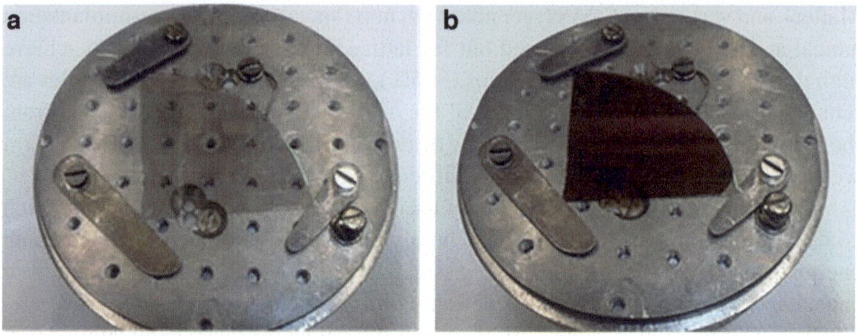

Fig. 3.10 (**a**) One-quarter of a SiC wafer piece prior to the onset of high-temperature ion implantation. The wafer is transparent due to the ultraviolet wavelength of the 4H-SiC bandgap. (**b**) The same SiC wafer piece after 400 °C heated ion implantation. The ion bombardment has degraded the SiC wafer crystallinity and the quarter piece is now opaque. (The photographs are courtesy of Dr. Roberta Nipoti)

degradation can be minimized [2]. Graphite, SiO_2, Si_3N_4, AlN, BN/AlN, and carbon-protective annealing cap layers are typically used [2, 3]. Carbon cap layers are known to give the best results. They can be formed by RF sputtering or graphitization of photoresist. No chemical reactions occur at the SiC/carbon interface during those processes. At completion of annealing, the carbon cap layer can be removed by plasma ashing (oxygen and fluorine are the most common reactive species) or 700–800 °C oxidation.

An SEM image of the surface of a wafer that underwent 1675 °C post-implantation furnace annealing for 30 min, in the presence of argon, *without a protective cap layer* is shown in Fig. 3.11. Extensive surface damage is observed. An SEM image of another SiC wafer that was capped with a graphite layer and underwent identical post-ion implantation thermal annealing at the same time and in the same chamber (as part of the same fabrication lot) is shown in Fig. 3.12. Apart from the absence of protective annealing capping on the surface of the wafer in Fig. 3.11, the two wafers contained SiC JFETs of identical design that underwent identical processing as part of the same fabrication lot, including both wafers simultaneously loaded in the anneal furnace to complete the same fabrication step. At completion of fabrication, the "uncapped" wafer yielded no functional JFETs in sharp contrast to the "capped" wafer that had a relatively high functional JFET yield of 85% [4–6].

The signature of the implantation's ion bombardment is visible in the dotted SiC surface texture in the SEM images of Figs. 3.11 and 3.12. High furnace-temperature tolerance and the ease of removal of the capping layer at completion of thermal annealing are the primary considerations in choosing a suitable capping layer material. After completion of ion implantation annealing, and capping layer removal, thermal and deposited dielectrics are employed to passivate the surface and improve device reliability. Combinations of thermal and high-temperature-densified deposited oxides, as well as sacrificial oxidations, are commonly performed. The effectiveness of the surface passivation is evidenced by low gate-to-source leakage

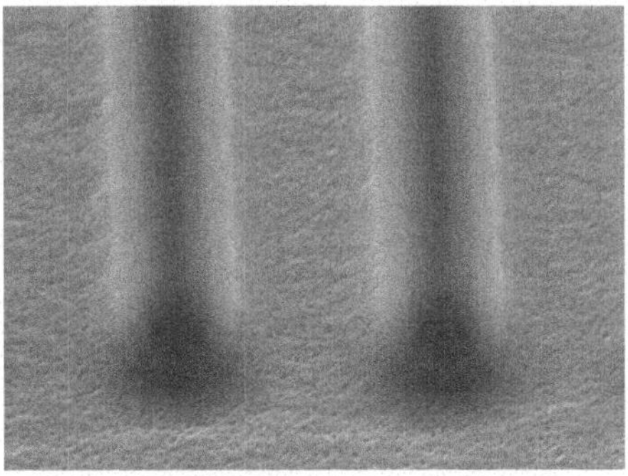

Fig. 3.11 SEM image of wafer surface after completion of 1675 °C post-implantation furnace annealing for 30 min in the absence of a protective cap layer

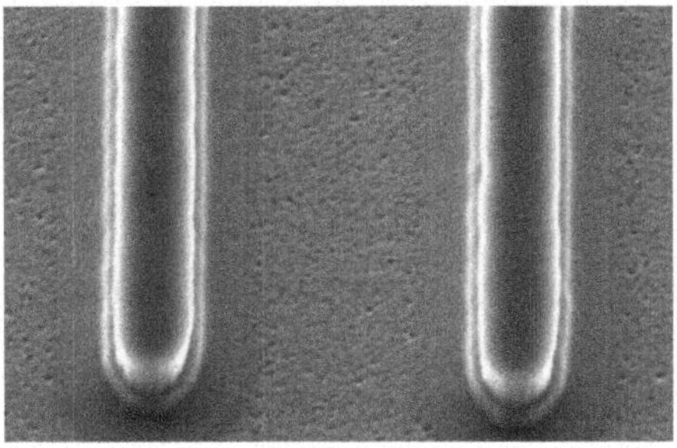

Fig. 3.12 SEM image of wafer surface after completion of 1675 °C post-implantation furnace annealing for 30 min in the presence of a protective cap layer

currents and sharp onsets of voltage breakdowns that contribute to long-term device reliability. Orders-of-magnitude reduction in the *pn* junction leakage current can be achieved by optimizing the thermal and sacrificial oxidations that passivate the SiC surface [4–7].

Ion implantation and subsequent high-temperature annealing can generate defects that degrade device performance. These include complex point defects, localized levels (shallow or deep), extended defects (basal plane dislocation loops and stacking faults), and new dislocations and/or movement of existing dislocations. These defects can degrade performance and present yield and reliability concerns. Thus, the ion implantation process must be improved. The goal is to optimize annealing ramp rates, temperature, annealing duration, and the overall implantation process to minimize defect generation. In addition, quality room-temperature heavy-dose ion implantation is highly desirable for low-cost volume manufacturing.

A cross-sectional SEM image of heavily doped *p*+ implanted gate regions in an *n*-channel is shown in the SEM image in Fig. 3.5. Ion implantation induces a small lateral straggle that undercuts the mask and is noted in the image of Fig. 3.5. Implantation profiles with a "boxlike" configuration can be achieved through targeted multiple-dose/energy implantation events (Fig. 3.13).

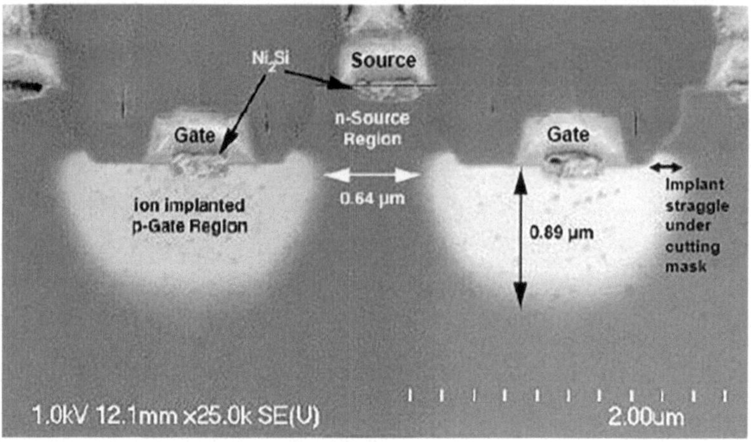

Fig. 3.13 Cross-sectional SEM image of heavily doped $p+$ implanted gate regions in an n-channel. Ion implantation induces a small lateral straggle that undercuts the mask and is noted in the image

3.4 Etching and Metallization

Dario Tenaglia and Paolo Badalà

The planar MOSFET structure is fabricated by starting with an N+ substrate in which an *N*-type epitaxial layer has been grown. A thick field oxide is first grown for the formation of the appropriate edge termination. A dedicated photoresist mask is used to open the active area for the floating field rings. The doping concentration in the JFET region can be enhanced by the addition of an *N*-type ion implant in the active area at this step process. Then, a photoresist mask is used to define the location of the P+ implant within the planar MOSFET cell window. Also, the same implant is utilized to form the P+ region for the edge termination. After that, the gate oxide is grown on the substrate followed by the deposition of the poly-Si gate electrode. The P-base region is implanted using polysilicon as a mask. A photoresist mask is used to define the location of the N+ source region. The inter-metal dielectric is deposited to cover the poly-Si gate. A mask is used to etch windows in the inter-metal dielectric film to open contacts to the N+ source and P-base regions. The same mask is used to form the contact with the gate electrode at the gate pads.

The source metal is deposited and patterned to define the source electrode as well as gate pads to complete the structure. The thickness of the N+ substrate is reduced by grinding and polishing processes to decrease its contribution to the device resistance, followed by backside ohmic contact formation by means of laser annealing. Finally, deposition of a solderable metal stack on the back surface is performed. Figure 3.14 shows the cross-sectional schematical drawing of a SiC planar MOSFET and of a SiC trench-gate MOSFET.

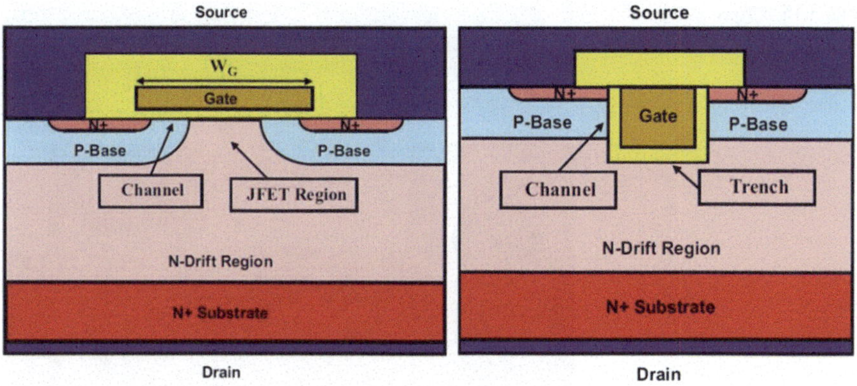

Fig. 3.14 (*Left*) Cross section of the planar MOSFET basic cell structure; (*right*) cross-section of the trench-gate MOSFET basic cell structure

3.4.1 Etching

Manufacturing of a device is a sequence of steps to define the structures according to the features required and the materials exposed.

Photoresist mask is commonly used for many etch processes, and thickness, critical dimension (CD), and lithography conditions depend on the further etch process.

Wet etch is an isotropic process to use for large features, for example, HF-based solutions are widely used for etching TEOS or SiO_2 as well as HNO_3-based solutions for etching aluminum or aluminum alloy.

Thanks to its chemical resistance, selectivity with SiC is not a problem for all typical manufacturing wet processes.

Dry etch is required in case of critical features because the dimension of the structures, shape, and process uniformity can affect the electrical performance of the device.

Looking at the doped area of a MOSFET, the quality of the process required needs a dry etch approach to guarantee a good uniformity in terms of profile and CD.

Figure 3.15 shows an example of anisotropic dry etch to define an array of thick TEOS with RIE process with plasma based on fluorinated chemistry (CHF_3, CF_4) assisted by Ar and O_2.

Even if the SiC is a hard-to-etch material, the consumption and the quality of the substrate exposed to the plasma must be considered especially to ensure the best condition for the other manufacturing steps.

Selectivity with SiC improves tuning the process parameters such as power, pressure, and gas mixture, but shape and CD can be modified as well.

Good results are also achieved by adding CO to the gas mixture to passivate the exposed material.

The effect of the plasma damaging is well visible on the sidewall of the trench.

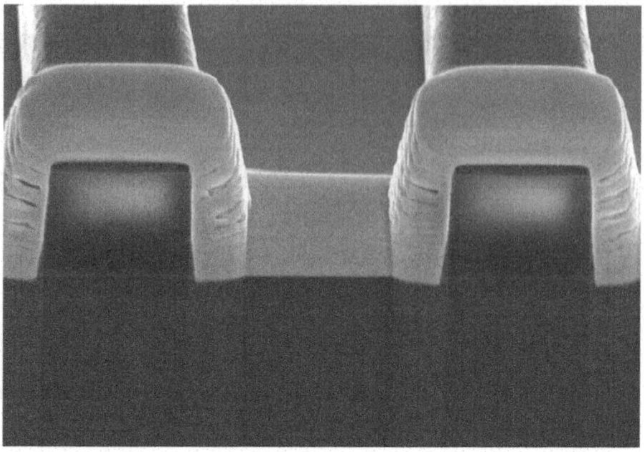

Fig. 3.15 Cross section of TEOS stripes etched on SiC

Fig. 3.16 (*Left*) Plan view of trenches on SiC; (*right*) cross section of trenches on SiC

Figure 3.16 shows a typical trench performed on SiC with SF_6 plasma.

The main feature of the trench MOSFET configuration is the presence of a U-shaped trench, produced by a dry etching process, which extends from the upper surface of the structure through the N+ source and P-base regions into the N-drift region.

In this configuration, the profile and the quality of the sidewalls of the trench are critical parameters.

3.4.2 Metallization

The typical fabrication flow of a SiC power device includes several metallization processes, devoted to contact formation and to electrode definitions and to ensure connections with package. The first metallization step in the process flow of a SiC MOSFET is the formation of the front side ohmic contact, which is usually made of Ni silicide, obtained through the sputtering deposition of a Ni layer of thickness in the range between 30 and 100 nm, followed by a rapid thermal annealing (RTA) process in N_2 ambient at a temperature between 950 and 1050 °C. Then the front side anode electrode is formed, by depositing a Ti-based barrier layer followed by a thick layer (between 4 and 6 μm) of an Al alloy, containing small amounts of copper and silicon, added to increase the resistance to electromigration and to inhibit diffusion processes, respectively. A wettable metal stack is deposited on an Al electrode, in order to favor the connection with package, by the sintering or soldering process, during the assembling phase. For this step, two different approaches can be adopted. In the first case, a Ti/Ni/Ag stack is deposited by sputtering and is defined through common lithographic and etching processes. In some cases, a small amount of vanadium can be added to Ni, and Ag can be replaced by Au. In the second case, a Ni/Pd/Au stack is plated by e-less deposition. When this approach is adopted, the stack is selectively grown on an Al layer, through a sequence of chemical processes aimed to activate the surface, and lithographic and etching steps are not required. Then the SiC wafer is usually thinned to the final desired thickness, and a Ni silicide-based ohmic contact is formed on the backside of the device, through the deposition of a Ni layer by sputtering and a laser annealing process, which makes the nickel react with SiC, to form a Ni silicide layer with the desired electrical, mechanical, and structural properties. Once the ohmic contact is formed, a final metallization stack, suitable for the sintering or soldering process, is deposited by sputtering on the backside of the wafer. The most adopted stacks are Ti/Ni/Ag and Ti/Ni/Au, with the optional introduction of a small amount of vanadium in the Ni layer.

Among the mentioned metallization processes, in the last years, the formation of backside ohmic contact by means of laser annealing has attracted increasing attention [8–14]. In SiC power devices of previous generations, in fact, the ohmic contact was formed by RTA in N_2 ambient at a temperature in the range between 950 and 1050 °C, with consequent thermal constraints that forced to perform thinning and annealing before the anode electrode metallization, with important limitations for wafer thickness reduction. On the contrary, laser annealing is considered an enabling process for new generations of silicon carbide power devices, since it allows to postpone thinning and ohmic contact formation at the end of the process flow, making the fabrication of power devices on very thin wafers possible [14]. Wafer thinning, indeed, is very important to reduce the total ON resistance of a device. The electrical, structural, and morphological properties of reacted layers have been deeply investigated, as a function of laser features, contact layer thickness and thinning process. By considering the sheet resistance evolution under irradiation, it is possible to get some interesting information about the reaction process. It has been observed that the typical sheet resistance curve, as a function of laser fluence, shows an increase of R_S values for low fluences, due to the initial intermixing between Ni and Si, and then a rapid drop of R_S to a final plateau at a fluence Φ_C [15]. This typical trend has been observed under several conditions (e.g., different thickness of Ni layer, different number of laser pulses, and different roughness of substrates), with a shift of the curve toward higher or lower values of fluence, upon the conditions. When the R_S curve is shifted, the value of Φ_C, corresponding to the drop of sheet resistance, is also shifted in the same direction. Figure 3.17 shows the evolution of Φ_C as a function of laser pulses, for a 100-nm-thick Ni layer. As reported, when the

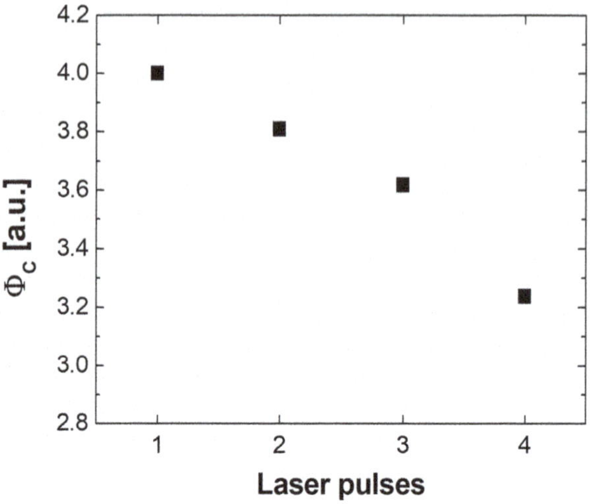

Fig. 3.17 Evolution of Φ_C as a function of laser pulses, for a 100-nm-thick Ni layer. When the number of laser pulses increases, the laser fluence at which the R_S drop is observed progressively decreases

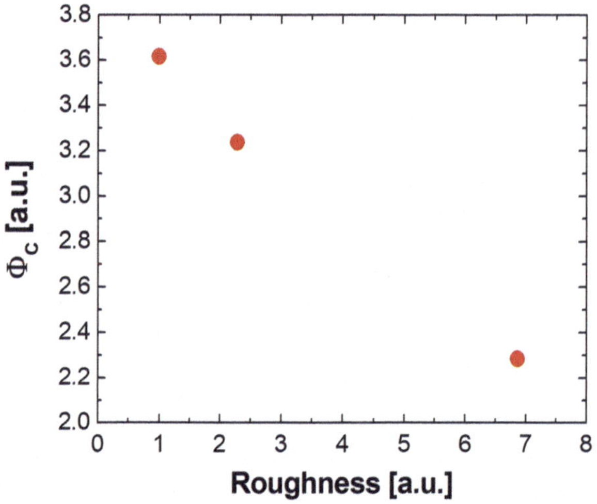

Fig. 3.18 Evolution of Φ_C as a function of surface roughness for a 100-nm-thick Ni layer. When the roughness increases, the laser fluence at which the R_S drop is observed progressively decreases

number of laser pulses increases, the laser fluence at which the R_S drop is observed progressively decreases.

The influence of the thinning process on silicide formation has been investigated. First of all, it has been observed that different thinning processes induce different surface roughness and different amounts of subsurface damage. These differences are reflected over the electrical properties of reacted layers. In fact, as reported in Fig. 3.18, for a 100-nm-thick Ni layer, with the increase of surface roughness, the decrease of laser fluence Φ_C at which the R_S drops toward a plateau is observed.

The reaction process has been also studied as a function of Ni layer thickness, observing a shift of the sheet resistance drop toward higher fluence with the increase of thickness.

3.5 Device Processing

Simone Rascuna

Silicon carbide (SiC) has emerged as a highly sought-after semiconductor material in recent years due to its exceptional properties, such as high thermal conductivity, high breakdown voltage, and high switching frequency capabilities. Its diverse applications span across power electronics, electric vehicles, and renewable energy systems. The fabrication of SiC electronic devices involves a series of processing steps, including ion implantation for doping, oxidation, and metallization, which contribute to the final structure, akin to Si technology. However, owing to the unique physical and chemical properties of SiC, several distinct technologies with specific requirements are also necessary.

3.5.1 Ion Implantation

The process of ion implantation is a critical step in the fabrication of various SiC devices [16], as it enables wide-range doping control for both *n*- and *p*-type conductivity. However, there exist significant differences between ion implantation technologies for SiC and Si. For instance, selective doping by a diffusion process is not feasible in SiC due to its extremely low diffusion constants. Moreover, lattice recovery after near-amorphous-level lattice damage poses a challenge, necessitating elevated temperature implantation.

According to A. Hallen et al.'s report [17] at fluences below 10^{14} cm^{-2}, the damage caused by implantation mainly consists of isolated point defects. At higher fluences, these point defects combine to form larger defects, such as dislocations and crystal imperfections. At very high fluences, the lattice collapses, leading to the formation of an amorphous material. Early on, it was recognized that amorphization during implantation doping of SiC must be prevented, as restoring the lattice through post-implantation thermal annealing to recreate a single-crystalline structure of the original polytype is challenging. The solution is to use implantations at elevated temperatures, as the thermal energy supplied by heating the target increases the dynamic annealing and helps to eliminate the point defects during implantation, thus avoiding amorphization. Figure 3.19 shows Rutherford backscattering spectroscopy channeling (RBS-C) results from 100 keV Al implantation in 4H-SiC at different temperatures for implantation fluence. Regardless of implant dose and temperature, high-temperature annealing (>1500 °C) is necessary for high electrical activation. An example of the use of ion implantation is the formation of junction termination extensions (JTE) or floating ring terminations on SiC power devices (see Fig. 3.20).

The efficiency of edge termination is particularly important in high-voltage SiC devices, where the electric field can be very high. In these devices, the

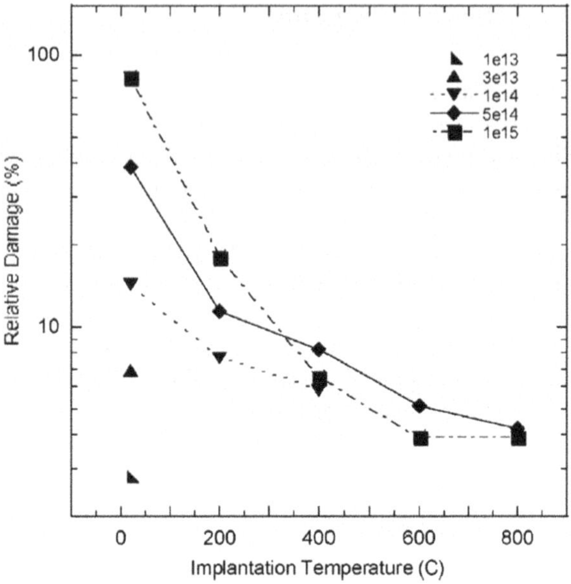

Fig. 3.19 RBS-C data from 100 keV Al implantation in 4H-SiC at different fluences as a function

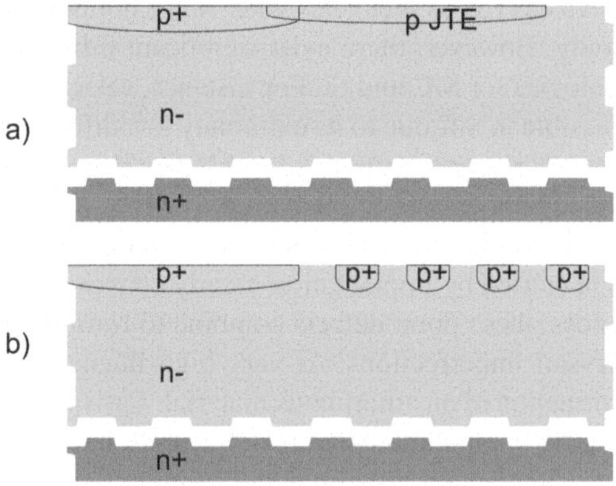

Fig. 3.20 (**a**) Cross-section of a single JTE structure. (**b**) Cross-section of a four-floating ring termination of implantation temperature

electric field at the edge of the device can be several times higher than the electric field in the active area. This can cause premature breakdown and reduce the device's overall reliability. The use of JTE or floating ring edge termination can help to reduce the electric field at the edge of the device and improve its overall reliability.

JTE edge termination (see Fig. 3.20a) works by creating a graded doping profile at the edge of the device. The doping concentration is highest at the edge of the device and gradually decreases toward the active area. This creates a smooth transition in the electric field, which reduces the electric field at the edge of the device and improves the breakdown voltage. The typical implantation fluence for the JTE is lower than 10^{13} cm^{-2}; thus, the process can be performed at room temperature resulting in very small damage.

In Fig. 3.3, the simulated breakdown voltage of a 650 V SiC diode device is depicted by varying the total ring dose of a single JTE. The edge termination's maximum efficiency is achieved when its breakdown voltage surpasses that of the active area, as shown in the plateau region of the image.

Floating ring terminations (see Fig. 3.20b) are concentric p + rings that surround the main junction and are typically formed in the same processing step as the main junction, eliminating the need for additional processing. The number of rings, the width of each ring, and the spacing between each ring are critical design parameters. The performance of the floating ring system is not significantly affected by the activated implant dose due to the heavily doped nature of the rings. Since the implantation fluence in this case will be, typically, close to 10^{15} cm^{-2}, the process must be performed at temperatures higher than 200 °C to avoid detrimental amorphization.

Ion implantation processes are also used in silicon carbide MOSFETs to realize the body, the source, and the channel region. The channel region is the region of the device that conducts current between the source and drain electrodes. The gate electrode controls the flow of current through the channel region.

3.5.2 Oxidation and Oxide/SiC Interface

Oxidation of silicon carbide is a key process for the fabrication of SiC-based devices. The oxide layer formed on the SiC surface acts as a passivation layer, protecting the underlying SiC material from further chemical reactions and electrical degradation. The oxide layer also serves as a dielectric layer in SiC-based devices, such as MOSFETs.

The gate oxide is the layer of an insulating material that separates the gate electrode from the channel region in the device (see Fig. 3.21). In SiC MOSFETs, the gate oxide is typically made of silicon dioxide (SiO_2), which is grown on the SiC surface using a thermal oxidation process. However, the thermal oxidation process can result in defects in the gate oxide layer, which can reduce the performance and reliability of the device.

To address this issue, post-oxidation annealing (POA) is used in the manufacturing of SiC MOSFETs to improve device performance. The process involves heating the MOSFET after the gate oxide has been formed, which helps to remove defects and improve the quality of the oxide layer.

According to T. Hosoi et al.'s report [18], the use of POA in a NO ambient is a common technique to enhance the channel mobility of SiC MOSFETs. However, it has been observed that this improvement in channel mobility can lead to threshold voltage instability, possibly due to the increased hole trapping [19–21]. An analysis using scanning X-ray photoelectron spectroscopy (XPS) to study the distribution of N atoms near the SiO_2/SiC interface revealed that N atoms were introduced on both

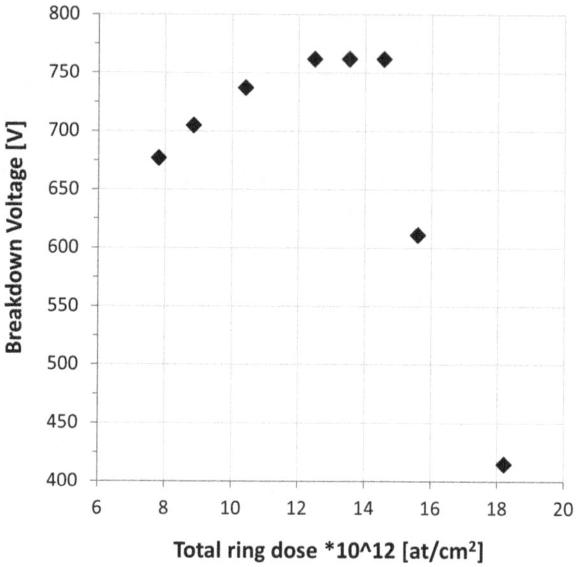

Fig. 3.21 Simulations of breakdown voltage values for a 650 V SiC diode by varying the total ring dose of a single JTE

the SiC and SiO_2 sides of the interface [22, 23]. The N atoms on the SiC side are believed to play a crucial role in improving channel mobility [22]. However, N atoms on the SiO_2 side can enhance charge trapping. To improve threshold voltage stability, selective removal of N atoms on the SiO_2 side at the SiO_2/SiC interface is necessary. This requires replacing N atoms with O atoms without causing SiC oxidation. This research study has shown that CO_2-POA at temperatures below 1300 °C is effective in improving both SiO_2/SiC interface properties and immunity to charge trapping [24].

Alternative gate oxide materials, such as aluminum oxide (Al_2O_3) and hafnium oxide (HfO_2), are being investigated for use in SiC MOSFETs. These materials have a higher dielectric constant than SiO_2, which can improve the device's performance. Additionally, these materials can be deposited using atomic layer deposition (ALD), which can result in a high-quality gate oxide layer with a low defect density.

3.5.3 *Metallization*

Metal/semiconductor contacts are commonly classified into two categories: ohmic and Schottky contacts [25]. Ohmic contacts exhibit a linear and symmetric current-voltage characteristic for both positive and negative applied bias. In contrast, Schottky contacts are rectifying, with current flow favored under forward bias and inhibited under reverse bias. The Schottky barrier height (SBH) is the most important parameter that describes a metal/semiconductor contact.

When a metal and a semiconductor are brought into close contact, if the semiconductor work function $q\Phi_s$ is lower than the metal work function $q\Phi_m$ (which is the case for most metals on SiC), electrons will flow from the *n*-type semiconductor into the metal. This leaves behind a region of positively charged donors over the depletion width W. The charge transfer continues until thermodynamic equilibrium is reached, and the two Fermi levels are aligned. At this point, the energy level of the electrons in the semiconductor will be raised near the metal/semiconductor interface by an amount qV_{bi}. The amount of band bending V_{bi} is often referred to as the built-in potential. The Schottky barrier height $q\Phi_{Bn}$ is defined as the difference between the metal work function $q\Phi_m$ and the semiconductor electron affinity $q\Phi_s$ [26].

The Schottky barrier height is a critical parameter that determines the electrical behavior of a metal/semiconductor contact. The Schottky barrier $q\Phi_{Bn}$ can be viewed as the energy required for electrons in the metal to enter the semiconductor, while qV_{bi} is the barrier observed from the electrons in the semiconductor side.

According to the classical theory, the current transport mechanism at a metal/semiconductor contact depends on the doping level of the semiconductor N. For lightly doped semiconductors ($N < 10^{17}$ cm^{-3}), the primary conduction mechanism is thermionic emission (TE), where carriers with sufficient thermal energy can pass from one material to the other by surmounting the Schottky barrier. For intermediate semiconductor doping levels (N in the range of 10^{17}–10^{19} cm^{-3}), the current transport through the barrier is governed by thermionic field emission (TFE), which involves carriers that lack sufficient thermal energy to surmount the barrier (as in the case of TE) but have enough thermal energy to tunnel at an energy higher than the Fermi level, where the barrier is thinner. Finally, for heavily doped semiconductors ($N > 10^{19}$ cm^{-3}), the small depletion layer W results in a very thin barrier. In this condition, the transport through the barrier is ruled by the field emission (FE) mechanism, as carriers can easily tunnel through this thin barrier.

Typically, *n*-type SiC is used to fabricate Schottky diodes because of the high mobility of electrons compared to holes and the low SBH values obtained in *n*-type materials compared to *p*-type materials.

The specific contact resistance, which is the resistance of the contact per unit area, is the primary characteristic of an ohmic contact. Ohmic behavior is typically achieved by using tunneling current through the thin potential barrier in SiC. Therefore, the key strategy for achieving low contact resistivity is to increase the surface doping density and select a metal that forms a low barrier height.

In this sense, the choice of metal and sintering after metal deposition are critical factors in the formation of good ohmic contacts. In the case of SiC, it is important to be aware of the reactions that can occur between the metal and Si or C as a function of temperature. The high temperature can cause reactions between the metal and the SiC, leading to the formation of unwanted phases and degradation of the device performance. Therefore, it is essential to carefully select the metallization materials and optimize the processing conditions.

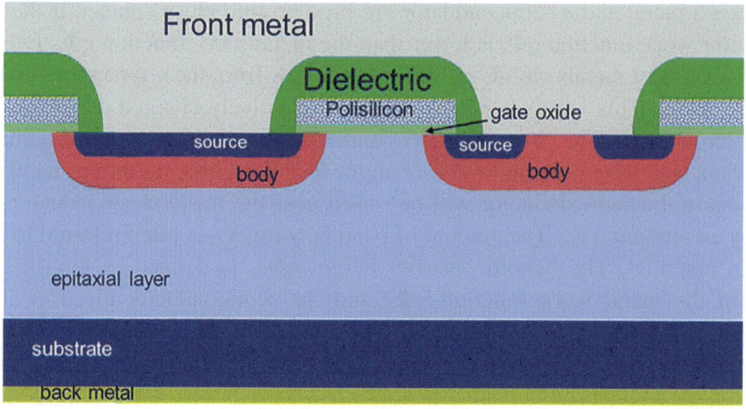

Fig. 3.22 Schematic of a metal-oxide-semiconductor field-effect transistor (elementary cells)

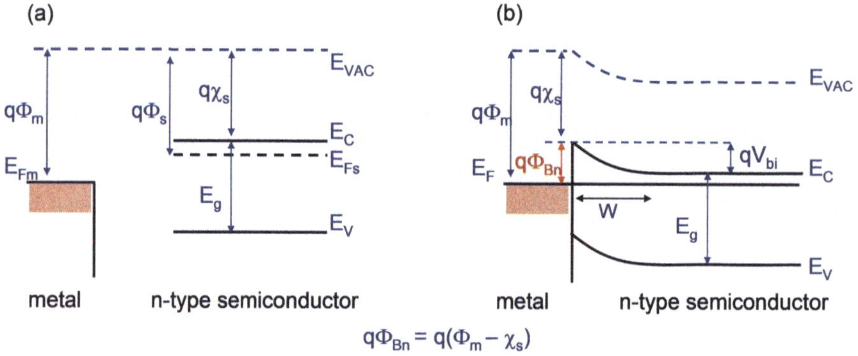

Fig. 3.23 Schematic representation of the energy band diagram for a metal contact onto a *n*-type semiconductor, before (**a**) and after (**b**) they are brought into intimate contact, showing the formation of the Schottky barrier height qΦBn

In the standard manufacturing process flow of a Schottky diode, a high-temperature rapid thermal process (900–1000 °C) is typically required for the formation of the nickel silicide (Ni_2Si) backside ohmic contact [27]. This process, however, must be carried out before defining the front side metal (e.g., Ti and Mo) to prevent undesired interface reactions and electrical degradation of the Schottky barrier. Hence, this standard silicidation process today represents a technological bottleneck when manufacturing brittle thinned wafers.

In this sense, laser annealing represents an alternative solution for achieving the silicidation with a limited heat transfer and, if applied to SiC technology, can enable to complete the diode front side process first and then process the backside contact without detrimental effects for the Schottky barrier. Indeed, in the last years, some works [28, 29] reported on the formation of ohmic contacts to SiC using laser annealing processes, suitable both for diode and MOSFET technology (Figs. 3.22 and 3.23).

References

1. S. Ortolland, C. Raynaud, J.P. Chante, M.L. Locatelli, A.A. Lebedev, A.N. Andreev, N.S. Savkina, V.E. Chelnokov, M.G. Rastegaeva, A.L. Syrkin, Effect of boron diffusion on high-voltage behavior of 6H-SiC *p+nn+* structures. J. Appl. Phys. **80**, 5464–5468 (1996)
2. E. Oliviero, M. Lazar, H. Vang, C. Dubois, P. Cremillieu, J.L. Leclercq, J. Dazord, D. Planson, Use of graphite cap to reduce unwanted post-implantation annealing effects in SiC. Mater. Sci. Forum **556–557**, 611–614 (2007)
3. K.A. Jones, M.C. Wood, T.S. Zheleva, K.W. Kirchner, M.A. Derenge, A. Bolonikov, T.S. Sudarshan, R.D. Vispute, S.S. Hullavarad, S. Dhar, Comparison of graphite and BN/AlN annealing caps for ion implanted SiC. Mater. Sci. Forum **556-557**, 575–6578 (2007)
4. V. Veliadis, L.S. Chen, E. Stewart, M. McCoy, T. McNutt, S. Van Campen, C. Clarke, G. DeSalvo, 2.1 mΩ-cm^2, 1.6 kV 4H-Silicon Carbide JFET for power applications, in International Semiconductor Device Research Symposium, presentation TP2-04, December 2005
5. V. Veliadis, M. Snook, T. McNutt, H. Hearne, P. Potyraj, A. Lelis, C. Scozzie, A 2055-V (at 0.7 mA/cm2), 24-A (at 706 W/cm2) normally-ON 4H-SiC JFET with 0.068-cm2 active area and blocking voltage capability of 94% of the SiC material limit. IEEE Electron Dev. Lett. **29**(12), 1325–1327 (2008)
6. V. Veliadis, M. McCoy, E. Stewart, T. McNutt, S. Van Campen, P. Potyraj, C. Scozzie, *Exploring the design space of rugged seven lithographic level silicon carbide vertical JFETs for the development of 1200-V, 50-A devices*, in International Semiconductor Device Research Symposium, pp. 1–2, 2007
7. X. Guo, A.L. Beck, X. Li, J.C. Campbell, Study of reverse dark current in 4H-SiC avalanche photodiodes. IEEE J. Quantum Electron. **41**, 562–567 (2005)
8. S. Sanzaro, C. Bongiorno, P. Badalà, A. Bassi, G. Franco, P. Vasquez, A. Alberti, A. La Magna, Appl. Surf. Sci. **539**, 148218 (2021)
9. S. Sanzaro, C. Bongiorno, P. Badalà, A. Bassi, I. Deretzis, M. Enachescu, G. Franco, G. Fisicaro, P. Vasquez, A. Alberti, A. La Magna, Materials **14**, 4769 (2021)
10. C. Berger, D. Alquier, J.F. Michaud, Electronics **13**, 217 (2024)
11. C. Berger, D. Alquier, M. Bah, J.F. Michaud, Mater. Sci. Semicond. Process. **151**, 106983 (2022)
12. P. Badalà, I. Deretzis, S. Sanzaro, F.M. Pennisi, C. Bongiorno, G. Fisicaro, S. Rascunà, G. Bellocchi, A. Bassi, M. Boscaglia, D. Pagano, P. Vasquez, M. Enachescu, A. Alberti, A. La Magna, Solid State Phenom. **344**, 15 (2023)
13. P. Badalà, E. Smecca, S. Rascunà, C. Bongiorno, E. Carria, A. Bassi, G. Bellocchi, S. Castorina, C. Tringali, A. La Magna, A. Alberti, Mater. Sci. Forum **1062**, 417 (2022)
14. S. Rascunà, P. Badalà, C. Tringali, C. Bongiorno, E. Smecca, A. Alberti, S. Di Franco, F. Giannazzo, G. Greco, F. Roccaforte, M. Saggio, Mater. Sci. Semicond. Process. **97**, 62 (2019)
15. P. Badalà, S. Rascunà, B. Cafra, A. Bassi, E. Smecca, M. Zimbone, C. Bongiorno, C. Calabretta, F. La Via, F. Roccaforte, M. Saggio, G. Franco, A. Messina, A. La Magna, A. Alberti, Materialia **9**, 100528 (2020)
16. T. Kimoto, J.A. Cooper, *Fundamentals of Silicon Carbide Technology* (Wiley, Singapore, 2014)
17. A. Hallén, M. Linnarsson, *Ion Implantation Technology for Silicon Carbide*

18. T. Hosoi et al., Appl. Phys. Exp. **15**, 6 (2022)
19. J. Rozen et al., J. Appl. Phys. **103**, 124513 (2008)
20. Y. Katsu et al., Mater. Sci. Forum **858**, 599 (2016)
21. M. Hayashi et al., Jpn. J. Appl. Phys. **57**, 04FR15 (2018)
22. K. Moges et al., Appl. Phys. Express **11**, 101303 (2018)
23. T. Nakanuma et al., Jpn. J. Appl. Phys. **61, SC1065** (2022)
24. T. Hosoi et al., Appl. Phys. Express **14**, 101001 (2021)
25. F. Roccaforte, G. Brezeanu, P.M. Gammon, F. Giannazzo, S. Rascunà, M. Saggio, *Schottky contacts to Silicon Carbide: physics, device technology and applications* (2018)
26. E.H. Rhoderick, R.H. Williams, *Metal-Semiconductor Contacts* (Oxford Science Publications, Oxford, 1988)
27. F. Roccaforte, M. Vivona, G. Greco, R. Lo Nigro, F. Giannazzo, S. Rascunà, M. Saggio, Proceedings of ICSCRM2017. Mater. Sci. Forum (2018)
28. T. Tabata, S. Halty, I. Toqué-Trèsonne, F. Mazzamuto, K. Huet, Y. Mori, Proceedings of the 21st International Conference Ion Implantation Technology (IIT2016) (2016) https://doi.org/10.1109/IIT.2016.7882917
29. F. Mazzamuto, S. Halty, H. Tanimua, Y. Mori, Mater. Sci. Forum **858**, 565 (2016)

John R. Silk has over three decades of industry experience in a variety of Southern California semiconductor companies. He has effectively fulfilled roles in both engineering and management in the areas of product engineering, operations, and quality assurance.

Dr. Victor Veliadis is the executive director and CTO of PowerAmerica, a member-driven consortium of industry, universities, and national labs accelerating the commercialization of energy-efficient silicon carbide and gallium nitride power semiconductor chips and electronics. At PowerAmerica, he has managed a budget of $156 million that he strategically allocated to over 210 industrial and university projects to catalyze SiC and GaN semiconductor and power electronics manufacturing, workforce development, and job creation. His PowerAmerica educational activities have trained 430 full-time students in collaborative industry/university WBG projects and engaged over 7000 attendees in tutorials, short courses, and webinars. Dr. Veliadis is currently negotiating a $64 M US Department of Energy PowerAmerica renewal to further catalyze WBG power technologies.

Dr. Victor Veliadis is an ECE professor at NC State and an IEEE fellow and EDS distinguished lecturer. He has 27 issued US patents, 6 book chapters, and over 150 peer-reviewed publications to his credit. He has given over 150 keynote/tutorial/invited presentations including keynotes at ICSCRM, APEC, ECCE, and WiPDA. Prior to entering academia and taking an executive position at PowerAmerica in 2016, Dr. Veliadis spent 21 years post-PhD in the semiconductor industry where his work included design, fabrication, and testing of SiC devices, GaN devices for military radar systems, and financial and operations management of a commercial semiconductor fab. Dr. Veliadis has a Ph.D. degree in electrical engineering from John Hopkins University (1995).

Dario Tenaglia has a Bachelor's Degree in Materials Science from the University of Catania in 2004. He joined ST in January 2005 and, after 13 years in R&D, moved to the manufacturing organization as process development engineer in 2018. Currently, he is involved in the new process and technology development especially in the etching, cleaning, and removal areas.

Paolo Badalà received his M.Sc. Degree in Physics from the University of Catania (Italy) in 2004. He joined STMicroelectronics in 2007. His current research interests include the development of thin-film deposition and laser annealing processes, mainly for wide bandgap semiconductor technology. He is co-author of several patents and papers published in international journals and conference proceedings.

Simone Rascuna has been working in research and development at STMicroelectronics for high-power devices since 2002. He has extensive experience in developing silicon super-junction MOSFET technologies and shifted his focus to compound semiconductors for high-power applications in 2010. As an advanced research senior manager, he leads the design and development group for silicon carbide diodes and has numerous patents in the field of SiC devices, including the fifth generation of the MDmesh super-junction technology.

Chapter 4
SiC Technology

Rudolf Elpelt, Andrea Piccioni, Jake Choi, Byeongyeop Lee, Sandol Lee, Paul Salmen, Olaf Zschieschang, Sebastian Fahlbusch, Ryo Takeda, Takamasa Arai, Llew Vaughan-Edmunds, and Victor Veliadis

This chapter of this insightful book delves deeply into the realm of silicon carbide (SiC) technology, elucidating critical aspects of design, operation, and reliability of SiC power devices. Authored by a distinguished panel of experts from leading companies and academic institutions, this chapter encapsulates a comprehensive overview of SiC device advancements. Rudolf Elpelt from Infineon Technologies initiates the discussion by dissecting the design intricacies and material selection for vertical power devices. Following this, Andrea Piccioni, also from Infineon, sheds light on SiC metal-oxide-semiconductor field-effect transistor (MOSFET)

R. Elpelt · P. Salmen
Infineon Technologies, Erlangen, Germany

A. Piccioni
Infineon Technologies, Villach, Austria

J. Choi
onsemi, South Portland, ME, USA

B. Lee · S. Lee
onsemi, Bucheon-si, Gyeonggi-do, Republic of Korea

O. Zschieschang
onsemi, Aschheim, Germany

S. Fahlbusch
Nexperia, Hamburg, Germany

R. Takeda · T. Arai
Keysight Technologies, Hachioji-shi, Tokyo, Japan

L. Vaughan-Edmunds
Navitas, Torrance, CA, USA

V. Veliadis (✉)
PowerAmerica, Raleigh, NC, USA
e-mail: jvveliad@ncsu.edu

M. Di Paolo Emilio (ed.), *SiC Technology*,
https://doi.org/10.1007/978-3-031-63418-5_4

switching cycles, highlighting operational nuances and performance optimization strategies. The subsequent sections authored by engineers from onsemi, Infineon, Nexperia, Keysight Technologies, Navitas, and PowerAmerica/NC State University delve into topics ranging from reliability challenges and packaging innovations to the dynamics of SiC FETs and junction field-effect transistors (JFETs). Each contribution underscores the transformative impact of SiC technology on power electronics, addressing key issues such as thermal management, reliability enhancement, and efficiency improvements. This collaborative effort by industry pioneers provides valuable insights into harnessing SiC's potential for high-voltage applications, renewable energy integration, and electrified transportation.

4.1 Structure and Design

Rudolf Elpelt

In this chapter, we focus on vertical power devices in which the flow of load current is from an active cell structure on one main surface (frontside) of the wafer to the opposite surface (backside) through the substrate and active layers. However, it should be mentioned that also in silicon carbide lateral devices for example, CMOS, integrated circuits are possible for special applications and conditions.

4.1.1 Vertical

Silicon carbide exists in a variety of polytypes—different crystalline structures that differ in basic properties, such as band structure, carrier mobility, and anisotropy of parameters along certain crystal directions. For power devices, the 4H polytype is generally the material of choice due to its high bandgap (~3.26 eV) and relatively low anisotropy of electron mobility. 4H means that the atoms are arranged in layers in a hexagonal grid. The stacking sequence of subsequent layers has a periodicity of 4, i.e., the crystal structure is repeated every fourth layer. Typically, after processing the active structures on the frontside of the wafer, the substrate is thinned from the backside, for example, by grinding, to limit the parasitic influences on electrical and thermal resistance of the device.

The buffer layer that epitaxially grows on the substrate prior to the active drift layer is typically highly doped (~1e17–~1e19/cm^3). The buffer thickness can range from a few hundreds of nanometers to several micrometers. Penetration of the electric field through the buffer during operation is limited or inhibited to protect the deeper layers/interfaces that have a higher concentration of defects. Due to the high doping, the buffer contributes only marginally to the total on-state resistance.

The active device drift layer that subsequently grows on the buffer layer accommodates the electric field in the blocking state and determines the static and dynamic losses during conduction and switching. Doping and thickness of the drift layer depend on the desired voltage blocking class. Due to the ~10 times higher electrical breakdown field, the resistance of a silicon carbide layer can be approximately a factor of 1000 lower compared to that of a silicon layer for the same breakdown voltage if mobility is considered the same for the sake of simplicity. This is why unipolar silicon carbide devices are alluring for high-voltage applications where otherwise bipolar silicon insulated-gate bipolar transistor (IGBT) devices would be used.

However, in unipolar devices, there is a trade-off between on-state losses and voltage class: higher voltage → lower doping/greater thickness → increase in R_{DSon}.

In unipolar devices, the current transport is partly governed by the electron mobility of the drift layer. Therefore, the resistance is dependent on the temperature. At higher blocking voltages, the dependence of R_{DSon} on the drift layer parameters and temperature can become dominant. To overcome or further push the resistance limits of the unipolar drift layer, it is possible, in principle, to build a charge-compensated drift layer—a so-called super junction as it is known in unipolar high-voltage silicon power devices. The drift layer, or part of the drift layer, is built by pillars of n- and p-type doping. The net doping of two adjacent pillars is effectively zero, i.e., intrinsic. The doping levels of n- and p-pillars can therefore be increased well over the standard drift zone level to yield a significantly lower resistance. However, if the charge balance is maintained, the blocking capability will still be determined by the thickness of the quasi-intrinsic drift layer. Of course, the realization of super junction pillars in silicon carbide poses additional challenges compared to silicon (Fig. 4.1).

4.1.2 Cell

Similar to conventional silicon, the basic power device types in vertical silicon carbide are unipolar and bipolar switches and diodes. However, due to the large bandgap, pure bipolar devices typically only become attractive at blocking voltages beyond ~10 kV.

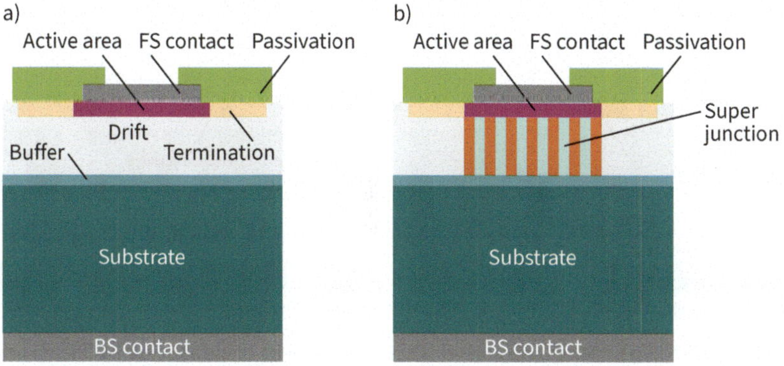

Fig. 4.1 (**a**) Conventional structure of a vertical power device. (**b**) Super junction structure

4.1.3 Switches

Field-effect transistors (MOSFETs and IGBTs) based on metal oxide semiconductors have become the most common type of switching device in the world of silicon. Research on silicon carbide MOSFET structures started in the 1990s. However, it has matured and made its way to large-scale industrialization only within the last decade.

Due to the material properties and process specialties of silicon carbide, formation of the gate oxide to silicon carbide interface is much more critical here than in silicon technology. SiO_2 is the native oxide for silicon carbide as it is for silicon. However, the excess carbon at the interface leads to defects and dangling bonds that reduce the interface electron mobility and increase the interface state density. Carriers can get trapped at the interface depending on the current condition and applied gate voltage. These trapped carriers contribute to scattering or the gate voltage hysteresis effect. The growth process of gate oxide or deposition and subsequent passivation of the interface are therefore crucial for the electrical properties and long-term reliability of MOS devices. The thermal budget of subsequent processes such as ohmic contact formation also needs careful consideration.

The electrical breakdown strength of silicon carbide is roughly ten times higher than that of silicon (2–3 MeV/cm). The field stress on the gate oxide in silicon carbide devices is therefore significantly higher than in silicon devices. Due to the difference in dielectric permittivity between the oxide and the semiconductor (SiO_2/SiC, 3.9:10), the electric field in the oxide can easily reach several MeV/cm. Electric field stress is important for device reliability and needs to be considered carefully when designing the device. The gate oxide structures in silicon carbide are typically designed with a p-type shielding structure, whereas this is not necessary in silicon MOS devices (Fig. 4.2).

There are two main ways to build a basic MOSFET cell structure. One way is by creating the channel laterally on the surface of the semiconductor. Such a device is normally referred to as a DMOS (double-diffused MOS). The naming stems from silicon technology, although the diffusion of dopants is generally negligible in silicon carbide. The advantages of the DMOS structure are as follows:

- Relatively low technological complexity.
- Can be realized with rather shallow implantation profiles.

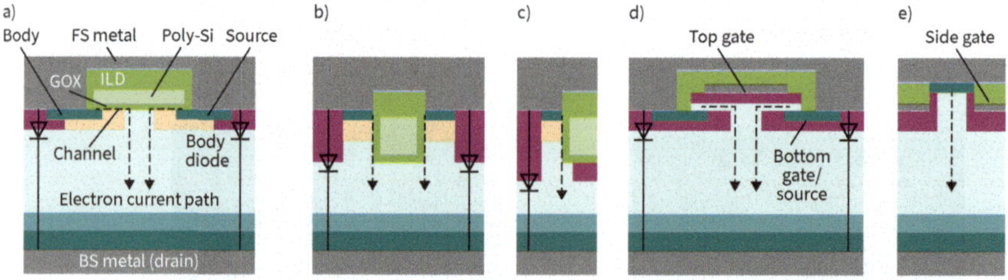

Fig. 4.2 (**a**) DMOS, (**b**) double-channel trench MOS, (**c**) single-channel trench MOS, (**d**) lateral-channel vertical JFET, and (**e**) vertical-channel JFET (static induction transistor—SIT)

The drawbacks are as follows:

- Limited shrink capability due to the lateral surface channel.
- Lower electron mobility at the gate oxide interface due to surface roughness.
- Higher interface state density compared to other crystal planes.

The second possibility is by using the trench MOSFET structure (also called U-MOS), where a vertical channel is created by etching a trench groove into the silicon carbide.

The advantages of U-MOS are as follows:

- Possibly increased channel density by pitch shrink due to the vertical channel.
- Possible utilization of the crystal planes with higher electron mobility in silicon carbide.

The disadvantage in this case is that the gate oxide at the trench bottom typically needs a deep p-type shielding structure to limit the electric field in the blocking state, and this can increase the process complexity and cost. Careful design optimization for a trade-off between the on-state performance and reliability is thus necessary.

Junction field-effect transistor (JFET) devices utilize the voltage-controlled space charge region of a pn-junction to control a conductive channel. They do not possess a critical gate oxide interface since the electron channel is formed in the bulk of the semiconductor. This can be beneficial in terms of reliability. However, properties of the electrical device, such as R_{DSon} and gate pinch-off voltage, V_{po} (the equivalent of V_{GSth} in the MOSFET), depend on the precise control and homogeneity of the doping profiles in the channel region. This can create some challenges in the designing and manufacturing of JFET devices.

JFETs are usually designed as normally-on devices. This means that the channel conducts at 0 V applied gate voltage and depletes and closes when negative gate voltage is applied. This is in contrast to the industry standard normally-off MOSFET devices and requires special measures depending on the application. A common method is to combine the normally-on JFET with a low voltage normally-off silicon MOSFET in series—a so-called cascode circuit. In the on-state, the low-voltage MOSFET only contributes marginally to the total resistance. In the off-state, the low-voltage MOSFET blocks and provides the pinch-off voltage to the JFET, which takes over the major part of the blocking voltage. In its simplest form, this cascode circuit appears as a standard, three-pin MOSFET device on the outside and can be integrated, for example, in standard TO packages.

4.1.4 Diodes

Diodes are devices with a rectifying junction that conduct load current in one polarity and block current in the opposite polarity. In silicon carbide semiconductors, mainly two types of diodes exist—the bipolar pn-diode and the unipolar Schottky barrier diode (SBD).

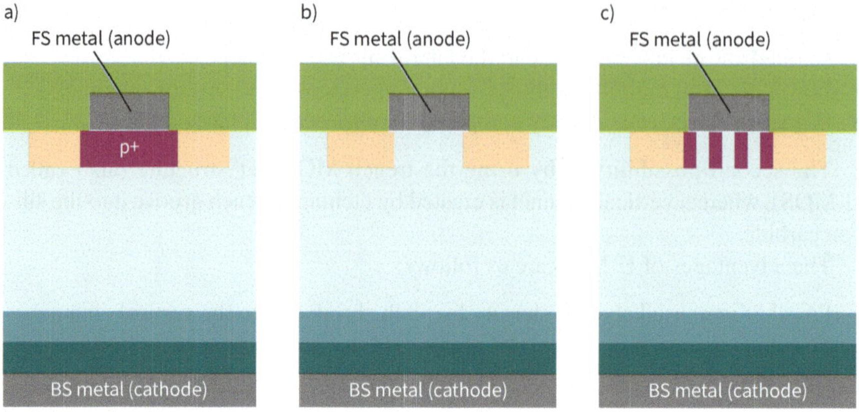

Fig. 4.3 (**a**) Plain pn-diode, (**b**) plain Schottky barrier diode (SBD), and (**c**) merged pn-Schottky diode (MPS)

A plain pn-diode exhibits a built-in voltage barrier for conducting a forward current of approximately 2.7–2.8 V at room temperature. Therefore, these devices are only attractive in terms of conduction loss at high voltages—in the range of 4 kV and above. Though bipolar silicon carbide devices have not been industrialized so far, there is ongoing investigation and research on their use in special high-voltage applications.

A Schottky barrier diode utilizes the rectifying junction of the metal semiconductor interface with typical barrier values, in the range of 0.9–1.2 eV, depending on the work function difference between the metal and the semiconductor. Examples of Schottky barrier metals with silicon carbide are molybdenum and titanium.

Schottky barrier diodes offer low losses in the forward direction. However, the temperature-dependent leakage current in plain Schottky barrier diodes limits the diode performance as very high electric fields must be avoided in the blocking state. The merged pn-Schottky diode (MPS) provides a significant improvement to this as it combines pn-junctions with metal junctions. The p-wells shield the electric field from the Schottky barriers, which lowers the leakage current in the blocking state. Secondly, for large surge currents during forward conduction, the p-wells can inject holes and, thus, enable the device to sustain high currents in the bipolar mode (Fig. 4.3).

4.1.5 Border

In the active area and in most parts of the device, the voltage applied in the blocking state is well accommodated within the drift layer. At the edge of the active area, the direction of the voltage drop is not only in the vertical (as it is in the active area) but also in the lateral direction. This makes edge termination in a power device a two-dimensional and, sometimes, even a three-dimensional problem.

The termination structure facilitates the decay of voltage on the active frontside (source or anode) over a certain lateral distance and avoids or attenuates field peaks where early avalanche breakdown could happen. Except boron, all the common dopants such as nitrogen, phosphorus, and aluminum exhibit an almost negligible diffusion in silicon carbide. This implies that implanted doping profiles stay as implanted and cannot be diffused by thermal processing. Therefore, implanted termination structures always exhibit rather sharp pn-junctions with distinct field peaks. The field peaks need to be tuned by the implanted doping density and by the geometric structure of the implantation, i.e., distance, depth, and overlap. In general, the lateral size of the termination structure depends on the voltage class or the desired blocking voltage and may range from a few micrometers for very low-voltage (<400 V) to several 100 μm for ultrahigh-voltage devices (>6.5 kV).

The standard termination structure is the junction termination extension (JTE), which is a single, shallow, low-doped p-type implantation adjacent to the main, high-doped p-type junctions in the active area. The lower-doped JTE layer depletes partially during blocking and distributes the electric field between the inner p+/p− junction and the outer p−/n-drift junction. This simple type of termination structure has the advantage of relatively low complexity and cost as it is only one additional mask layer and implantation. However, balancing the electric field and stability depends strongly on the implanted doping concentration, and therefore a single JTE might be more prone to reliability issues under harsh environmental conditions than other, more sophisticated termination schemes. To counter this, the single-step JTE can be enhanced by two or more lateral adjacent p-implantations with decreasing doping, for example (Fig. 4.4).

Another common way to distribute the electrical potential laterally across the termination is through field guard rings (FGRs). These are a sequence of p-implanted stripes surrounding the active area. The single rings are not fully depleted when the

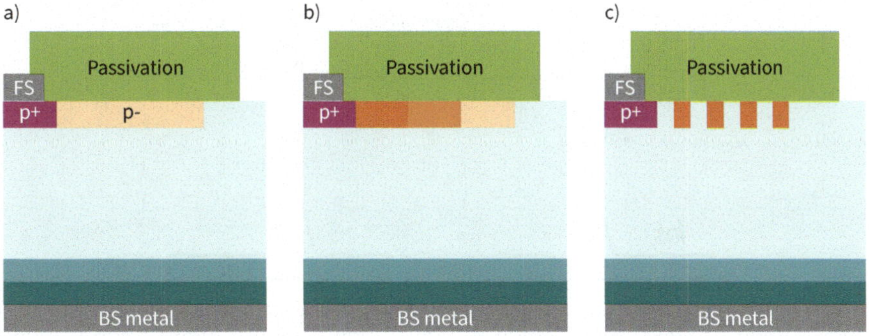

Fig. 4.4 (**a**) Single-step junction termination extension (JTE), (**b**) multistep JTE, and (**c**) field guard ring (FGR) termination

device is in the blocking state. The electric potential then decays stepwise from ring to ring. The total size and lateral dimension depend on the desired blocking voltage, as mentioned before. FGRs can, for example, also be formed with or combined with trench-etched structures—single or multistep JTEs.

Last but not the least, the dielectric passivation layers need to be designed based on the requirements of the package and application. A simple passivation, for example, a molded package, can consist of an organic layer such as polyimide. A more advanced passivation might use an inorganic and organic multilayer stack, for example, with oxide, nitride, and polyimide.

4.2 Cracking the Code of SiC MOSFET Switching Cycles: Understanding the Variable and Parameter Driving Performance and Efficiency

Andrea Piccioni

The shift from Si-based to wide bandgap devices, such as SiC MOSFETs, has brought many benefits that are now widely recognized in the power electronics field. SiC MOSFETs have enabled non-multilevel, unipolar power conversion in applications where conventional bipolar-based converters were once the only option (e.g., renewable energy and traction systems). This has improved efficiency in low- to mid-load operations, significantly reduced turn-off energy, and eliminated the need for separate diode chips in switching cell designs. What was once only possible in less than 1000 V applications is now achievable in more than 1000 V applications, thanks to SiC MOSFETs and SiC Schottky diodes. To further understand the benefits of a SiC MOSFET-based converter, a breakdown of a typical hard-switching cycle is mandatory, especially with parameters that affect performance and their influence in operation based on design and boundary conditions (see Fig. 4.5).

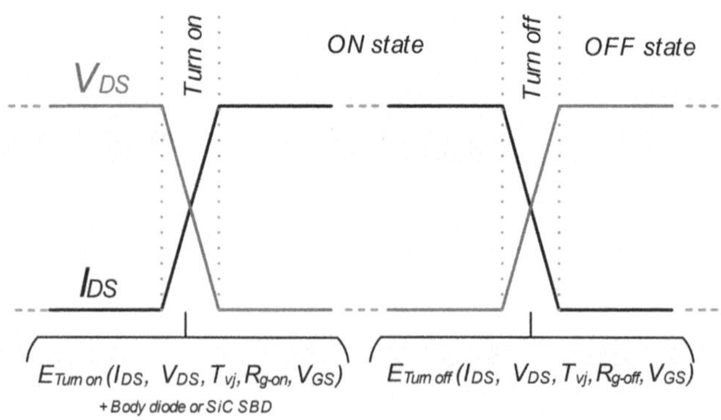

Fig. 4.5 Ideal hard-switching cycle decomposition and energy loss dependency with the boundary and design conditions

4.2.1 *Conduction Loss:* On-*State*

Conduction losses and, thus, the $R_{DS(on)}$ of SiC MOSFETs have several contributors, as defined in Eq. 4.1:

$$R_{DS(on)} = R_{ch} + R_{JFET} + R_{epi} + R_{sub} \tag{4.1}$$

Equation 4.1 Decomposition and different structure contribution of the $R_{DS(on)}$.

Based on the cell design and voltage class, some contributors are more pronounced than others. Of particular importance is the channel resistance. It has a negative temperature coefficient, i.e., $V_{GS(th)}$ decreases with increasing temperature, as defined in Eq. 4.2:

$$R_{ch} = \frac{L_{ch}}{\mu_n C_{ox} W_{ch} \left(V_{GS(on)} - V_{GS(th)}\right)} \tag{4.2}$$

Equation 4.2 Channel resistance approximated formula, where L_{ch} is the length of the channel, W_{ch} is the width of the channel, μ_n is the free electron mobility, C_{ox} is the gate oxide capacitance, $V_{GS(on)}$ is the positive on-state gate voltage, and $V_{GS(th)}$ is the threshold voltage of the device.

The higher the contribution of channel resistance in a design, the lesser the dependence of conduction loss on temperature (also known as the RDS(on)HOT / RDS(on)COLD ratio). Generally, in all commercially available SiC MOSFETs, the total SiC MOSFET RDS(on) shows a positive temperature coefficient for temperatures above 25 °C. This enables static current balancing in paralleled devices. Another important contributor to RDS(on) is the positive gate supply, VGS(on). It affects conduction losses directly. The higher the VGS(on), the lower the channel resistance. Due to the strong electric field in silicon oxide, a trade-off between performance and gate oxide reliability must be considered. Manufacturers provide the recommended positive gate voltage for long-term reliable operations in their datasheet.

4.2.2 *Off-State*

The off-state is loss-free since there is no current flowing into the device while blocking the drain-to-source voltage. The zero gate voltage drain current, also known as leakage current, is usually specified in the datasheet in the range of the μA but is usually in the range of the tens of nA at the given blocking voltage of the device, e.g., for 1200 V, the dissipation in on-state is in the order of tens of μW. In wide bandgap devices due to their disruptive commutation speed, a hidden risk emerges even under the "safe" *off*-state. It is the so-called parasitic turn-on (PTO).

PTO represents an unwanted turn-on caused by the complementary switch, positive dV_{DS}/dt. Due to Miller capacitance (C_{GD}) coupling, this affects the stability of the *off*-state causing a half-bridge short as defined in Eq. 4.3:

$$\frac{\mathrm{d}V_{GS}}{\mathrm{d}t} = \frac{1}{C_{GS}}\left(\frac{\mathrm{d}V_{DS(\text{complementary }sw)}}{\mathrm{d}t}C_{GD} - \frac{V_{GS}}{R_{GOFF}}\right) \tag{4.3}$$

Equation 4.3 Gate-to-source voltage variation due to Miller capacitance coupling.

Several methods, such as using a negative V_{GS} supply, increasing the R_{gon}/R_{goff} ratio, and using an active Miller clamp, can be implemented to prevent or mitigate the V_{GS} spikes that can result in a PTO.

4.2.3 Turn-On Process and Influence of Diode Performance

Figure 4.6 shows the turn-on process with its different phases, from phase 1 (initial still off) to 4 (fully turned on), of a 40 mΩ CoolSiC™ MOSFET in a TO247–4 package (part number of the DUT IMZA120R040M1H). During phase 1 (turn on delay), no current flows in the chip and the device can be considered to be in the

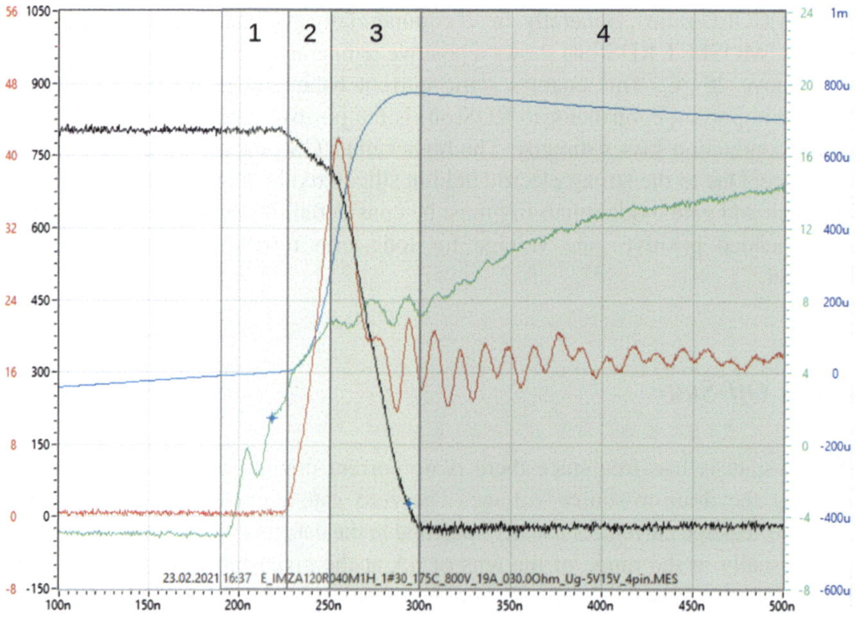

Fig. 4.6 Turn-on of a 40mΩ four-pin TO247 CoolSiC™ MOSFET with explication of the different phases from off-state to on-state

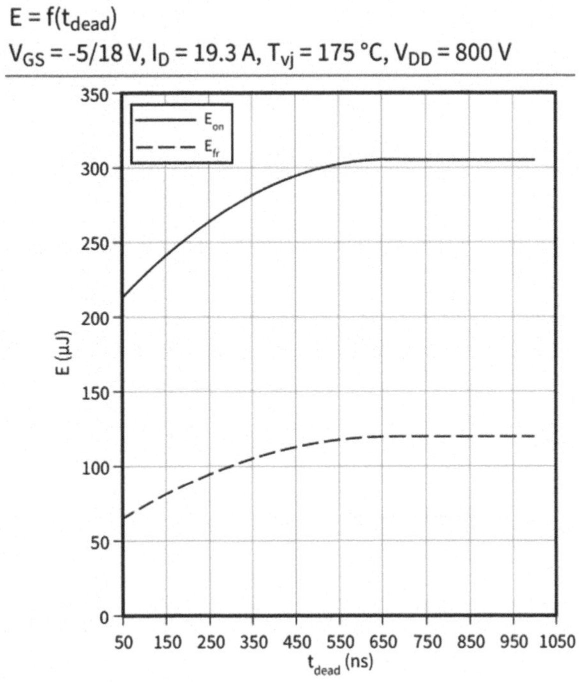

Fig. 4.7 Depiction of the turn-on and forward recovery energy function of the dead time as per IMZA120R040M1H datasheet

off-state. The gate driver supplies the charging process of CGS till V_{GS} reaches ~4.1 V, i.e., $V_{GS(\mathrm{th})}$. Hereafter, the current starts flowing toward the chip, beginning phase 2 (current rise). In phase 2, as soon as the current flows toward the MOSFET channel, C_{oss} slowly discharges. This causes a drop in V_{DS} due to parasitic inductances. Here, phase 3 (Miller plateau) starts, where the complete discharge of C_{oss} causes voltage transition and exit of saturation region. After this, the device can be considered as fully on, i.e., in phase 4. However, during phase 2, the SiC MOSFET's turn-on performance is influenced by the Q_{rr} performance of the diode, which is switching against, in contrast to their wide bandgap counterpart, the GaN devices. Consequently, there is a significant loss difference when using the internal bipolar body diode or a unipolar SiC SBD that has almost negligible reverse recovery. Dead time also plays a pivotal role in increasing the turn-on performance. Decreasing the duration of the dead time means that the third channel of the SiC MOSFET will conduct for a longer time than the body diode, as shown in Fig. 4.7.

4.2.4 Turn-Off Process and Overvoltage Spikes

Figure 4.8 shows the turn-off process with its different phases, from phase 1 (initial still on) to 4 (fully turned off), of the same DUT under the same conditions. During phase 1 (turn-off delays), the C_{GS} is discharged till V_{GS} reaches the Miller plateau. In

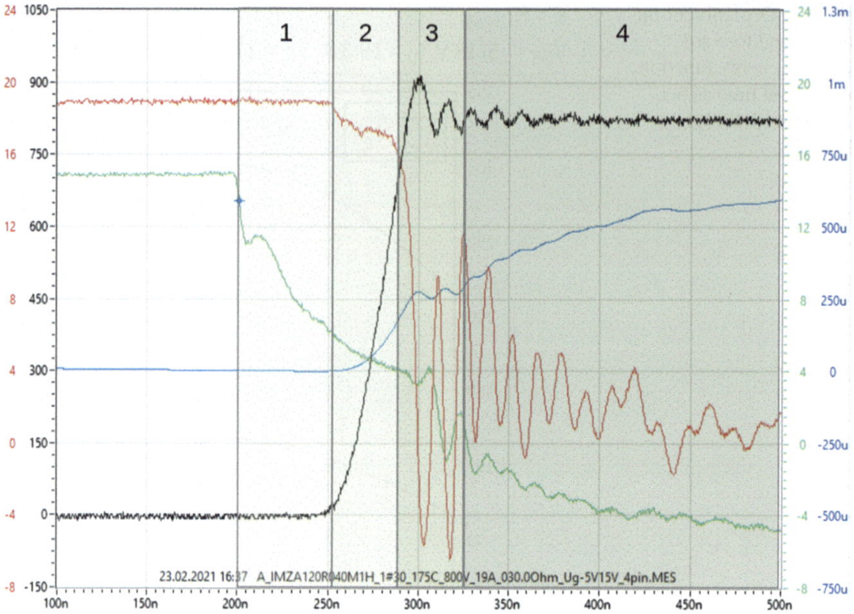

Fig. 4.8 Turn-off of a 40mΩ four-pin TO247 CoolSiC™ MOSFET with explication of the different phases from on-state to off-state

phase 2, the voltage transition begins, and C_{oss} starts getting charged and enters the saturation mode. Then, the MOSFET channel moves the current to the diode, thereby ending phase 3 and the current commutation when $V_{GS} < V_{GS(th)}$. Finally, in phase 4, the device is fully turned off and V_{DS} oscillations are dumped.

Voltage ringing and its peak value during the turn-off phase are of particular interest to researchers and designers. These overvoltage values by themselves cannot even remotely damage the semiconductor (usually the breakdown voltage is 15% higher than the datasheet value). However, the insulation of the load (e.g., electrical machine stator winding) could deteriorate significantly if the overshoots are not considered or mitigated, especially in low- and medium-voltage drive applications, which usually include cables that act as amplifiers for the abovementioned voltage overshoot.

The groundbreaking speed of SiC MOSFET commutations represents a major blocking point, after short circuit withstand time, for penetration of wide bandgap devices into electric drive application markets. Here, SiC MOSFETs face a significant challenge from bipolar devices, such as Si diodes and IGBTs. Afterall, these devices also target to electrical drive applications in which softness of the diode, controllability, and acceptable saturation current are required.

4.2.5 Challenges of Parallel SiC MOSFETs

In the pursuit of targeting high-power applications that use SiC-based converters, the design is constrained by the necessity of employing parallel dies (within power modules) or devices (discrete components). This is because commercial single-die SiC MOSFETs generally cap at around 100 amperes. This limitation in achieving high current using individual SiC MOSFET dies arises due to its manufacturing process, specifically defect density and wafer yield. Designers and researchers are, therefore, increasingly getting interested in the idea of using multiple SiC MOSFETs in parallel to collectively achieve higher current ratings. This concept also, however, faces challenges caused due to the spread of parameters and difficulties in achieving design symmetry that is vital for effective parallel operation. Despite assuming an ideal symmetric layout, the spread of parameters ($V_{GS(th)}$ and $I_{D(VGS)}$) during dynamic events (especially for turn-on) among the parallelly connected SiC MOSFET devices can lead to unfavorable effects on the electrothermal equilibrium of the system. Figure 4.9 shows turn-on events with DUTs in a parallel operation, highlighting the current and consequently ΔE_{on} imbalances that arise hand in hand with $\Delta V_{GS(th)}$ between the two DUTs. As shown in Fig. 4.10, during a continuous PWM

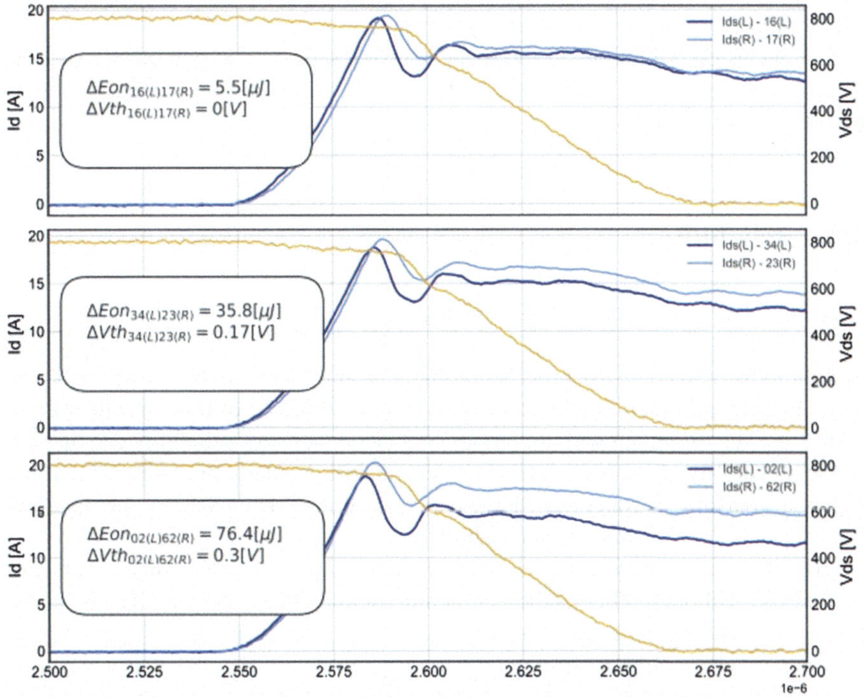

Fig. 4.9 Double-pulse tests with different combinations of samples having different ΔVGS(th)—VDC = 800 V Rg = 42 Ω—DUTs 2xIMZ120R060M1H 2xIDW40G120C5B

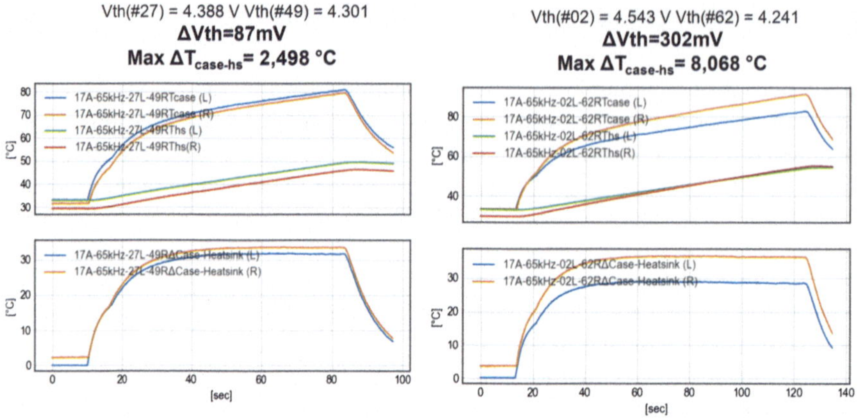

Fig. 4.10 Application test DC-DC boost with thermal camera, Vin = 400 V Vout = 800 V, decoupled heatsink with forced air cooling—DUTs 2xIMZ120R060M1H 2xIDW40G120C5B

operation, the temperatures of the DUTs also differ from one another, and ΔT_{case} becomes proportional to $\Delta V_{GS(th)}$. If no countermeasures are taken (e.g., device screening and clustering and using active gate drivers), this imbalance of temperature can affect the total output current. Improper use of SiC MOSFET devices can lead to degradation in the reliability of the "hotter device."

In conclusion, variations in the manufacturing process and parameters such as $V_{GS(th)}$ and $I_{D(VGS)}$ play a pivotal role in the reliable operation and full utilization of high-power SiC-based converters.

4.3 Reliability

Jake Choi, Byeongyeop Lee, and Sandol Lee

Reliability is defined as the probability that a product, system, or service will perform its intended function adequately for a specified period of time or will operate in a defined environment without failure. Stated another way, reliability can be seen as the probability of success, durability, dependability, quality over time, and availability to perform a function. (defined by ASQ, American Society for Quality, https://asq.org/quality-resources/reliability).

4H-SiC (silicon carbide) has wide bandgap of 3.26 eV than Si (silicon) bandgap of 1.12 eV. Although all crystalline structure yield a wide bandgap, 4H-SiC is the preferred electronic material because it has the widest bandgap (3.26 eV) among the common polytypes and because it has a high electron mobility that is nearly isotropic which has small lattice constants, the strong bond strength. It helps for a high electric breakdown field and thermal conductivity.

SiC semiconductors give many benefits for industrial and automotive applications. And it is facing more and more stringent reliability constraints coming from the increased lifetime expectation and safety requirements. The industrial and energy sectors are also following the same trend, so that strong efforts are made to improve the reliability of power electronic systems with cost-effective and sustainable solutions. In the meanwhile, wide bandgap (WBG) semiconductor devices are capturing increasingly new power applications, as they are quite promising in terms of switching speed, efficiency, thermal management, and high-voltage capability, even if their reliability performance is still far to be demonstrated [1, 2], as a challenge of wide bandgap SiC, high electric field, gate instability, package interaction, and bipolar degradation for the reliability.

4.3.1 Testing

Nowadays, high-power discrete modules are widely used in various applications that suffer from stressful operation conditions. It has many challenges to the reliability of the power system. Silicon carbide (SiC) has good characteristics such as high breakdown electric field, high thermal conductivity, and wide bandgap. The extrinsic and intrinsic reliabilities of SiC are fundamental for addressing the escalating demands of advanced electronic systems, particularly in power electronics. SiC is known for its superior material properties and leverage extrinsic and intrinsic reliability to meet critical performance benchmarks and ensure the longevity of electronic devices. Several key factors from the intrinsic reliability test, including gate oxide integrity, time-dependent dielectric breakdown (TDDB), unclamped inductive switching (UIS), and instability in breakdown voltage (BVDSS), and from the extrinsic reliability test, including die-to-package interaction (environmental, endurance, device test), underscore the imperative need for reliability in SiC MOSFET technology.

Gate Oxide Integrity (GOI)

Gate oxide integrity is a pivotal aspect of SiC MOSFET intrinsic reliability, emphasizing the robustness of the insulating layer between the gate electrode and the channel. SiC MOSFETs rely on gate oxide stability to control the flow of current through the device. Intrinsic reliability ensures that the gate oxide maintains its structural and electrical integrity over time, resisting factors such as high electric fields and voltage overstress. A stable gate oxide is essential for sustaining device performance, preventing malfunctions, and guaranteeing the overall reliability of SiC MOSFETs. The gate oxide reliability can be investigated by gate oxide integrity GOI/VRAMP measurements at room temperature in Figs. 4.11 and 4.12. Figure 4.11 shows gate leakage current at room temperature. This reflects mature, reliable, and manufacturable process control. The breakdown voltages, VBD, are extracted and fitted with Weibull distribution in Fig. 4.12.

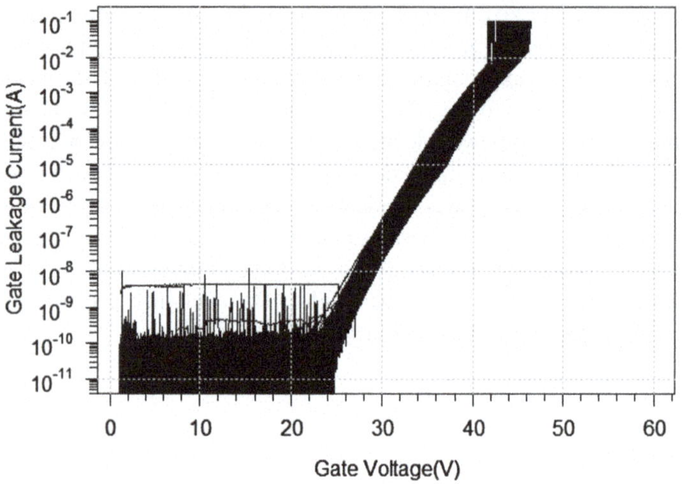

Fig. 4.11 Gate leakage current at room temperature

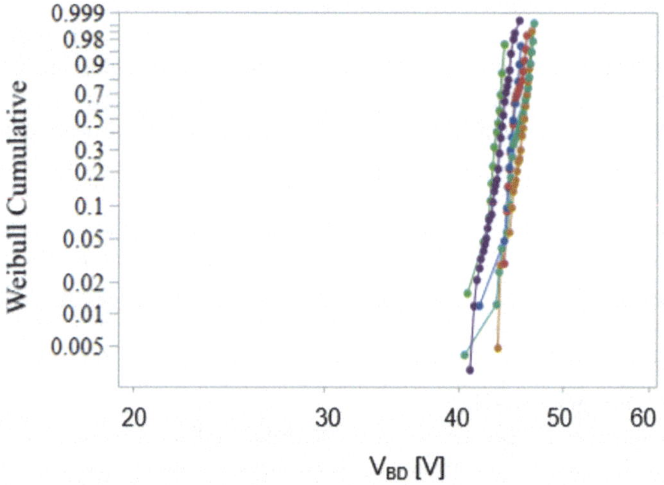

Fig. 4.12 Gate-source breakdown voltage distribution

Time-Dependent Dielectric Breakdown (TDDB)

Intrinsic reliability plays a crucial role in mitigating the impact of time-dependent dielectric breakdown (TDDB), a phenomenon that can compromise the insulating properties of the gate oxide over the device's operational lifetime. TDDB is a time-related degradation mechanism, and SiC MOSFETs with strong intrinsic reliability resist the gradual breakdown of the gate oxide. Enhanced materials and fabrication processes contribute to the prevention of TDDB, ensuring that SiC MOSFETs maintain their insulating properties and overall reliability over extended periods.

Unclamped Inductive Switching (UIS)

Unclamped inductive switching (UIS) is a reliability concern, particularly in power electronic applications where SiC MOSFETs are extensively deployed. UIS occurs during rapid switching, leading to voltage spikes and potential stress-induced failures. SiC's intrinsic properties, such as high breakdown voltage and wide bandgap, effectively mitigate the impact of UIS. The intrinsic reliability of SiC MOSFETs ensures their resilience in high- power switching scenarios, contributing to the robustness and dependability of these devices in power electronics applications.

Breakdown Voltage (BVDSS) Instability

Breakdown voltage (BVDSS) instability is a challenge that can affect the reliability of SiC MOSFETs. Intrinsic reliability becomes imperative in addressing BVDSS instability, ensuring that the breakdown voltage remains consistent and predictable over the device's lifetime. BVDSS instability can lead to unpredictable device behavior and potential failures. SiC MOSFETs with enhanced intrinsic reliability provide stable and consistent breakdown voltages, contributing to the overall reliability of these devices in various applications. From the traditional HTRB (hot temperature reverse breakdown) measurement, there was no changes of BVDSS with no fail, but it was not possible to check whether the charge balance of the drift region was good or not. The new avalanche stability measurement is very similar to HTRB stress, except that the stress time is much shorter and that a constant current avalanche injection is used instead of a constant voltage stress. It could check BVDSS robustness and characterization of walk-in or walk-out with charge balance. To confirm the avalanche stability, BVDSS was measured versus time using constant current injection to the drain electrode at room temperature and high temperature for hundreds of seconds. Figure 4.13 shows that MOSFET exhibits a walk-in capability ranging from 0.6 to 12.0 V. It showed high performance at breakdown voltage without any workout issues. This high performance can be attributed to the optimization of MOSFET, which is designed to have balanced charges.

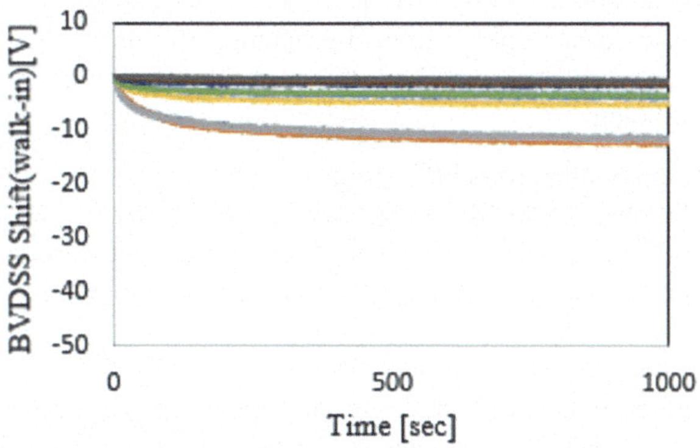

Fig. 4.13 BV_{DSS} (walk-in) with constant current injection in drain

4.3.2 Extrinsic (Product/Package Level) Reliability

Preconditioning
This test is to simulate the board mounting process according to moisture sensitive level. The target is a surface-mount device assuming that it can cause reliability failure and issue due to moisture absorption and elevated temperature during board mounting.

Gate Stress Test (Gate Bias)
It is a test to check gate integrity via a gate bias stress under a high-temperature condition. The test ensures that the gate leak and threshold voltage instability mechanism are triggered by oxide trap filling and oxide trap activation. The HTGB test biases gate or other oxides of the device samples. The devices are normally operated in a static mode or dynamic mode at, or near, maximum-rated oxide breakdown voltage levels.

Breakdown Voltage (BVDSS) Instability (Reverse Bias)
It is configured to reverse bias major power handling junctions of the device samples. The devices are characteristically operated in a static/dynamic operating mode at, or near, maximum-rated breakdown voltage and/or current levels. The test could ensure the breakdown voltage and leakage instability mechanism.

Humidity Endurance Test
The steady-state bias or non-steady-state bias lifetime test is performed for the purpose of evaluating the reliability of non-hermetic packaged devices operating in humid environments. This test is a standard for accelerated life testing with respect to corrosion, discoloration, and other humidity penetration-driven degradation mechanisms. It employs severe conditions of temperature, humidity, and bias which accelerate the penetration of moisture through the external protective material (encapsulant or seal) or along the interface between the external protective materials and the metallic conductors passing through it. When moisture reaches the surface of the die, the applied potential forms an electrolytic cell, which can corrode the aluminum, affecting electrical parameters through its conduction, and eventually causes catastrophic failure by opening the metal. The presence of contaminants such as chlorine greatly accelerates the reaction as does excessive phosphorus in the layers.

Temperature Shock/Cycling
The temperature shock/cycle test is conducted for the purpose of determining the resistance of devices to alternating exposures at extremes of high and low temperatures. Permanent changes in electrical characteristics and physical damage produced during temperature cycling result principally from mechanical stress caused by thermal expansion and contraction from CTE mismatch. The effects of temperature cycling include cracking and delamination of packages, cracking or cratering of die, cracking of passivation, delamination of metallization, and various other

changes in the electrical characteristics resulting from thermo-mechanically induced damage.

Power Cycle
The power cycle test is performed to investigate package wear-out from the thousands of on/off operations. The repetitive heating/cooling effect caused by multiple on/off cycles can lead to fatigue cracks and other degrading thermal and/or electrical changes in devices which generate significant internal thermal heating under maximum load conditions. From the test result by different delta Tj, it can be used to support in the creation of the lifetime of package.

Temperature Storage Endurance Test
The high- and low-temperature storage is employed for the purpose of determining the effects of storing devices at elevated temperatures without the application of electrical stresses.

The need for reliability in SiC MOSFETs arises from the critical role these devices play in power electronics, high-temperature environments, and demanding applications. As electronic systems continue to evolve, the reliability of SiC MOSFETs becomes a key for advancing technology, meeting stringent performance requirements, and ensuring the durability of devices in diverse operational conditions. Researchers and engineers are continually innovating to enhance the intrinsic reliability of SiC MOSFETs, driving progress in the field of power electronics and semiconductor technology.

4.4 Reliability and Lifetime

Paul Salmen

When switching from silicon-to-silicon carbide devices, one of the biggest concerns is the possibility of a decreased lifetime in applications when using the wide bandgap semiconductor. The root cause for this concern is the changed base material properties and the increased crystal defect density in SiC, which lead to new or changed degradation phenomena for SiC devices compared to Si. In the past decade, a lot of effort has been put into addressing and understanding the failure mechanisms that have their origin in either of these two reasons. A comparison of changed material properties and application parameters is shown in Table 4.1.

To enable the usage of SiC without reliability issues, new stress tests such as a gate-switching stress (GSS) [3, 4] have been developed, and old tests such as the time-dependent dielectric breakdown (TDDB) [5, 6] and the high-temperature reverse bias (HTRB) test [7, 8] have been adapted. Using these tests, devices can be qualified, in a reliable manner, to guarantee a lifetime without unexpected deviations. Besides these more electrically driven mechanisms, the changed mechanical properties of SiC can also lead to a decreased lifetime in thermomechanical tests

Table 4.1 Comparison of SiC and Si properties

Parameter	Typical for Si	Typical for SiC
Short-circuit withstand time	0–10 μs	0–3 μs
Frequency	< 20 kHz	< 200 kHz
dU/dt	~25 kV/μs	~100 kV/μs
Internal electric field	kV/cm	MV/cm
Young's modulus	162 GPa	501 GPa
$V_{GS,max}$	20 V	20–25 V
$V_{GS,min}$	−20 V	−10 to −4 V

such as power cycling. To address all of this, many papers and articles have been published showing how to verify device reliability.

Additionally, organizations such as JEDEC or ECPE started releasing first guidelines on device reliability, for example, JEP184, JEP190, JEP194, and JEP195 [3, 6, 9, 10] or AQG324 [11].

This chapter will provide a short overview of the different failure mechanisms that need to be understood to design a reliable system using SiC devices. Except in the case of gate oxide-driven failure mechanisms, the mechanisms described in this chapter are valid for both SiC MOSFETs and SiC diodes.

When transitioning from Si to SiC, a first intuitive step to understand device reliability can be to look at all the tests known from Si (-IGBT) devices of the same voltage class. In most of these tests, SiC devices have been highly stable resulting in very long lifetimes even in conditions that are very demanding for Si. For example, high-voltage humidity tests without failure and stress times of up to 3000 h have been reported by multiple groups [12–14]. Due to this, not much literature is available about humidity-driven failure mechanisms except some discussions on a new failure mode under dynamic switching in humidity [11, 13, 15–17]. Also results from well-known tests such as HTRB have not shown fails, e.g., in edge termination and in SiC devices.

Things are different when it comes to the SiO_2/SiC interface that is strongly influenced by the material properties, especially the higher (compared to Si) defect density in a SiC crystal. The higher defect density leads to a higher number of extrinsic defects in the gate oxide that might end up in an earlier breakdown of the device and, therefore, result in higher failure rates even under normal use conditions, especially if vendors do not take proper measures, such as using an effective gate oxide screening [5, 18]. As such defects are incorporated into the SiC raw material available currently, this is something that chip manufacturers need to address.

To calculate the influence of these defects on a device's lifetime, classical tests, such as the TDDB, are also used for SiC [6, 19]. These tests can be performed on SiC just as they are on Si—different levels of positive bias high enough to induce fails are applied. Time until failure under extreme conditions, obtained from these experiments, can be extrapolated to normal use conditions. This well-known test works for SiC with one drawback: impact ionization starts driving the breakdown at the extremely high fields required to induce failure in the thick gate oxide (within a feasible time). This leads to a voltage acceleration that is much higher than in any application-relevant low field region [5, 6]. The different slopes for intrinsic low-voltage and intrinsic high-voltage breakdown can be seen in Fig. 4.14 as contours of the green-colored area.

As known from Si, TDDB results can therefore be used to compare two different distributions, or if a field acceleration factor lower than that from the experiments is used for extrapolation. However, if caution is not maintained in calculations, the results from TDDB tests can give projected lifetimes higher than they really are. This is, in principle, the same for Si [20], but due to different gate oxide thicknesses and screening capabilities, it is not considered a big issue in Si devices. To avoid overestimation of lifetime and to address the influence of extrinsic defects, one group proposed a so-called marathon test [5]. A high number of devices are stressed at voltage levels above usage conditions but with the same field acceleration as that at usage conditions. This does not allow the estimation of intrinsic lifetime but enables the calculation of field failure rates due to extrinsic defects in the on-state. See Fig. 4.17 for a comparison between the different test approaches and their relation to the usage conditions.

Unlike Si, due to the high field and permittivity of SiC, the gate oxide of a SiC MOSFET is also critically stressed in the blocking mode. As shown by different

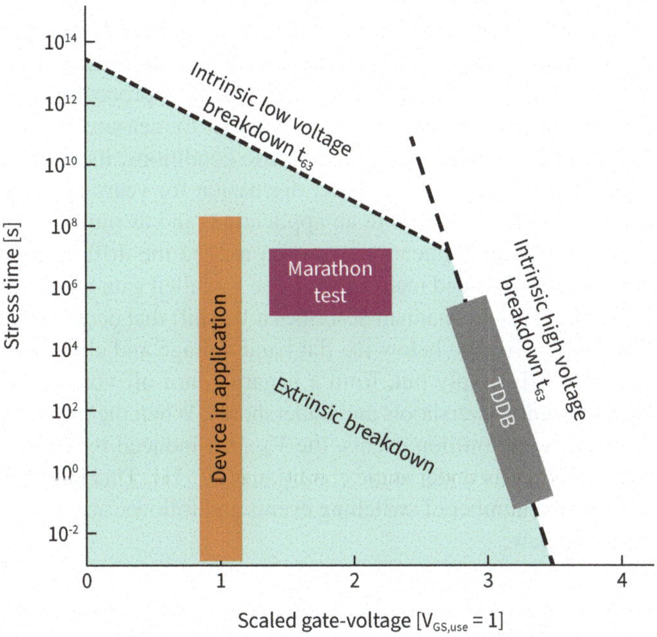

Fig. 4.14 Overview of an application and the different on-state stress tests depending on stress/use voltage and stress time. The dotted lines mark the intrinsic breakdown robustness, while breakdowns inside this area are caused by extrinsic defects

research groups, gate oxide can fail under increased drain source voltages even without an applied gate bias [7, 8, 19]. These fails happen due to an increase in the field of the gate oxide, E_{ox}, under high drain bias. To prevent this, device manufacturers must shield the gate oxide by adapting the JFET region of the device. If this is done properly, the device gets sufficient high lifetime even if the failure rates are driven by gate oxide defects, as shown in [21] for capacitors with a device design based on SiC MOSFETs.

To investigate the intrinsic device lifetime under these conditions, a new test method has been proposed—the accelerated lifetime testing-HTRB (ALT-HTRB) [7]. Similar to the TDDB test, in this test too, the device is stressed with a voltage above the maximum rated voltage, but with a drain source bias. Fails occurring under these extreme conditions can have different root causes, e.g., cosmic radiation. Through failure analysis, it must be verified that only the fails located in the gate oxide [7, 8] are considered. From this, the intrinsic gate oxide lifetime under reverse bias can be calculated. Till now, this method has worked for DMOSFETs with a gate oxide on the 0001-plane. No reports are available for trench devices.

Besides the topic of gate oxide breakdown induced by high gate-source or drain-source fields, another aspect of gate oxide reliability has also been heavily discussed in literature in the past years. At the beginning of SiC MOSFET development, a strong focus was on investigating and reducing the drift of the threshold voltage V_{th} under applied gate bias [22–24]. This mechanism, the bias temperature instability (BTI), is well known in Si but enhanced in SiC [25] because a larger number of defects and dangling bonds are present at the atomic bond level. Besides higher V_{th} drifts, this property also causes a larger hysteresis in SiC MOSFETs. The hysteresis creates the necessity for more sophisticated measurement procedures to determine the V_{th} during readouts [9, 22, 26] and makes it harder to measure the drift in a reliable way. As this drift is mainly seen under static conditions, its impact on device reliability in an application has been under discussion for years. In contrast to the assumption that the drift is reduced in an application, and as multiple groups have shown now, the drift in an application can even exceed the drift seen under static conditions due to a newly found mechanism—the so-called gate-switching instability (GSI) [3, 27–30]. This mechanism describes a V_{th} shift that occurs when a device is switched between a voltage below the flat band voltage and above the threshold voltage of a device [3], simply put, from a negative turn-off voltage to a positive turn-on voltage including overshoots and undershoots. When the number of switching cycles is above some million events, the V_{th} drift induced by GSI exceeds the drift known from BTI tests under static conditions [27, 31]. This GSI drift depends not on time, but on the number of switching events and follows, according to [3, 29], the following power law:

$$\Delta V_{th} = \mathrm{a}\mathrm{N}^{\mathrm{b}}$$

This V_{th} shift leads, due to the reduced overdrive $V_{gs,on} - V_{th}$, to an increase in the R_{DSon} during the lifetime of a device. To quantify the impact of this drift, the GSS test as described in [3, 4] is performed. Here, $V_{DS} = 0$ V is used to simplify the test.

During the test, the gate is switched at a high frequency against the drain source. As shown in Fig. 4.15, this results in the same drift as when the device is actively switched, including V_{DS} and I_{DS}. From this test data, the drift until the end of application profile can be derived and then used to calculate the necessary, but not overdimensioned, safety margins for SiC inverters [3]. An example of the necessary safety margins based on temperature and gate voltage is shown in Fig. 4.16. Here, it is clearly visible that the relative drift of the R_{DSon} depends, in contrast to the V_{th} drift, on several parameters such as V_{GS} and T_{vj}. The reason is that the total R_{DSon} of the devices is the sum of many different contributors such as the channel, drift layer, and JFET. Since only the R_{DSon} of the channel is influenced by GSI, its effect on the overall R_{DSon} can be complex [3, 28].

Considering this drift mechanism is not necessary for Si. However, in SiC it enables a more reliable design since all the effects that originate from the SiO_2/SiC interface of this new technology are included.

Additionally, not only the changed electrical properties of the wide bandgap device and the much more complicated and defect density-driven SiO_2/SiC interface but also the mechanical properties of the SiC material influence the reliability and lifetime estimates. Due to its much higher Young's modulus, the mechanical stress on the interconnection technology is higher compared to that in silicon chips [32]. This leads to an accelerated solder aging (see Fig. 4.17) and, thereby, to a factor 3–4 earlier end of life in power cycling tests with power modules [32]. In contrast to power modules with Si IGBTs, the end of life in SiC power modules is not governed by a degradation of the bond wires but mostly by a degradation of the (soft) solder layer [32]. Other parameters known to influence degradation, mentioned in [33], seem to be stable, even though the validity for very small temperature swings is under discussion [13, 34–36]. This lifetime disadvantage compared to Si must be considered when using SiC modules. This, besides the higher cost, also prevents

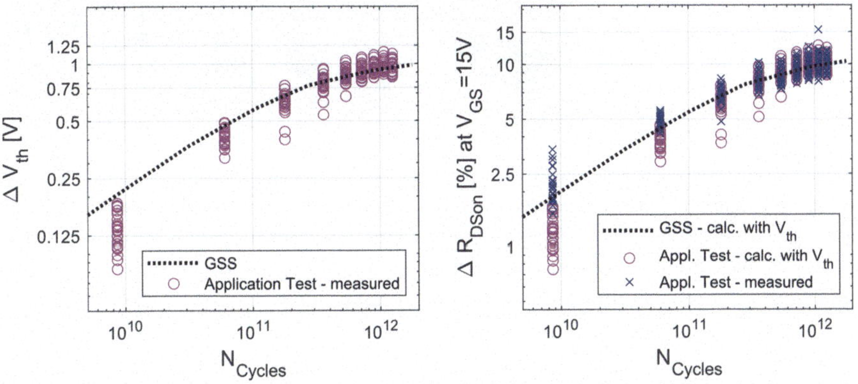

Fig. 4.15 Comparison between Vth drift in a GSS and an application test under similar conditions (left) and corresponding R_{DSon} drift (right) [4]. The R_{DSon} drift after a GSS test is mostly calculated from the Vth drift. Here it is compared with the calculated drift after an application test (magenta) and with the test that is measured on an application test bench (blue) [4]

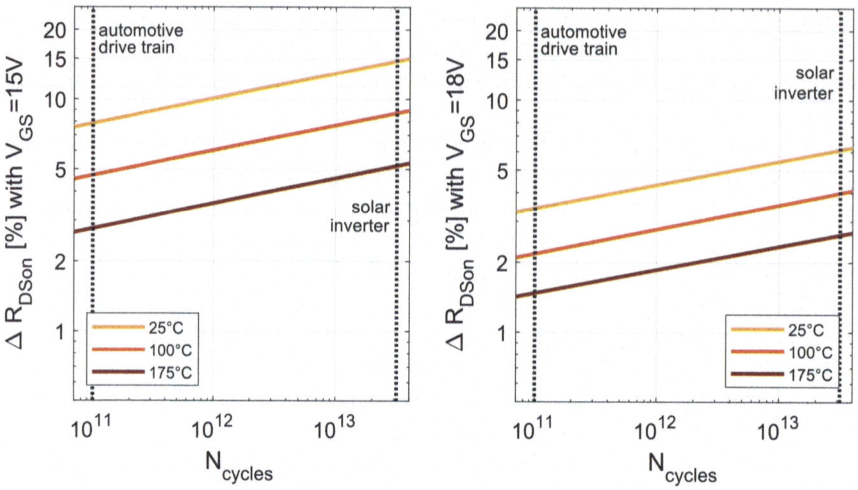

Fig. 4.16 R_{DSon} drift for different temperatures and gate voltages depending on the number of cycles in an application. The dashed black lines show typical numbers of cycles for an automotive drive train and a solar inverter

Fig. 4.17 A mechanical cut through a SiC chip after power cycling in a module. Solder degradation is clearly visible and the reason for the end-of-life detection. Additionally, a small crack at the foot of the bond wire is visible

some applications from transitioning to SiC. To overcome this disadvantage, improved interconnection technologies that lead to longer lifetimes are under evaluation [32, 37]. Interestingly, this disadvantage is not seen in molded discrete packages where the overall lifetime seems to be comparable to that of Si chips packaged in the same way [38].

To sum it up, SiC devices are not worse than Si devices in terms of reliability. Due to the changes in material properties, some additional failure mechanisms come into play that need to be considered. If these challenges are properly taken into account in the device design, SiC devices can be even more reliable and exceed the lifetimes seen so far in Si devices.

4.5 Packaging for SiC

Olaf Zschieschang

Power semiconductors are playing an important role in our life. They can be found everywhere where energy is generated, conducted, or converted.

While early devices were made of germanium (Ge), silicon (Si)-based semiconductors for many years dominated the applications and are still widely used. Over the last decade, another semiconductor material, silicon carbide (SiC), a so-called wide bandgap (WBG) material, came into the focus of applications. Compared to Si, the fabrication of SiC devices is more complex and significantly more costly. But as the material properties of SiC provide significant performance improvements, such semiconductors were able to gain access into many areas, such as automotive and renewable energy applications. SiC allows higher operating temperatures and faster switching at lower switching losses.

For the use in an application, the semiconductor devices need to get assembled into a semiconductor package. Discrete packages typically contain one function only, while modules may consist of basic circuit configurations and can use multiple devices in parallel to increase the power rating. External connectors of the package provide the electrical connection of the electrodes in the application, which can be realized, for example, by means of soldering, welding, or screw connections. The package also needs to enable good thermal dissipation to extract heat generated by losses and to protect the devices from environmental influences and mechanical damage.

The enhanced characteristics of SiC devices allow higher power densities which lead to a size reduction of the devices. This, in turn, results in smaller contact areas for heat dissipation and current flow. To conquer negative impacts of the size reduction on power ratings and reliability, improvements of the packaging technology are required.

4.5.1 Substrate Materials

Power modules are typically using ceramic-based substrates to provide isolation against the heatsink. Typically, copper layers are attached to both sides of the ceramic sheet. Those enable a good thermal performance due to superior thermal conductivity and heat spreading. In the past mainly Al_2O_3 ceramic-based direct bonded copper (DBC) substrates were used for modules with Si devices. The availability of AlN-based substrates and active metal brazed (AMB) substrates using a Si_3N_4 ceramic with their higher thermal conductivity enables improvements of the thermal path (Table 4.2).

Although AlN exhibits the highest thermal conductivity, it is known to be brittle, causing reliability limitations and problems in the assembly. The ceramic layer

Table 4.2 Thermal conductivity of ceramics used as substrate materials

Material	Al_2O_3	Si_3N_4	AlN
Thermal conductivity [W/(m*K)]	24	90	180

needs to remain relatively thick (>0.5 mm). Si_3N_4 ceramics are very robust and thicknesses down to 0.25 mm (or even 0.15 mm, depending on isolation requirements) are possible. In the AMB process, those can be combined with very thick copper sheets (up to 0.8 mm) where the typical DBC is using 0.3 mm copper sheets only. This allows better heat spreading in the copper layers. Although more expensive, the superior characteristics of Si_3N_4 are making it a valuable choice for modules using SiC devices.

4.5.2 Die Attach

In traditional modules and in discrete packages, the Si devices are assembled to the substrate or lead frame by means of soldering. Tin-based, or for higher reliability requirements even lead-based, solders have been used. In combination with SiC devices, it was seen that the power cycling capability of the solders as an expression of the expected lifetime of the product is very poor.

The introduction of silver sintering for the assembly of the devices into the module showed a 3× improvement in cycling capability, compared to soldering. In the assembly process, an Ag-based paste is applied on the substrate and the devices are placed and the connection is made in a process using elevated temperatures and pressures (e.g., ~240 °C and 15 MPa). As a result, a highly reliable Ag joint is created which exhibits the high melting temperature of Ag (960 °C). Due to the reduction of bond line thickness compared to a solder joint, the sinter die attach also enables an improvement of the thermal resistance Rth. For a TO-247 package, it provided a 30% reduction of the R_{thjc}.

4.5.3 Top Side Connection

Aluminum wires with diameters between 150 μm and 500 μm are still state of the art for the electrical connection of the electrodes on top of the power devices to the substrate or lead frame. The wires are attached by means of an ultrasonic welding process, called wire (wedge) bonding. With the size reduction of SiC devices, the number of wires gets limited, while currents and temperatures increase. As a result of the higher stresses, the lifetime of such wire connections is reduced.

As an alternative, copper wires can be used for the top side connection of power devices. Copper has a higher hardness compared to aluminum and requires higher forces in the ultrasonic bonding process. Therefore, an adapted top surface of the device is needed. This can be done by thick copper plating or, alternatively, by applying the copper as an interface layer. The so-called Die Top System® (DTS) [Heraeus Electronics GmbH, Document Number: HET61002-0822-9 | Version: 2022] uses copper pads that are attached in a process such as the aforementioned Ag sintering die attach to the die top, followed by copper wire bonding.

High-power diode and thyristor modules are using copper clips, soldered to the top surface of the die instead of wire bond connections already since many years. This technique was rediscovered for the connection of modern semiconductors but combined with a silver sintering process instead of soldering. As thermomechanical stresses are significantly higher with SiC devices and their high temperatures, special attention must be paid to the source clip design. At the same time, the clip solution is also reducing the parasitic inductance of the package, leading to lower switching losses and efficiency improvements in the application (Figs. 4.18 and 4.19).

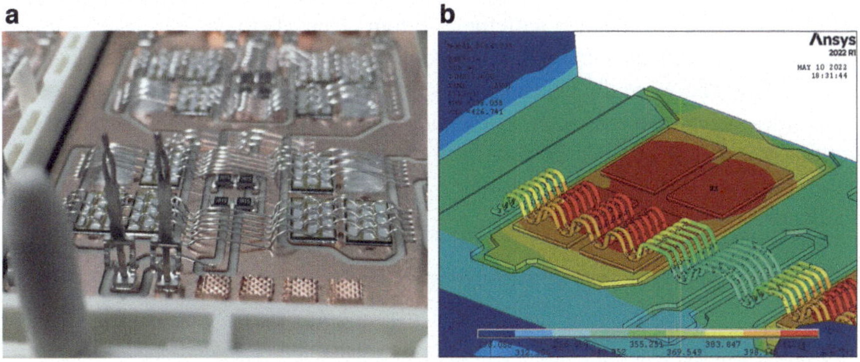

Fig. 4.18 (**a**) SiC MOSFETs with wire bond connections, (**b**) temperature distribution, simulated

Fig. 4.19 Example of a clip design for the die top side connection

4.5.4 Environmental Protection

There are different ways used to protect the devices and their connections from environmental influences and mechanical damages. Plastic housings and a soft silicone gel for potting are often used for power modules. Depending on the design of the package, such modules exhibit great flexibility to realize multiple different circuit configurations at low initial costs.

Alternatively, the package can be formed by a hard epoxy. This is done in a so-called transfer molding process where an epoxy material is injected into a mold cavity in liquid form at high pressures and high temperatures. Similar to known discrete packages such as TO-247, this is also done for more complex modules. Tooling is more expensive and inflexible to modifications or variants, but this technology shows an improved power cycling capability.

To improve the thermal performance of a module using SiC devices further, packages can be designed to use dual side cooling as an option. In such case, typically, the devices are attached with their bottom electrode to one substrate. The top side of the device is then connected to another substrate either directly or via spacers. The space between the two substrates is finally filled with epoxy.

Besides better cooling, such a dual side cooled package in addition allows to improve the design of the current path in the module, reducing its parasitic inductance. This in turn reduces the switching losses and increases the efficiency of the application.

4.5.5 Cooling

In industrial applications, power modules are traditionally attached to a heatsink by means of a thermal interface material. Such materials are filling air gaps due to unevenness, pores, etc. of the contact materials, improving the thermal conductivity compared to air voids, but the thermal resistance of such materials is still high. Such applications typically use a forced air-cooling solution. In applications that are using a liquid cooling, SiC-based power modules can be equipped with a baseplate with a pin-fin structure which is then placed directly into the flow of the coolant. Alternatively, modules can be attached directly to a closed cooler. In both designs, the module substrates are attached to the baseplate or cooler by means of soldering. Such large-area solder joints are prone to exhibit solder voids, or thickness differences, and delaminations may occur. Besides specifically designed solder preforms that provide distinct spacing to prevent tilting, alternatively Ag sintering is used to attach the modules to the cooling structure and improve the performance and reliability of this joint (Figs. 4.20 and 4.21).

Fig. 4.20 Epoxy transfer molded module, for dual side cooling

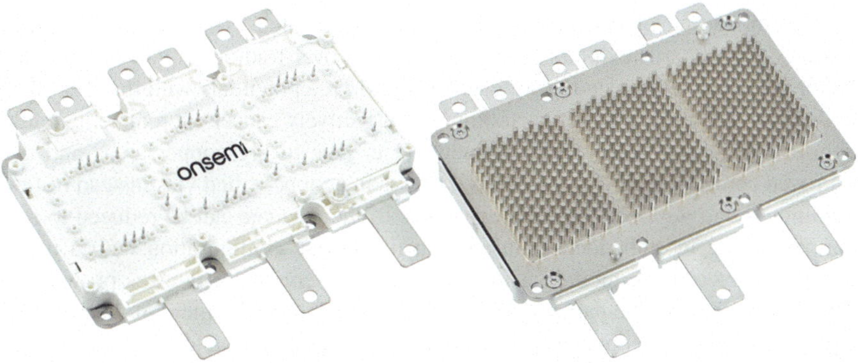

Fig. 4.21 Power module with pin-fin style baseplate

4.5.6 External Connections

For fast switching devices such as SiC MOSFETs based on the standard TO-247 package, a modified version was designed with four leads. This enables a return path for the driver signal via a separate kelvin source pin, while at the standard three-lead package, the signal needs to share the same path with the high current source potential. Separating the driver signal from the main current path reduces the parasitic inductance and enables higher switching speeds and reduces the switching losses up to 35% (Fig. 4.22).

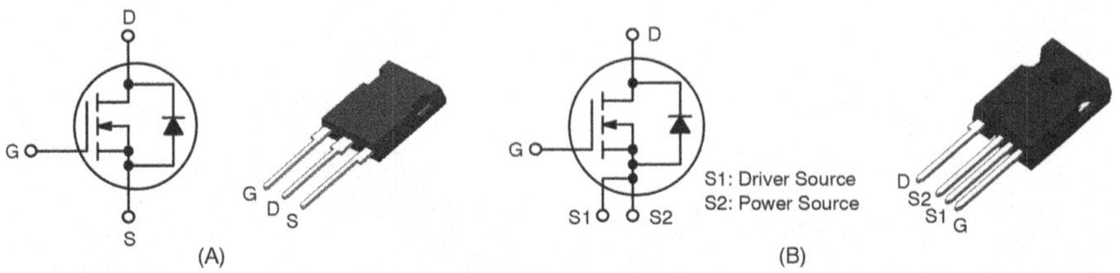

Fig. 4.22 (**a**) TO-247-3 L, (**b**) TO-247-4 L

4.5.7 Embedding

Embedding of active and passive components into printed circuit boards (PCB) is a known technology for low-power applications. Meanwhile, with the availability of improved PCB materials, this technology is evolving into high-power, high-voltage applications.

The embedding of fast switching power devices such as SiC MOSFETs into printed circuit boards exhibits several advantages over traditional power semiconductor packages. Optimized routing of DC+ and DC- potentials with the short distance of PCB layers on top of each other can effectively cancel out magnetic fields, significantly reducing the parasitic inductance. Due to big copper areas and the improved connection of the power device by many copper-filled vias instead of few aluminum bond wire contacts, the resistance of the package can be reduced too. A comparison of an embedded solution of a full-bridge configuration with a traditional transfer molded module using the same semiconductor devices provided a 10× improvement of the parasitic inductance and a 3× improvement of the package resistance for the power path as well as for the gate loop path (Table 4.3). Meanwhile, in a next-generation design, the parasitic inductance was further reduced and a L_S of 0.4nH was achieved. This enables faster switching with lower over-voltages and lower switching losses. As switching losses of the SiC MOSFETs are accountable for about 70% of the losses in a conventional drive inverter, any reduction of such losses is significantly increasing the system efficiency.

The embedded structure allows a more symmetrical stack-up and exhibits lower thermomechanical stresses, leading to improvements in the cycling capability and improving the lifetime of such a semiconductor package, while potential new failure mechanisms such as copper migration and aging effects on the PCB materials and their impact on reliability are still subject of evaluations.

Another advantage, especially for automotive applications, is on the reduction of size and weight and, from environmental perspective, the reduction of material content. The power embedded PCB allows to design additional, low-power layers on top. This allows to integrated the gate driver and control circuitry very close to the devices.

Table 4.3 Comparison of a standard transfer molded module versus the same die in an embedded design concept, simulation results

	DC+ to DC- inductance (nH)	Gate loop inductance (nH)	DC+ to DC- resistance (mohm)	Gate loop resistance (mohm)
Refence module	20	10	75	55
Embedded demonstrator	2.5	1	25	17

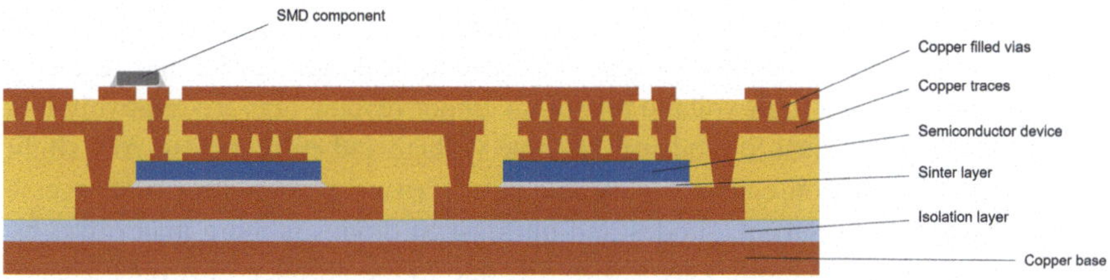

Fig. 4.23 Draft example of a PCB embedded half-bridge configuration using SiC MOSFETs

PCB embedded structures can alternatively also be used for subassemblies, enabling the parallel connection of several smaller SiC MOSFET devices, providing bigger contact areas on top of the PCB and the possibility to pre-test such pre-packaged devices before integrating those into a bigger module (Fig. 4.23).

4.6 Diodes

Sebastian Fahlbusch

The use of silicon carbide (SiC) Schottky diodes can provide significant efficiency and other performance improvements over silicon (Si) P-N diodes in power conversion applications. Since their introduction almost two decades ago, these devices have evolved from pure Schottky diodes to the advanced merged PiN Schottky (MPS) diode structure that is widely used today. This chapter describes the structure of SiC MPS diodes and explains their advantages over their silicon counterparts. It also discusses improvements in diode performance due to advances in process tech nology, namely, thin SiC.

4.6.1 SiC Diodes' Operating Characteristics

SiC diodes are unipolar Schottky metal-semiconductor devices where conduction is only through majority carriers, i.e., electrons, resulting in negligible charge stored in the junction depletion layer during forward bias. In contrast, P-N junction diodes are bipolar and store charges that must be removed during the transition to reverse bias.

This causes a spike in reverse current and leads to higher-power losses in the diode, switching transistors, and snubbers, which worsens as the switching frequency increases. The SiC diode also exhibits a reverse current spike due to the discharge of its self-capacitance under reverse bias, but this can be an order of magnitude less than in a P-N junction diode, meaning less power dissipation not only in the diode itself but also in the corresponding switching transistor. The maximum forward voltage drop of SiC diodes is comparable with the best ultrafast silicon types and is still being improved, with minor differences for higher blocking voltage ratings. Despite being a Schottky type, the reverse leakage current and resulting power dissipation in a high-voltage SiC diode under reverse bias are relatively low and similar to an ultrafast Si diode of the same voltage and current class. Any slight difference in dissipation caused by variations in forward voltage drop and reverse leakage current between SiC diodes and ultrafast Si diodes is more than offset by the improvement in dynamic losses with SiC due to the absence of reverse charge recovery effects. This is further elaborated in the section on MPS diode reverse recovery characteristics.

A silicon diode's recovery current and time vary widely with temperature changes, making circuit optimization difficult, but significantly, with SiC, there is no such variation. In some circuits, such as "hard-switched" power factor correction stages, a silicon diode acting as the boost rectifier can dominate losses, going from forward bias at high current to reverse bias at typically 400 V for a typical single-phase AC input. The characteristics of SiC diodes can bring significant efficiency gains to this application.

An additional advantage of SiC diodes over Si is that they can be connected in parallel because their forward voltage drop has a positive temperature coefficient (in the application-relevant area of the I-V curve), which helps correct any current imbalances. This compares with the negative temperature coefficient of Si P-N diodes which can produce thermal runaway when devices are connected in parallel, requiring heavy derating or additional active circuitry to force current sharing.

A further benefit of SiC diodes' comparable soft-switching behavior is a significant reduction in EMI. When Si diodes are used as switching rectifiers, the potentially snappy spike in reverse recovery current, with its broad frequency spectrum, can cause conducted and radiated emissions. These create system disturbances (via various coupling paths) that may result in system EMI limits being exceeded. At these frequencies, filtering can be complex due to this stray coupling. In addition, EMI filters designed to attenuate switching fundamental and low harmonic frequencies (often below 1 MHz) typically have high self-capacitance, making them less effective at higher frequencies. A snubber can be used across a silicon fast recovery diode to limit edge rates and damp oscillations, reducing stress on other components and EMI. However, the snubber has to dissipate significant energy, adversely impacting system efficiency. Another often-overlooked source of loss is forward recovery. During transition from off- to on-state temporarily, the diode voltage drop increases in silicon diodes which produces overshoot, ringing, and additional loss.

This effect relates to the initial lower conductivity of the junction of Si-based P-N diodes and is absent in SiC diodes.

The absence of minority carriers in SiC diodes enables their superior efficiency. Consequently, they require less cooling, meaning they can be used to provide higher-power density in smaller and lighter products. Due to the significant reduction in switching losses (in both the device itself and the corresponding transistor), SiC diodes enable higher switching frequencies. The dissipated heat is therefore reduced, meaning additional cost savings can be made through less complex EMI filtering and smaller magnetic components. Additionally, device junctions can withstand higher temperatures for more rugged performance and better reliability without risking thermal runaway.

4.6.2 Merged PiN SiC Schottky

Compared to silicon fast recovery diodes, some characteristics of the basic SiC Schottky diode can be further improved, including leakage current (caused by imperfections at the metal-semiconductor interface) and surge current rating (which relates to the device's unipolar nature and relatively high drift layer resistance). Leakage current can be improved with a thicker drift layer, which increases its ohmic and thermal resistance, but this is a disadvantage in power applications. Merged PiN Schottky (MPS is a hybrid of a SiC Schottky and P-N junction diode) overcome this constraint while providing maximum robustness in the presence of surge currents.

P-doped areas are implanted in the drift zone of a conventional Schottky arrangement forming a P-ohmic contact with the metal at the Schottky anode and a P-N junction with the lightly doped SiC drift or epi-layer. Under reverse bias, the P-wells "push" the general area of maximum field strength downward into the almost defect-free drift layer, away from the metal barrier with its imperfections, reducing the overall leakage current. The effect is shown in Fig. 4.24.

The P-wells' physical placement and area, compared with the size of the Schottky diode, and doping concentration all affect the end characteristics, with forward voltage drop traded against leakage and surge currents. Consequently, an MPS device can operate at a higher breakdown voltage with the same leakage current and drift layer thickness. The improvement in surge current rating comes from the fact that the differential resistance of a bipolar device is lower than that of a unipolar type. Although the Schottky area has a lower voltage drop (at low current) and can conduct the whole current (even high surges), this voltage increases beyond the P-N diode cut-in voltage, which starts to conduct with lower differential resistance. This diverts current, limiting dissipated power and relieving the MPS diode from thermal stress (Fig. 4.25).

Under normal conditions, the P-N junction does not conduct, the overall effect is that of a unipolar diode, and there are almost no reverse recovery losses. Without the P-N diode, the Schottky diode alone would have to be significantly over-dimensioned

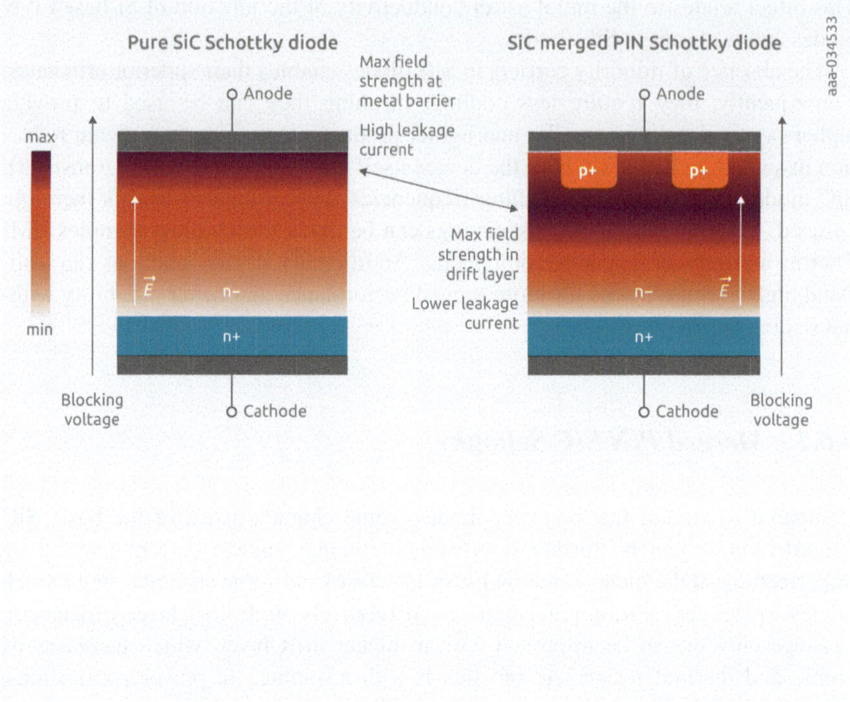

Fig. 4.24 Standard SiC Schottky diode structure (left) and Nexperia's SiC MPS diode structure (right)

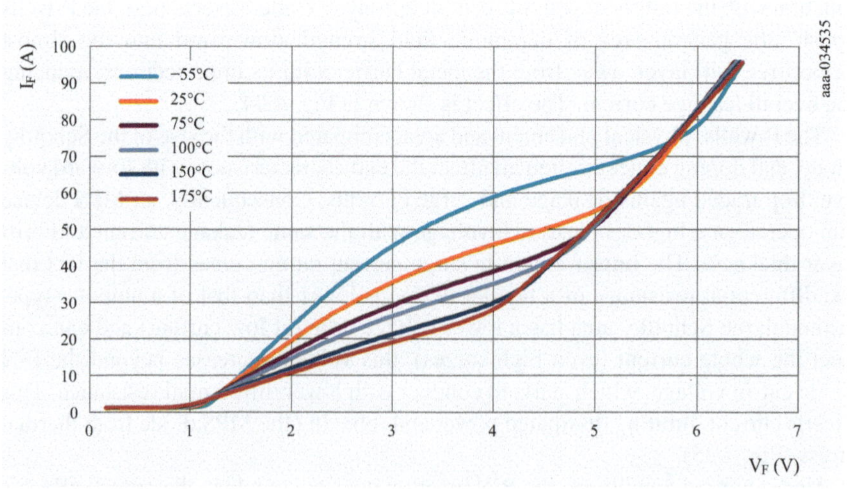

Fig. 4.25 Static I-V behavior of Nexperia's SiC MPS diode PSC1065K, including overcurrent

if transient overcurrent events could occur in the target application. Parts can be paralleled (or extra circuitry added) to limit over current, but this increases cost. Again, the dimensions and doping of the P-wells create a trade-off between forward voltage drop (in regular operation) and surge withstand capability. The best choice is application-specific, and Nexperia offers diodes to suit a wide range of hard- and soft-switching applications.

4.6.3 MPS Diode Reverse Recovery Characteristics

Besides the static advantages, SiC MPS diodes offer benefits during dynamic operation in switching mode. One significant advantage over silicon-based P-N diodes relates to the reverse recovery behavior. As previously described, SiC MPS diodes behave like Schottky diodes under nominal conditions. Unlike conventional Si fast recovery diodes, only the majority carriers contribute to the overall current for SiC diodes. As a result, these show a purely capacitive switching behavior, resulting in a lower reverse recovery charge than a Si fast recovery diode with the same electrical rating. The reverse recovery charge is one of the main loss contributors and thus adversely affects converter efficiency. Figure 4.26 depicts the effects of different parameters affecting the reverse recovery behavior, such as diode turn-off currents and junction temperatures. SiC exhibits almost constant behavior in the presence of these variations and little of the nonlinear behavior of Si fast recovery diodes. Hence, its behavior is easier to predict for power designers because they do not need to consider a variety of ambient temperatures and load conditions.

4.6.4 "Thin SiC" Diodes

Nexperia's "merged PiN Schottky" exhibits improved leakage and surge current ratings but also provides other advantages derived from reduced die thickness during manufacture. The unprocessed SiC substrate is N-doped, and SiC epitaxial layers are grown to form the drift region. The substrate starts with a thickness of up to 500 μm, but after epitaxy, this adds unnecessary electrical and thermal resistance in the current and heat-flow path to the backside metal. This increases forward voltage drop and junction temperature for a given current. A solution is to "thin" the underside of the substrate by grinding. During this process step, material quality is vital, along with the precision of the grind, to avoid non-uniformities and consequent performance degradation of the diode and potential field failures. In addition, an advanced manufacturing capability is required for this process, not least because of the hardness of SiC (9.2 to 9.3 on the Mohs scale, compared with silicon at around 6.5). Figure 4.27 illustrates the effect of the process, with Nexperia's "thin SiC" technology achieving a reduction to about one-third of the original substrate thickness. As a result, the thermal resistance from the junction to the backside metal is

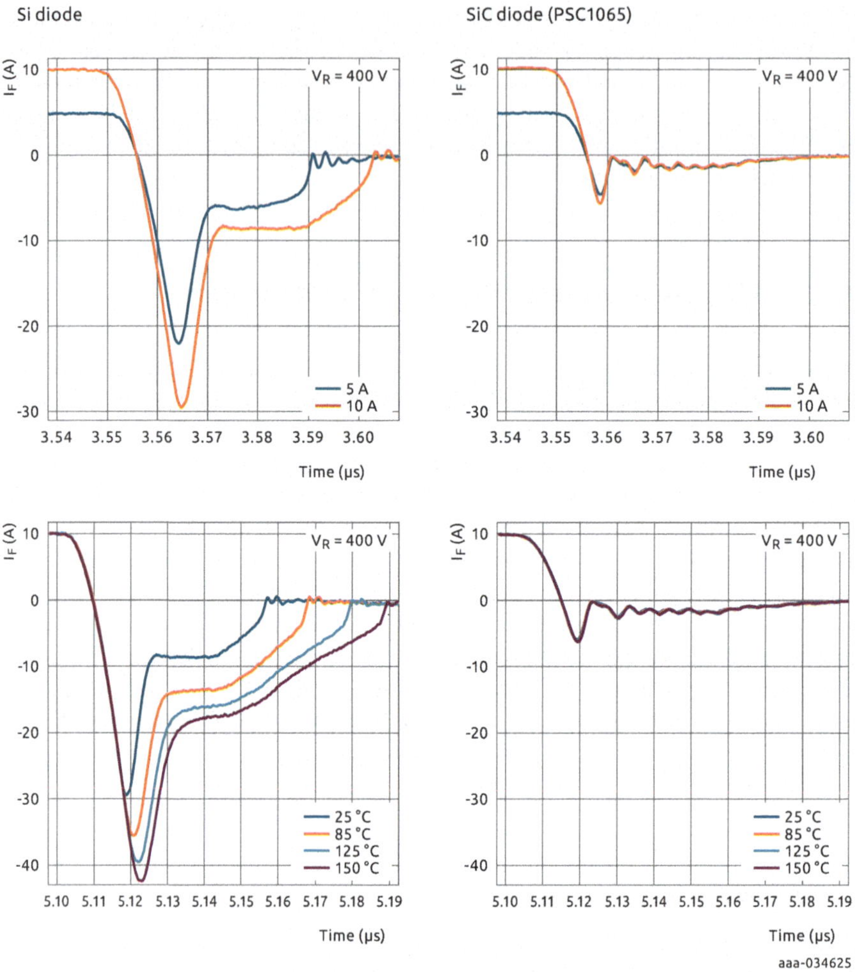

Fig. 4.26 Comparison of reverse recovery behavior of Si PN diode and the Nexperia SiC MPS PSC1065K under various operational conditions

dramatically reduced, resulting in lower operating temperature, higher reliability and lifetime, higher surge current capability, and lower forward voltage drop (Figs. 4.26 and 4.27).

4.6.5 Conclusion

Compared to silicon, SiC diodes provide superior technical advantages such as significantly better switching performance and allow system engineers to improve their power applications. SiC diodes enable higher efficiency and power density, lower EMI, better reliability, and lower system complexity. In addition, higher switching

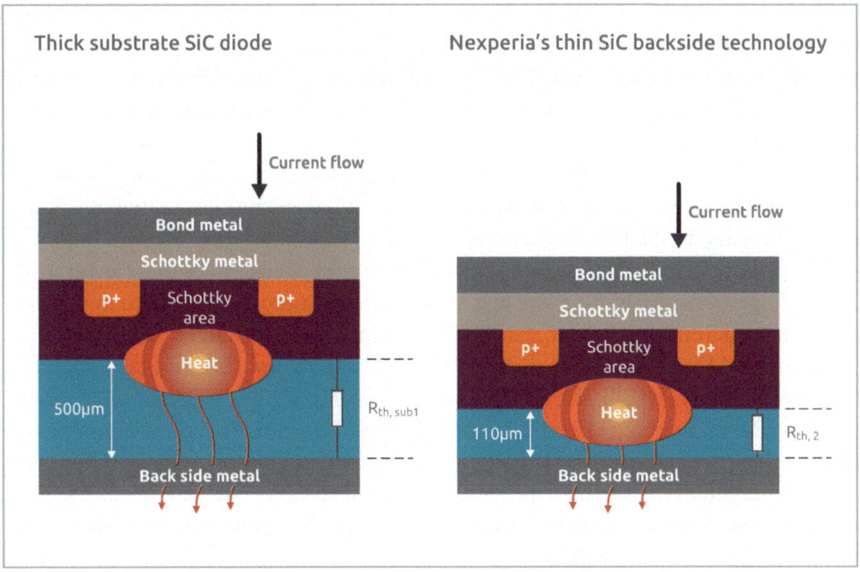

Fig. 4.27 The "thin SiC" process (right) improves diode electrical and thermal performance compared to standard SiC diode construction (left)

frequencies can be achieved without sacrificing output power or overall system efficiency. Consequently, smaller magnetics and other passive components can be used, reducing converter size and lowering the total cost of ownership.

4.7 SiC Power Semiconductor Device Dynamic Characterization

Ryo Takeda and Takamasa Arai

Silicon carbide (SiC) power semiconductor devices such as SiC MOSFET (metal-oxide-semiconductor field-effect transistor), are increasingly finding applications in high-power scenarios such as electric vehicles (EVs), solar power inverters, trains, and wind power generation. This is due to their ability to simultaneously achieve high breakdown voltage, low on-resistance, fast operation, low switching loss, and superior performance under high-temperature environments compared to silicon (Si) power devices.

However, owing to its faster operation speed compared to its Si counterpart, improper design of a power electronics circuit using SiC MOSFET can lead to undesired behaviors such as ringing or false turn-on. The very high-power density of SiC MOSFETs can result in a significant surge of current during accidental false turn-on, leading to a short circuit operation that may cause a complete breakdown of SiC power devices or the entire power electronics system.

Therefore, a detailed characterization of SiC power devices is the critical initial step in designing a power electronics circuit. The fast operational speed, accompanied by a high current slew rate (di/dt) and high-voltage slew rate (dv/dt), is a cause of unwanted behaviors such as ringing or false turn-on. Hence, dynamic characterization of SiC power devices is crucial. Switching characteristics with high voltage such as 1 kV, or high current, in the order of a few hundred amps, are necessary for measuring SiC power devices. Safety is especially important during dynamic characterization due to the high-power density of SiC power devices.

SiC power devices exist not only in discrete form but also in various power module configurations. Therefore, the test system needs to provide a way to make connections to such devices under test (DUT) and facilitate measurements.

Traditional Si devices, such as IGBT (insulated-gate bipolar transistor), lack fast operational speed, which allows the use of custom-built test systems or commercially available dynamic testers primarily designed for production testing in power semiconductor manufacturing sites. However, using such test systems for SiC power devices is challenging because the fast operational speed of SiC devices can lead to measurement errors caused by subtle differences, such as parasitic elements in the test circuit and skew of probes.

Differences in measurement results can occur in various instances, such as between device manufacturers and their customers or among different offices or departments within the same company. Identifying the cause of these differences is difficult and consumes valuable engineering time, as variations may arise not only from test system differences but also from various sources, including variations in device characteristics, degradation of device performance over time, and operational mistakes.

A robust test system, built by eliminating the influence of parasitic elements in the system through a deep understanding of measurement science and by using appropriate probing technologies, is necessary to obtain repeatable and reliable dynamic test results on SiC devices.

4.7.1 Basic Dynamic Characterization: Double-Pulse Test

Dynamic characteristics or switching characteristics are measured with the so-called double-pulse test. Using a half-bridge structure of field-effect transistors (FET), two pulses are applied to a DUT. Due to the load inductor between the power supply and the DUT, the current gradually increases inversely proportional to the load inductance value when the first pulse is applied. By controlling the width of the first pulse, you can manage the current value for the switching event.

When the first pulse ends, the current cannot flow through the DUT because it is turned off, but it continues flowing in the loop consisting of the load inductor and the freewheeling diode of the high-side FET. When the second pulse is applied, the DUT is turned on again, and the current flows through the DUT to the ground. With this sequence of two pulses, the waveform shown in Fig. 4.28 is obtained. The edge

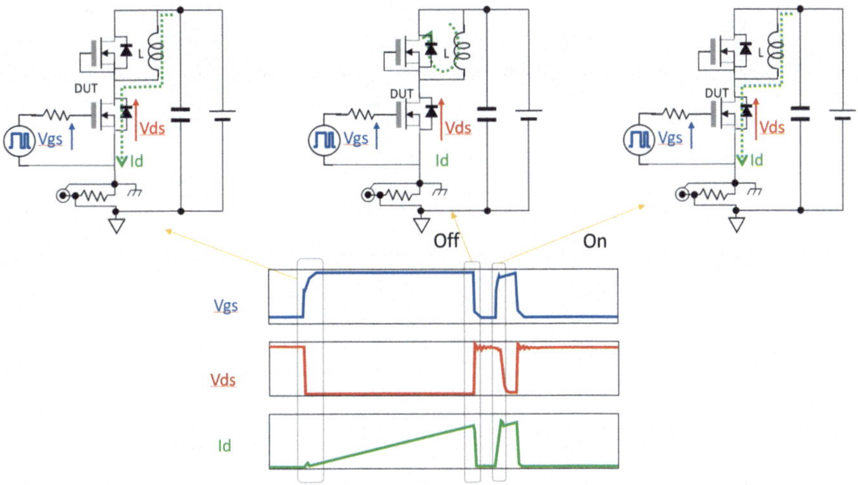

Fig. 4.28 Double-pulse test for switching characterization

at the end of the first pulse represents off-characteristics, and the edge at the beginning of the second pulse represents on-characteristics. In other words, you can conveniently measure turn-on and turn-off switching characteristics with a single measurement event.

Reverse recovery characteristics are also measured using the double-pulse test. In this case, the load inductor is connected between the neutral point of the half-bridge and the ground. The two pulses are applied to the high-side device. By controlling the width of the first pulse, the current reaches the target value. When the first pulse ends, the current continues to flow in the loop consisting of the load inductor and the freewheeling diode of the low-side FET. Carriers are formed in the PN or PIN junction while this current flows. Then, the second pulse is applied to the high-side FET, and the current commutates and stops flowing through the diode. The carriers in the junction flow instantaneously and are measured as reverse recovery current (Fig. 4.29).

4.7.2 Eliminating the Influence of Parasitics in the Test System

When performing a double-pulse test for SiC power devices, the tested waveform is likely to exhibit distortions or ringing due to parasitic elements in the test circuit interacting with the fast di/dt and dv/dt of the switching process, producing voltage (L * *di/dt*) and current (C * *dv/dt*) distortions. For example, stray inductance at the source terminal of the low-side device causes back-electromotive force (EMF) that affects the gate voltage of the DUT. Additionally, stray inductance at the drain

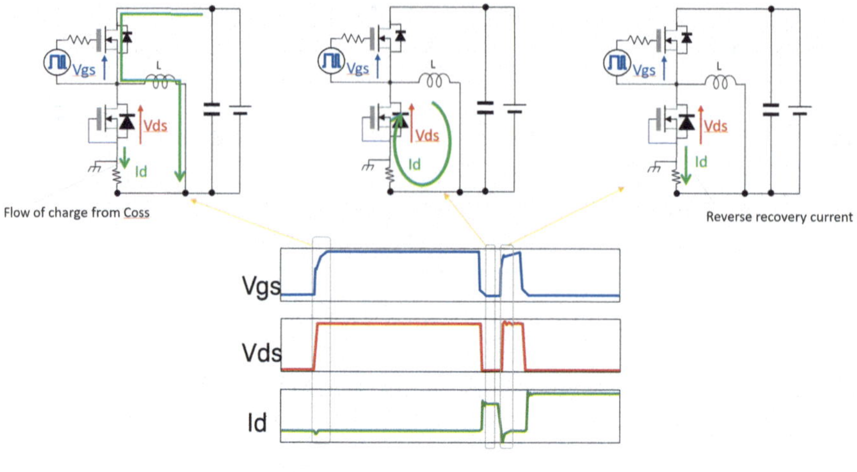

Fig. 4.29 Double-pulse test for reverse recovery characterization

terminal of the low-side device can resonate with the output capacitance of the DUT, leading to surge voltage or ringing. Careful design of the test board is crucial to minimize stray inductance and reduce its negative influence on measurement quality.

To cancel out the magnetic field generated by currents, it is recommended to overlap forward and return current traces across the DUT board if possible (see Fig. 4.30). Ground loops can also cause ringing. The slow ringing in the MHz range can be suppressed by incorporating common-mode chokes in the measurement probe, reducing ground loop currents.

One of the parameters indicating the performance of the test system is the power loop inductance, which is calculated using the following equation. Here, $V_{ds,droop}$ represents the dip at the drain voltage when the drain current (Id) ramps up. This parameter is easily calculated from turn-on characteristics, whereas other stray inductances, such as source inductance or drain inductance, are difficult to separate:

$$L_{\text{powerloop}} = \frac{V_{ds,droop}}{dI_d / dt}$$

Fig. 4.31 depicts measured waveforms obtained from two different test setups. The one with a smaller power loop exhibits significantly less ringing and smoother waveforms for both turn-on and turn-off.

Caution is necessary when assessing the power loop inductance of commercially available double-pulse test systems, as the device inductance is already subtracted from the actual power loop inductance.

Parasitics do not necessarily exist only on the DUT board. Components on the DUT board, such as current sensors, have insertion inductance. As described in the

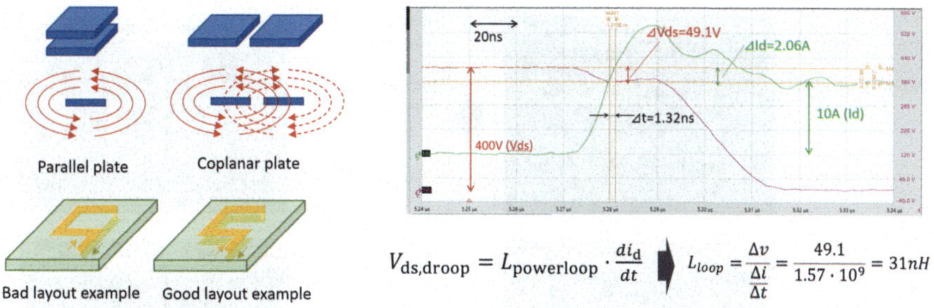

Fig. 4.30 Layout example and power loop calculation

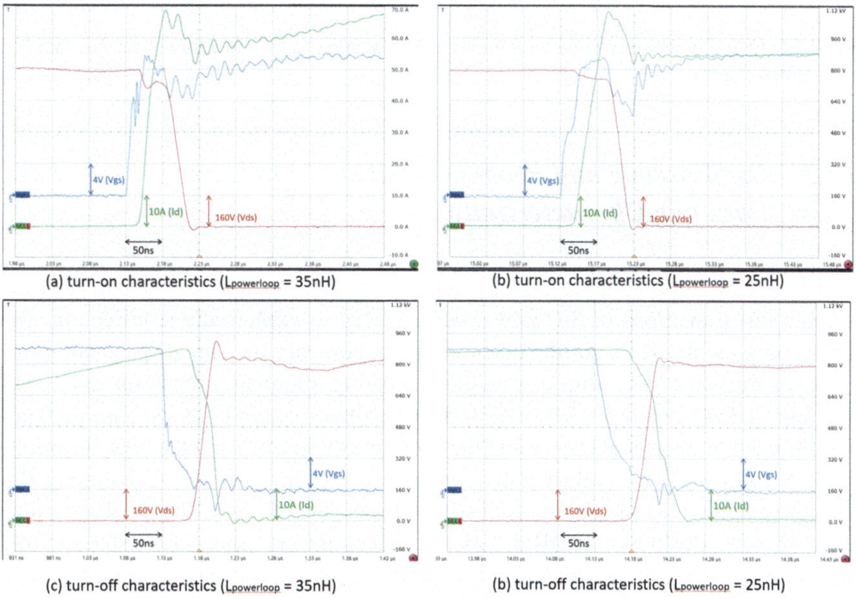

Fig. 4.31 Switching waveforms with two different test setups

next section, bandwidth is crucial for measuring SiC power devices, and therefore, a current sensor with high bandwidth (>100 MHz), such as a coaxial shunt resistor, is useful. However, coaxial shunts do not always have the bandwidth described in the datasheet due to production variations. In such cases, radio frequency (RF) compensation is beneficial, as shown in Fig. 4.32.

Additionally, compensations such as probe deskew, adjustment of gate voltage, and drain voltage are very important to obtain repeatable and reliable measurements.

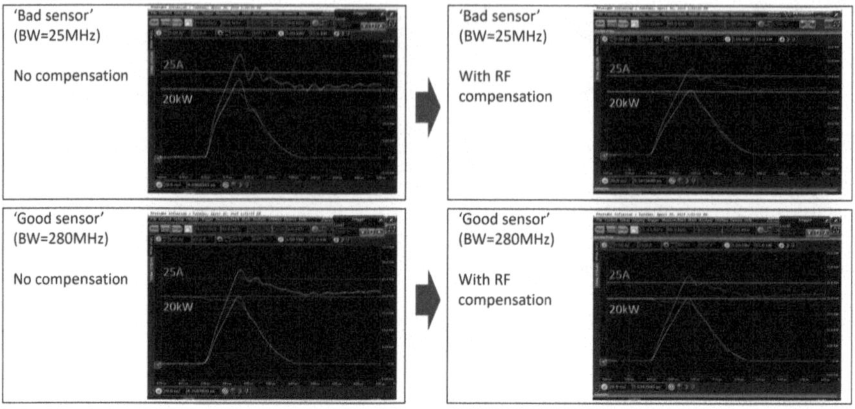

Fig. 4.32 Id and power measured with different BW sensors (left) and adjusted results with RF components (right)

4.7.3 Oscilloscope and Probes for SiC Dynamic Characterization

Oscilloscopes and probes are critical components for dynamic test systems to capture the fast transitions of SiC power devices. Let us explore the requirements for these components. To begin the discussion, we compare the signal spectrum of an ideal square wave (V2) with a waveform showing a realistic transition time (V1) (see Fig. 4.33a). V1 has a 10–90% transition time of about 1.8 ns. The markers highlight when the signal is crossing the 10% threshold (m1) and the 90% threshold (m2). V2 has an almost ideal transition time (i.e., time ~ 0 ns). Calculating the respective spectrums yields, the following results (Fig. 4.33b). The ideal spectrum of V2 shows a decline of 20 dB/decade. However, the spectrum of V1 starts to roll off at a certain frequency. The frequency at which the amplitude has dropped by half (6 dB) compared to the ideal spectrum is labeled as the "knee frequency" Fknee [39]. In this example, Fknee happens to be 278 MHz. We can relate Fknee and transition time (tr) with the equation in Fig. 4.33b. Because a square wave signal carries a significant amount of energy up to Fknee, it is important to capture all frequency components up to Fknee plus some headroom. The necessary amount of headroom is discussed after describing oscilloscope properties.

There are two types of oscilloscopes in the market: a traditional analog oscilloscope and a digital oscilloscope. An analog oscilloscope has many analog amplifiers in a chain from input to output and therefore exhibits a Gaussian response, while a digital oscilloscope has a flat response. Figure 4.33c shows the equation to calculate intrinsic transition time for these frequency responses. You cannot simply assume that an oscilloscope with a Gaussian frequency response yields results with higher accuracy by seeing these equations. On the contrary, using an oscilloscope with a flat response and sufficiently high bandwidth margin yields better results.

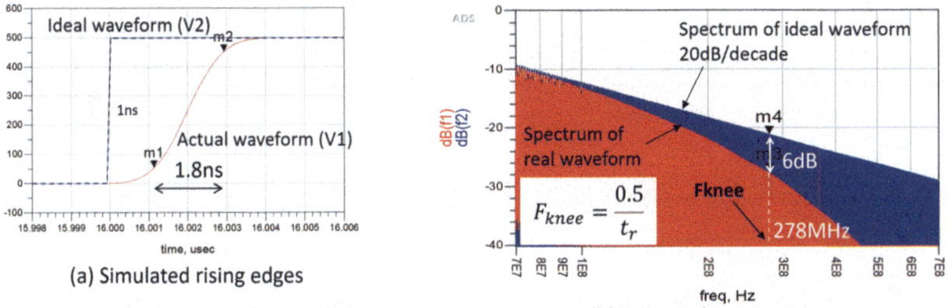

(a) Simulated rising edges

(b) Extracted spectrum

Gaussian response	Flat response
$t_r = \frac{0.35}{BW}$	$t_r = \frac{0.4 \ldots 0.5}{BW}$

(c) Intrinsic rise time of the step response with bandwidth (BW)

Meas. Error	BW Gaussian	BW Flat-response
20 %	$1.0 \cdot F_{Knee}$	$1.0 \cdot F_{Knee}$
10 %	$1.3 \cdot F_{Knee}$	$1.2 \cdot F_{Knee}$
3 %	$1.9 \cdot F_{Knee}$	$1.4 \cdot F_{Knee}$

(d) Required bandwidth

Fig. 4.33 Bandwidth to capture transition

Figure 4.33d shows some evaluation results down to a practical limit with an accuracy of 3% [40]. In summary, you can estimate the required bandwidth for a given transition time of a power device switching by determining the frequency response of the oscilloscope and the allowable measurement error. If the rise time or fall time of the SiC power device is 8 ns, then the necessary bandwidth is calculated to be about 90 MHz.

Another important characteristic to capture is the ringing. Figure 4.34b shows an equation to calculate the resonant frequency of an LC circuit. The capacitance value of a power device is typically found in the device datasheet as output capacitance (Coss). The inductance of the test system is calculated with the formula mentioned earlier. Then, the required bandwidth should be around the calculated resonant frequency.

Figure 4.34a shows the captured ringing of a SiC MOSFET turn-off characteristic with sufficiently high bandwidth (500 MHz). The measured ringing frequency is about 80 MHz. The resonant frequency calculated from the system loop inductance and Coss of the DUT is 103 MHz. Figure 4.34c shows the captured ringing of the same DUT by intentionally reducing the test system frequency to 103 MHz.

The measured ringing frequency and amplitude are the same as in Fig. 4.34a, indicating that a bandwidth of around 100 MHz is sufficient to capture the ringing in this case.

Finally, the sampling rate should be set sufficiently high. Figure 4.34d shows the guideline of the sampling rate dependent on the frequency response of an oscilloscope. Figure 4.34e provides the required bandwidth guideline for SiC MOSFET. The guideline for gallium nitride (GaN) FET is also shown as a reference.

SiC power devices exist not only in discrete form but also in various power module configurations. To build a test system for power modules, it is important to reduce cable connections for the power loop and gate loop area by designing the interface board with minimized parasitics. One of the challenging tests for the

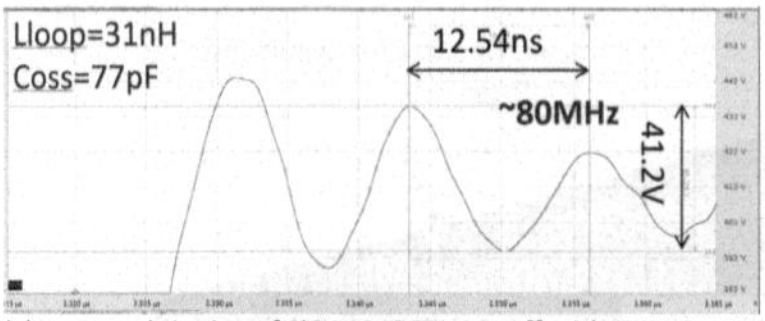

(a) Captured ringing of SiC MOSFET turn-off with 500MHz BW

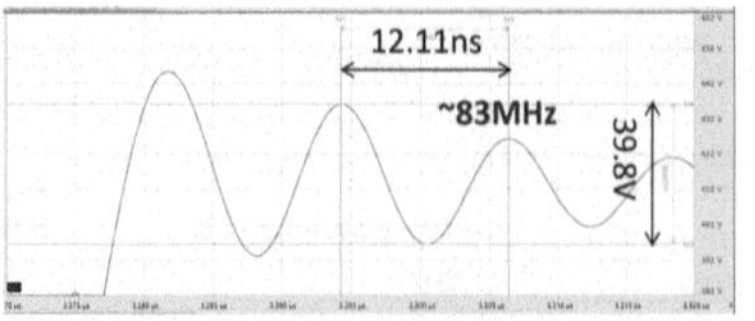

(c) Captured ringing of SiC MOSFET turn-off with 103MHz BW

$$f_{ringing} \approx \frac{1}{2\pi\sqrt{L_{loop}C_{oss}}}$$

(b) Resonant frequency of an LC circuit

Gaussian response	Flat response
> 4*BW	2.5*BW

(d) Sufficiently high sample rate from the system BW

DUT type	Typ. Transition times	Required BW (transition time)	Typical loop inductance	Typical C_{oss}	Required BW (ringing)
SiC (TO247)	8-15ns	50-90MHz	30nH	60pF	120MHz
GaN (SMD)	2-5ns	150-350MHz	4nH	25pF	500MHz

(e) Estimated required BW for SiC and GaN power devices

Fig. 4.34 Bandwidth to capture ringing

power module is high-side Vgs measurement because the common-mode noise is in the range of hundreds of volts, while the gate voltage is less than 20 V or so. Optically isolated probes are available for such measurements, providing sufficient common-mode rejection ratio (CMRR). It is also important to check not only the CMRR but also the noise performance when selecting the solution for high-side Vgs measurement.

4.7.4 Building a Safe Test System

Double-pulse test on SiC power devices requires substantial amount of energy to provide high voltage and high current simultaneously. Bank capacitors are necessary in the configuration to store such high energy. It is imperative to ensure safe discharge procedure with built-in safety redundancies. The discharge process should be linked to multiple events, such as the emergency-off switch, accidental power cuts, door unlocks, and completion of measurements. A robust mechanism is essential to prevent the user from contacting a part of the system where the voltage exceeds 42 V.

In the case of commercial test system, it is advisable to select a system that complies with CSA and UL safety regulations.

4.7.5 Summary

The dynamic characteristics of SiC power semiconductor devices require high-voltage, high-current, and high-speed transient measurements simultaneously. The test system should be built by minimizing parasitics in the system, employing appropriate system bandwidth and compensation, implementing safety with redundancy, and so on. Commercial double-pulse test systems covering the aforementioned requirements and challenges are available, such as Keysight PD1500A and Keysight PD1550A double-pulse testers.

4.8 No-Compromise "Trench-Assisted Planar-Gate" SiC FETs

Llew Vaughan-Edmunds

As the world moves from fossil fuels to renewable energy, power electronics take center stage, and legacy, performance-limited silicon FETs, and IGBTs cannot contribute fully to this critical transition. SiC's stronger covalent bonds are comparable to those of diamond and give SiC a bandgap (the energy required to free an electron from its nuclear orbit) of 3.26 eV, which is nearly three times that of conventional silicon. This higher bandgap allows SiC to provide a dielectric breakdown strength ten times that of conventional silicon and enables it to support higher speed operation, higher frequencies, and higher voltages. SiC's superior thermal conductivity and stability also make it more reliable in high-voltage, high-temperature applications.

In renewable energy generation and grid distribution, the higher efficiency of all-SiC inverters will revolutionize the integration of renewable energy supply, storage, and delivery. SiC is used, for example, in inverters that aggregate the power from strings of panels and convert the generated DC into AC. SiC MOSFETs and diodes in the inverter and boost stages are more efficient than silicon and again contribute to reducing overall system size, weight, and cost.

In transport, SiC is also improving the fuel efficiency, weight, and size of traction and other onboard power systems in trains, buses, planes, and other vehicles. And in motor control, replacing the antiparallel silicon diodes in motor current commutation with SiC reduces switching and conduction losses; replacing all the power silicon IGBTs and diodes with SiC is even more efficient.

While a SiC FET requires more energy to create than a Si IGBT—due to SiC boule growth—the system benefits in terms of dematerialization and efficiency upgrades for higher-power, longer-lifetime systems, far outweigh the initial footprint. In fact, each SiC FET shipped is estimated to save 25.2 kg of CO_2 versus legacy silicon IGBTs. SiC offers the potential to address a further reduction of 3.4 Gtons CO_2/year by 2050 [41].

4.8.1 Fabricating SiC Semiconductors

Planar or trench techniques have traditionally been used to fabricate SiC devices. Planar fabrication, depositing active device regions on the flat substrate surface, is relatively straightforward and familiar in the semiconductor industry through established use. For SiC, it is high yield and low cost and produces reliable devices but with slower switching and high drain-source on-resistance than devices made with trench techniques, which involve etching active device regions into the SiC substrate. However, traditional trench process tends to produce lower yields and lower reliability at a higher cost.

Navitas' patented, next-generation GeneSiC trench-assisted planar-gate design combines the two approaches into a process that eliminates the compromises inherent in each (Fig. 4.35). The process offers repeatable, high-yield manufacturing at a low cost. The design minimizes energy losses at high speeds, raising efficiency and performance levels. It produces reliable SiC MOSFETs with fast switching and enables the lowest on-state resistance, even at high temperatures. The trench-assisted, planar-gate approach also monolithically integrates a merged PiN Schottky (MPS) diode as shown in Fig. 4.36.

4.8.2 No-Compromise Device Performance

Both on-state and switching performances are critical for SiC FETs. A SiC technology report compared the $R_{DS(ON)}$ characteristics of 12 SiC FET technologies, and the novel GeneSiC technology outperformed all the others, including the trench-gate

	Planar	Trench	GeneSiC

Manufacturability	» Repeatable » High yield » Low cost	» Inconsistent trench etch » Lower yields » High cost	» Repeatable » High yield » Low cost
Performance	» High $R_{DS(ON)}$ / area » Slow switching » High $R_{DS(ON)}$ / Δ temp	» Lower $R_{DS(ON)}$ / area » Faster switching » High $R_{DS(ON)}$ / Δ temp	» Lower $R_{DS(ON)}$ / area » Fastest switching » Lowest $R_{DS(ON)}$ / Δ temp
Reliability	» Rugged gate oxide (stable V_{TH})	» Failures due to non-uniform gate oxide » Lower short-circuit capability	» Highest 100% tested avalanche » Long short-circuit withstand time » Rugged gate oxide (stable V_{TH})

Fig. 4.35 Navitas GeneSiC technology: trench-assisted planar-gate fabrication compared to traditional trench or planar processes

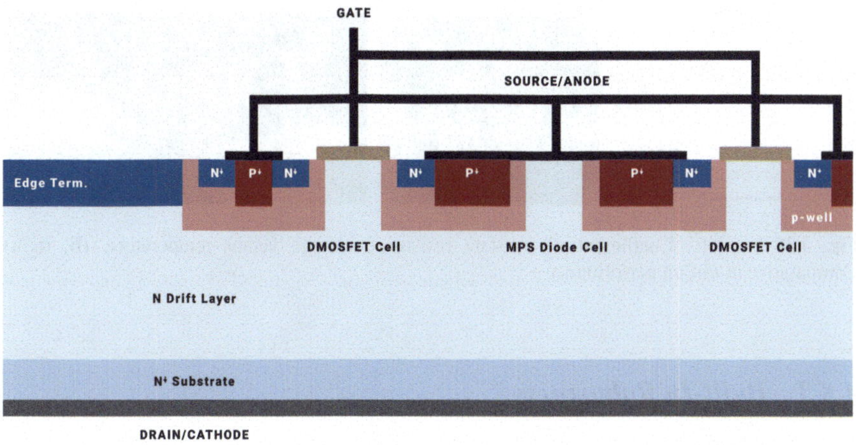

Fig. 4.36 Structure of the GeneSiC MOSFET with monolithically integrated MPS diode

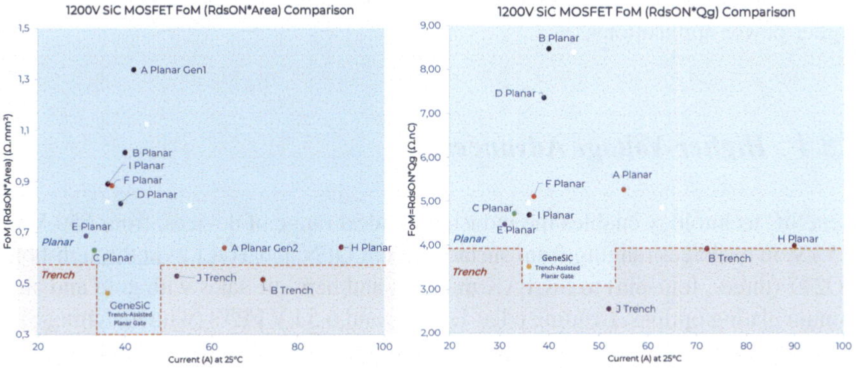

Fig. 4.37 Industry trench, planar, and trench-assisted planar-gate devices' figures of merit for on-state resistance per unit area ($R_{DS(ON)}$ Ω/mm^2) and on-state resistance with gate charge ($R_{DS(ON)}$ ΩnC) (Yole)

structures, while maintaining the ruggedness, short circuit capability, and simpler manufacturing process of planar-gate devices (Fig. 4.37) [41].

This is important for applications where the devices are required to operate in challenging thermal conditions, as outlined above. In such applications, ambient system temperatures can rise to 80 °C, with even higher junction temperatures due to power cycling. GeneSiC technology enables the lowest commercially available drain-source on-resistance temperature coefficient. Traditionally, "headline" rated at 25 °C, $R_{DS(ON)}$ increases dramatically for traditional devices (Fig. 4.38a).

In bench tests, the GeneSiC 1200 V, 40 mΩ SiC MOSFET in a D2PAK outperformed the comparable leading competition. Under equivalent conditions, GeneSiC MOSFETs exhibited a 25 °C cooler case temperature, reduced losses, higher system efficiency, and enhanced reliability, likely multiplying device lifetime threefold.

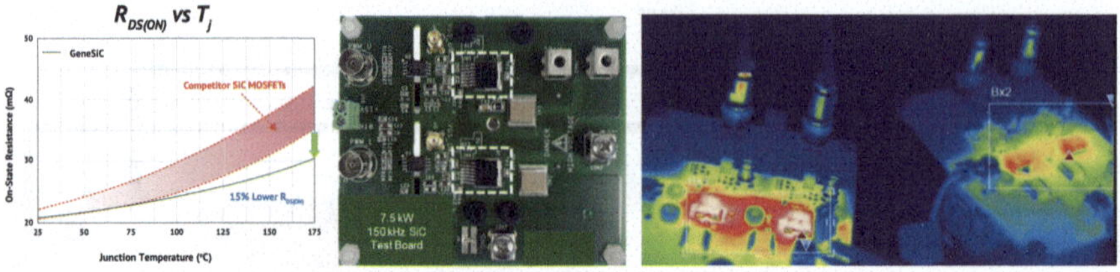

Fig. 4.38 (**a**, left) Coefficient of on-state resistance change versus temperature, (**b**, right) Comparative in-circuit performance

4.8.3 Built-In Robustness

Device- and system-level robustness and reliability are key for high-power systems. The trench-assisted planar-gate devices are 100% tested for avalanche and have a long "short-circuit withstand time" for robust performance in abnormal conditions. Predictable threshold voltages mean that devices can be paralleled more reliably for higher-power applications.

4.8.4 Higher-Voltage Advances

GeneSiC technology enables the industry's widest range of devices, from 650 V to 6.5 kV, in packages ranging from surface-mount QFN and TOLL, via through-hole TO247 (three-, four-pin) to SiCPAK modules and bare-die sales with gold and aluminum plating options. Leading edge 1.7, 3.3, and 6.5 kV FETs (with matching SiC Schottky diodes) enable simple, reliable high-power applications, avoiding complex topology-stacking techniques (Fig. 4.39).

Deploying SiC

SiC FETs dominate new designs across high-power applications, with a prediction that the SiC market will grow from around $2B in 2023 to between $11 and $14 billion by 2030 [41].

As power-hungry AI processors ramp in production, the need for higher-power density AC-DC conversion has opened the door for SiC. A new design for 220 V_{AC} to 48 V_{DC} conversion uses 650 V GeneSiC FETs in TOLL packaging for the high-speed switching bridgeless PFC design, achieving 8.4 W/cc (138 W/in^3) in the OCP standard CRPS185 form factor [42].

In solar, KATEK is a pioneer in the renewable energy industry and develops and produces high-end power electronics for grid inverters and energy storage as well as control technology for photovoltaic systems and fuel cell systems. The Steca

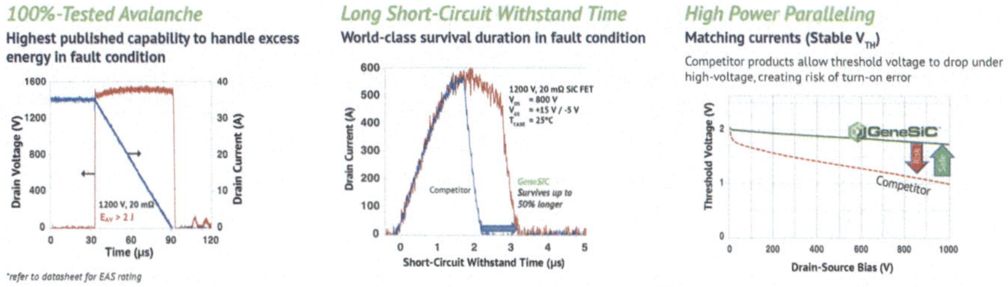

Fig. 4.39 Trench-assisted planar-gate ruggedness at device and system level: (**a**, left) avalanche, (**b**, middle) short-circuit, and (**c**, right) V_{TH} stability

coolcept fleX model solar inverter converts DC power from a string of solar panels into 4.6 kW AC power for use in the home, returning to the grid, or being stored locally for later use—to smooth demand and/or support power during an outage. Next-generation GeneSiC technology has enabled a major step in system performance without compromise. Each inverter uses 16× GeneSiC G3R75MT12J SiC MOSFETs [43]. These 1200 V, 75 mΩ-rated devices are used in a two-level converter, with bidirectional boost converters and an H4 topology for AC voltage output. Increased switching frequency shrinks the size and weight of "passive" components, which optimizes size and weight compared to legacy silicon-based inverters.

In industrial warehousing, materials-handling robots require reliable, fast charging to maintain productivity. For Exide, high-frequency chargers convert 220 V_{AC} to a battery-level voltage between 24 and 80 V for lead-acid and lithium-ion technologies. The 7 kW modules uses GeneSiC G3R60MT07D (750 V) MOSFETs and GD10MPS12A (1200 V) MPS Schottky diodes, with frequency-optimized architecture. The same platform can be upgraded to 10 kW, with four modules in parallel to provide 40 kW of reliable fast-charging power.

For EVs, upgrading from 400 V to 800 V batteries increases cuts transmission losses, heat dissipation, weight, and cost in the power cables. As the onboard charger and traction drive need to be designed for the higher voltage, 1200 V SiC MOSFETs, with their efficiency and performance, are a perfect fit. Benefits include smaller and lighter batteries, motors, and EVs as a whole, which lead to greater range, fuel efficiency, and lower cost [43].

In long-haul EV trucking, 27 nations, including Germany, the US, the UK, and France have signed up to the COP27 agreement to eliminate diesel-truck sales by 2035, and a new "Megawatt Charging System" specification (SAE J3271 [44]) has been developed to provide up to 3.75 MW, over a 1250 V cable [42]. Higher voltage reduces I^2R conduction loss, cost, and weight while increasing motor speed and torque. This is especially critical in aviation electrification, for example, EVTOL "flying cars" where most major players are using high-voltage architectures around 1 kV [43].

4.8.5 No-Compromise, Next-Generation SiC

Early planar and trench designs have exposed limitations in manufacturing, performance, and/or reliability. GeneSiC trench-assisted, planar-gate technology offers a no-compromise solution with the industry's broadest voltage range, from 650 V to 6.5 kV [41–50].

4.9 SiC Junction Field-Effect Transistors

Victor Veliadis

The SiC junction field-effect transistor (JFET) is capable of high-power and high-temperature switching as it only uses *pn* junctions in the active device area, where the high electric fields occur and can therefore fully exploit the high-temperature properties of SiC in a gate voltage-controlled switching device. Provided the gate-to-source junction of the JFET is biased below its built-in potential, negligible gate current is needed to drive the device and voltage-controlled switching is realized. JFETs are free of MOS native oxide problems such as low channel mobility, threshold voltage instability, and lack of reliability at elevated temperatures. They have demonstrated electrostatic discharge immunity to 16 kV [51] and as unipolar devices do not suffer from forward voltage degradation at the same degree as bipolar devices. SiC power JFETs are almost exclusively implemented in a vertical configuration and are native depletion mode or normally-*on* (n-on).

Several SiC JFET designs have been demonstrated over the years, schematically shown in Fig. 4.40. The first power 4*H*-SiC JFETs were reported by H. Mitlehner et al. in 1999 [52]. They were vertical JFETs (VJFETs) with lateral channel (Fig. 4.40a), and they have been further developed by D. Stephani et al. [53] and

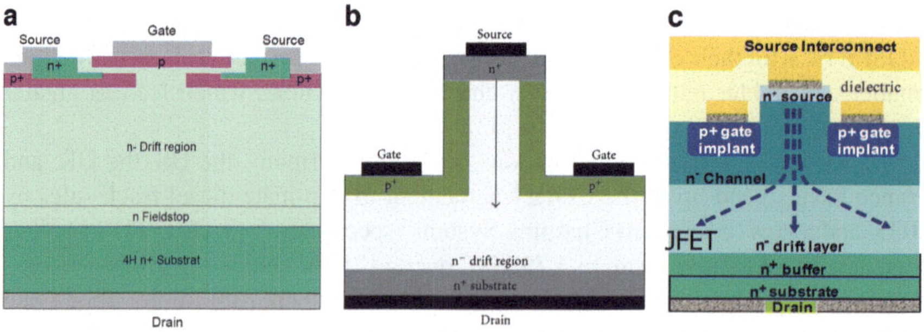

Fig. 4.40 Simplified cross sections of (**a**, left) a vertical JFET with a lateral channel (the *n*+ source is embedded in the *p*-well and reaches below the *p* gate, to minimize source resistance), (**b**, center) a trench and implanted vertical JFET, (**c**, right) a depletion mode ion-implanted SiC vertical-channel JFET

S-H. Ryu et al. [54]. This JFET is similar to a SiC double-implanted MOSFET (DMOSFET), with the oxide-controlled inversion channel having been replaced with a bulk channel. This eliminates the SiC/SiO2 interface, with its channel mobility and reliability drawbacks, and creates a bulk channel where the SiC mobility is fully utilized.

Zhao et al. have implemented a JFET design with relatively critical dimensions (Fig. 4.40b) [55]. The design requires three implantation events. To implant the lower portion of the sidewalls, the wafer must be tilted against the direction of the ion beam and rotated. No epitaxial regrowth is needed.

A simplified cross-sectional schematic of a trenched-gate vertical channel *p+* ion-implanted depletion mode (normally on) 4H-SiC JFET is shown in Fig. 4.40c [56]. This representative JFET design will be used to analyze electrical characteristics including on-state and blocking characteristics, thermal performance, and ruggedness. Numerous publications report on the detailed fabrication and the related electrical characteristics [57–59] of this SiC JFET design.

The inherent simplicity of the design shown in Fig. 4.40c, which does not require epitaxial regrowth and is fabricated in a single implantation event that forms the *p* gates and edge termination guard rings, allows for reliable JFETs with excellent yields and parameter uniformity [56]. Indeed, this JFET design has low process complexity [56–58]. And 1680 V SiC JFETs with an active area of 0.143 cm^2 (0.19 cm^2 total area) and on-state current capability of 54 A were fabricated in seven photolithographic levels, with a single masked ion-implantation event that simultaneously implanted the *p+* gates and guard rings [60]. Representative room-temperature *on*-state drain-current versus voltage characteristics are shown in Fig. 4.41a at a gate bias range of 0 to 2.5 V, in steps of 0.5 V. To maintain voltage-control capability (high I_D/I_G gain), the gate must be biased below its 2.7 V built-in potential value. If the gate bias increases in excess of 2.7 V, significant gate current injection occurs into the channel of the JFET, and its current gain I_D/I_G degrades. At a gate-to-source bias of 2.5 V, the JFET outputs 53.6 A at a forward drain voltage drop of 2.08 V. The specific *on*-state resistance is 5.5 $m\Omega{\cdot}cm^2$, and the transistor current gain is $I_D/I_G = 26{,}800$.

The blocking voltage characteristics of the 0.143 cm^2 active-area JFET at gate biases of −4 to −24 V, in steps of −2 V, are shown in Fig. 4.41b. At a gate-to-source bias of −24 V and a low drain-current density of 1 mA/cm^2, the JFET blocks 1680 V.

An important element of SiC device reliability is eliminating threshold voltage instability. In SiC MOSFETs, threshold voltage instability is primarily due to the oxide traps at the SiC/gate oxide interface.

To investigate the threshold voltage stability of JFETs, a JFET based all-SiC normally-off switch was implemented by combining a 1200 V normally-on JFET with a low-voltage normally-off (enhancement mode) JFET in the cascode configuration [61]. To evaluate threshold voltage shift with changes in temperature, the I_{DS}-V_{DS} characteristics of the all-SiC cascode switch were measured at junction temperatures of 25 °C, 100 °C, 200 °C, and 300 °C and are shown in Fig. 4.42a [62]. The cascode's *on*-state resistance is 6.2 $m\Omega{\cdot}cm^2$ and was extracted from the data of Fig. 4.42a at $V_{DS} = 0.5$ V. The increase in *on*-state resistance with temperature agrees

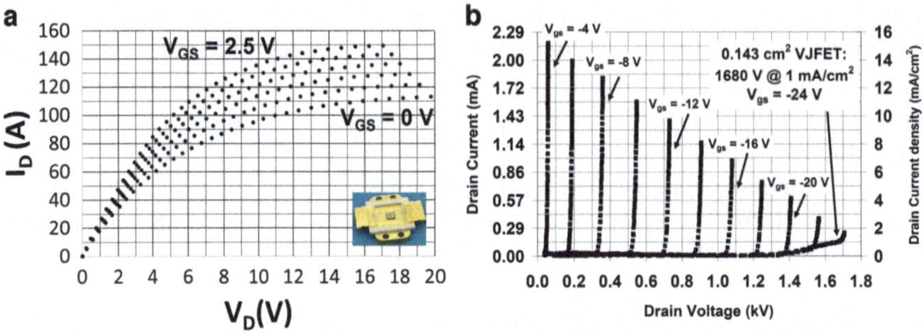

Fig. 4.41 (**a**, left) On-state drain current versus drain voltage and (**b**, right) blocking voltage characteristics of a single 1680 V, 0.143 cm² packaged SiC JFET

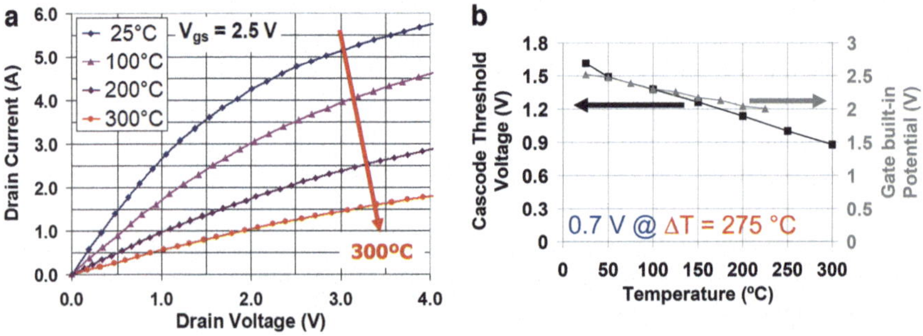

Fig. 4.42 Temperature dependence of the SiC cascode's (**a**, left) *on*-state drain current at $V_{GS} = 2.5$ V and (**b**, right) threshold voltage (left axis) and gate-junction built-in potential (right axis). At a temperature swing of 275°, the threshold voltage only shifts by 0.7 V

well with the theoretical reduction of the electron mobility in 4H-SiC. The cascode threshold voltage was extracted and is plotted as a function of temperature on the left axis of the graph of Fig. 4.42b.

The threshold voltage decreases from 1.6 V to 0.9 V as the temperature increases from 25 °C to 300 °C; the cascode switch remains normally-off at 300 °C. The cascode's gate-junction built-in potential variation with temperature was also extracted and was plotted on the right axis of the graph of Fig. 4.42b. As the temperature increases from 25 °C to 225 °C, the cascode's threshold voltage decreases by 0.54 V, while its gate-junction built-in potential decreases by 0.52 V. This excellent agreement confirms that the decrease in cascode threshold voltage with temperature stems from the reduction of its gate-junction built-in potential, as expected from solid-state physics. Thus, SiC JFETs have remarkably stable threshold voltages because they use *pn* junctions instead of gate oxides to control current flow.

To evaluate ruggedness, a SiC JFET was subjected to over 2.4 million 1200 V/115 A hard-switching events at 150 °C, at 13 times its 8.8 A 150 °C rated current (Fig. 4.43) [63]. The JFET drain voltage is plotted in black (left axis) in Fig. 4.43a, while the current through the JFET is plotted in gray (right axis). By multiplying

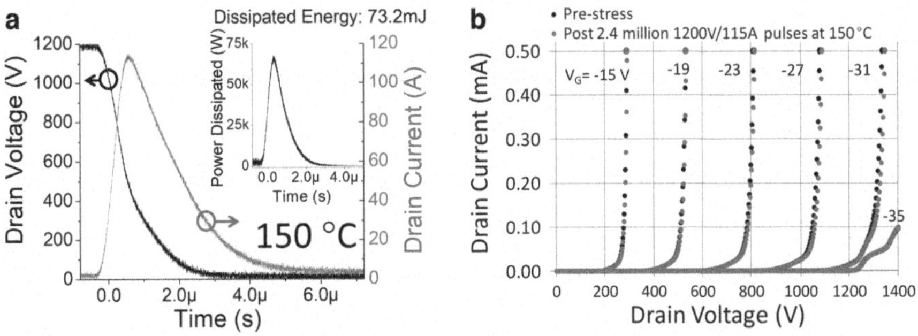

Fig. 4.43 (**a**, left) Representative 1200 V/115 A hard-switching waveforms of the SiC JFET testing at 150 °C. The energy dissipated by the JFET during each hard-switching event is 73.2 mJ (inset), and the peak dissipated power is 68.2 kW. (**b**, right) Blocking voltage JFET curves before (black circles) and after (gray circles) the over 2.4 million hard-switching events at 150 °C. The blocking voltage characteristics do not change

voltage by current, the power dissipated by the JFET is calculated and plotted in the inset of Fig. 4.43a. The energy dissipated by the JFET during each 1200 V/115 A switching transient is 73.2 mJ, and the peak dissipated power is 68.2 kW. The electrical conduction and blocking voltage characteristics are unchanged after over 2.4 million 1200 V/115 A hard-switching events at 150 °C, indicating the JFET is reliable and rugged. The before-and-after blocking voltage characteristics are shown in Fig. 4.43b.

The ruggedness of SiC JFETs at high temperatures has been demonstrated. Short-term operation of packaged 4*H*-SiC junction field-effect transistor (JFET) logic-integrated circuits (ICs) at temperatures exceeding 800 °C is discussed in reference [64]. SiC lateral JFETs with operating times of 6000 h at 500 °C, which was limited by the thermal degradation of the metal stack used for the formation of the ohmic contacts, are shown in references [65, 66].

Today, power SiC JFETs are commercially available as discrete components in the 650–1700 voltage range [67]. MOSFETs dominate SiC-based power electronics, while JFETs are utilized in smaller numbers. As SiC power electronics continue to gain ground, the JFET has the potential to be the device of choice for rugged high-temperature applications.

References

1. American Society for Quality, https://asq.org/quality-resources/reliability
2. F. Lannuzzo, M. Ciappa, Reliability issues in power electronics. Microelectron. Reliab. **58**, 1–2 (2016)
3. JEDEC, JEP195: Guideline for Evaluating Gate Switching Instability of Silicon Carbide Metal-Oxide-Semiconductor Devices for Power Electronic Conversion
4. P. Salmen, M.W. Feil, K. Waschneck, H. Reisinger, G. Rescher, I. Voss, M. Sievers, T. Aichinger, Gate-switching-stress test: Electrical parameter stability of SiC MOSFETs in switching operation. Microelectron. Reliab. **135**, 114575 (2022)

5. T. Aichinger, M. Schmidt, Gate-Oxide Reliability And Failure-Rate Reduction of Industrial SiC MOSFETs, 2020
6. JEDEC, JEP194: Guideline for Gate Oxide Reliability and Robustness Evaluation Procedures for Silicon Carbide Power MOSFETs
7. D.J. Lichtenwalner, S. Sabri, E. van Brunt, B. Hull, S.-H. Ryu, P. Steinmann, A. Romero, S. Ganguly, D.A. Gajewski, S. Allen, J.W. Palmour, Accelerated Testing of SiC Power Devices, South Lake Tahoe, 2020
8. P. Moens, Unique failure mode of SiC MOSFETs under accelerated HTRB, in *ICSCRM*, (2023)
9. JEDEC, JEP184: Guideline for Evaluating Bias Temperature Instability of Silicon Carbide Metal-Oxide-Semiconductor Devices for Power Electronic Conversion
10. JEDEC, JEP190: Guideline for Evaluating dv/dt Robustness of SiC Power Devices
11. ECPE, ECPE Guideline AQG 324: Qualification of Power Modules for Use in Power ElectronicsConverter Units in Motor Vehicles—Release 03.1/2021
12. F. Hoffmann, P. Friedrichs, N. Kaminski, H3TRB-Test on 1200 V SiC Schottky Diodes After Previous Operation, Germany, 2020
13. P. Salmen,P. Friedrichs, Qualifying a silicon carbide power module: reliability testing beyond the standards of silicon devices, in CIPS 2022; 12th International Conference on Integrated Power Electronics Systems, 2022
14. I.-H. Ji, A. Mathew, J.-H. Park, N. Oldham, M. McCain, S. Sabri, E.V. Brunt, B. Hull, D.J. Lichtenwalner, D.A. Gajewski, J.W. Palmour, High Temperature and High Humidity Reliability Evaluation of Large-Area 1200V and 1700V SiC Diodes, Monterey, 2023
15. I. Voss, T. Aichinger, T. Basler, P. Friedrichs, R. Rupp, Reliability and Ruggedness of SiC Trench MOSFETs for Long-Term Applications in Humid Environment, Nuremberg, 2018
16. Power Electronics for Photovoltaics—Actual Status and Future Trends, 2017
17. D.-P. Sadik, J.-K. Lim, F. Giezendanner, P. Ranstad, H.-P. Nee, Humidity Testing of SiC Power MOSFETs An Update, Warsaw, 2017
18. J. Berens, T. Aichinger, A Straightforward Electrical Method to Determine Screening Capability of GOX Extrinsics In Arbitrary, Commercially Available SiC MOSFETs, Monterey, 2021
19. D. A. Gajewski, B. Hull, D. J. Lichtenwalner, S.-H. Ryu, E. Bonelli, H. Mustain, G. Wang, S.T. Allen, J.W. Palmour, SiC Power Device Reliability, South Lake Tahoe, 2016
20. M. Beier-Moebius, J. Lutz, Breakdown of Gate Oxide of SiC-MOSFETs and Si-IGBTs Under High Temperature and High Gate Voltage, Nuremberg, 2017
21. E.V. Brunt, D.J. Lichtenwalner, J.H. Park, S. Ganguly, J.W. McPherson, Lifetime Modeling of the 4H-SiC MOS Interface in the HTRB Condition Under the Influence of Screw Dislocations, Monterey, 2023
22. T. Aichinger, G. Rescher, G. Pobegen, Threshold voltage peculiarities and bias temperature instabilities of SiC MOSFETs, vol. 80, pp. 68–78, 2018
23. A.J. Lelis, R. Green, D.B. Habersat, M. El, Basic mechanisms of threshold-voltage instability and implications for reliability testing of SiC MOSFETs. IEEE Trans. Electron Devices **62**(2), 316–323 (2015)
24. D.A. Gajewski, Challenges and Peculiarities in Developing New Standards for SiC, 2020
25. J. Berens, M. Weger, G. Pobegen, T. Aichinger, G. Rescher, C. Schleich, T. Grasser, Similarities and Differences of BTI in SiC and Si Power MOSFETs, 2020
26. JEDEC, JEP183: Guideline for Measuring the Threshold Voltage (VT) of SiC MOSFETs
27. H. Jiang, X. Zhong, G. Qiu, L. Tang, X. Qi, L. Ran, Dynamic gate stress induced threshold voltage drift of silicon carbide MOSFET. IEEE Electron Device Lett. **41**(9), 1284–1287 (2020)
28. Infineon, Application Note 2018–09: guidelines for CoolSiC MOSFET gate drive voltage window, www.infineon.com, 2018
29. P. Salmen, M.W. Feil, K. Waschneck, H. Reisinger, G. Rescher, T. Aichinger, A New Test Procedure to Realistically Estimate End-Of-Life Electrical Parameter Stability of SiC MOSFETs in Switching Operation, Monterey, 2021
30. D.B. Habersat, A.J. Lelis, AC-Stress degradation and its anneal in SiC MOSFETs. IEEE Trans. Electron Devices **69**(9), 5068–5073 (2022)

31. M.W. Feil, K. Waschneck, H. Reisinger, J. Berens, T. Aichinger, P. Salmen, G. Rescher, W. Gustin, T. Grasser, Towards Understanding the Physics of Gate Switching Instability in Silicon Carbide MOSFETs, Monterey, 2023
32. C. Herold, M. Schaefer, F. Sauerland, T. Poller, J. Lutz, O. Schilling, Power cycling Capability of Modules with SiC-Diodes, Nuremberg, 2014
33. R. Bayerer, T. Herrmann, T. Licht, J. Lutz, M. Feller, Model for Power Cycling lifetime of IGBT Modules—various factors influencing lifetime, Nuremberg, 2008
34. F. Hoffmann, N. Kaminski, S. Schmitt, Comparison of the Power Cycling Performance of Silicon and Silicon Carbide Power Devices in a Baseplate Less Module Package at Different Temperature Swings, Nagoya, 2021
35. J. Lutz, C. Schwabe, G. Zeng, L. Hein, Validity of Power Cycling Lifetime Models for Modules And Extension to low temperature swings, Lyon, 2020
36. P. Salmen, T. Methfessel, C. Kuenzel, O. Schilling, Impact of Load-Pulse Duration on Power-Cycling Capability of SiC Devices, Berlin, 2020
37. R. Schmidt, R. Werner, J. Casady, B. Hull, A. Barkley, Power Cycle Testing of Sintered SiC-MOSFETs, Nuremberg, 2017
38. C. Schwabe, N. Thoenelt, T. Basler, Reliability Investigation of SiC MOSFETs Under Switching Operation in Various Packages, Berlin, 2022
39. H. Johnson, M. Graham, *High-Speed Digital Design: A Handbook of Black Magic* (Prentice Hall, 1993)
40. Effect of Oscilloscope Frequency Response on Rise-Time Accuracy: Understanding Oscilloscope Frequency Response and Its Effect on Rise-Time Accuracy, Application Note, Keysight Technologies
41. Navitas sustainability report (GaN and SiC), https://navitassemi.com/sustainability-and-gan/
42. Yole SystemPlus SiC Transistor 2022
43. McKinsey, New silicon carbide prospects emerge as market adapts to EV expansion, October 2023
44. AI Power Roadmap, Navitas, https://navitassemi.com/navitas-gan-and-sic-technologies-to-enable-next-gen-ai-power-delivery/
45. Navitas and KATEK Accelerate Solar Adoption with Higher Efficiency and Lower Costs, Navitas, https://navitassemi.com/navitas-and-katek-accelerate-solar-adoption-with-higher-efficiency-and-lower-costs/
46. Exide 40kW Charging using GeneSiC, https://navitassemi.com/navitas-next-gen-sic-power-semis-adopted-in-industrial-chargers-from-exide-technologies/
47. Navitas GaNSafe and GeneSiC Double engine drive automotive power future, CPSSC 2023, https://navitassemi.com/articles/
48. SAE J3271 specification, https://standardsworks.sae.org/standards-committees/j3271-megawatt-charging-system-electric-vehicles-tf
49. US DoE/Argonne National Lab, https://www.anl.gov/reference/faq-charging-for-heavyduty-electric-trucks
50. EVTOL battery voltages: Archer Aviation 800V, https://archer.com/technologies, Lilium 900V, https://lilium.com/newsroom-detail/first-high-voltage-electrical-harnesses-roll-off-the-line-for-the-all-electric-lilium-jet, and Joby 1000V https://joby-site.cdn.prismic.io/joby-site/5f82ea34-645e-4468-8e3f-14a16e298941_Joby-Charging-GEACS-final.pdf
51. V. Veliadis, Private communication
52. H. Mitlehner, W. Bartsch, K. O. Dohnke, P. Friedrichs, R. Kaltschmidt, U. Weinert, B. Weis, D. Stephani, *Dynamic characteristics of high voltage 4H-SiC vertical JFETs*. Presented at the 11th International Symposium on Power Semiconductor Devices and ICs, ISPSD'99 Proceedings, 1999
53. D. Stephani, Friedrichs, Silicon carbide junction field effect transistors. Int. J. High Speed Electron. Syst. **16**(3), 825–854 (2006)
54. S.-H. Ryu, S. Krishnaswami, B.A. Hull, B. Heath, F. Husna, J. Richmond, A. Agarwal, J. Palmour, J. Scofield, A comparison of high temperature performance of SiC DMOSFETs and JFETs. Mater. Sci. Forum **556–557**, 775–778 (2007)

55. J.H. Zhao, K. Tone, X. Li, P. Alexandrov, L. Fursin, M. Weiner, 3.6 mΩ cm2, 1726V 4H-SiC normally-off trenched and-implanted vertical JFETs and circuit applications. IEE Proc. Circuits Devices Syst. **151**(3), 231–237 (2004)
56. V. Veliadis, Silicon carbide junction field effect transistors (SiC–JFETs), in *Wiley Encyclopedia of Electrical and Electronics Engineering*, pp. 1–37, Online ISBN: 9780471346081, https://doi.org/10.1002/047134608X.W8232, Dec 15, 2014
57. V. Veliadis, Chapter 7: 1200-V SiC vertical-channel-JFETs and cascode switches, in *Silicon Carbide: Volume 2: Power Devices and Sensors*, ed. by P. Friedrichs, T. Kimoto, L. Ley, G. Pensl, ISBN: 978–3–527-41002-6 (Wiley-VCH, 2009), pp. 157–191
58. V. Veliadis, Chapter 13: Silicon carbide vertical junction field effect transistors and cascode switches for 1200 V power conditioning applications, in *Advanced semiconductor Materials and Devices Research—SiC and III-Nitrides*, ed. by Ho-Young Cha, ISBN: 978–81–7895-371-7 (Research Signpost, 2010), pp. 407–446
59. V. Veliadis, M. Snook, T. McNutt, H. Hearne, P. Potyraj, A. Lelis, C. Scozzie, A 2055-V (at 0.7 mA/cm2), 24-A (at 706 W/cm2) normally-ON 4H-SiC JFET with 0.068-cm2 active area and blocking voltage capability of 94% of the SiC material limit. IEEE Electron Dev. Lett. **29**(12), 1325–1327 (2008)
60. V. Veliadis, T. McNutt, M. Snook, H. Hearne, P. Potyraj, C. Scozzie, A 1680 V (at 1 mA/cm2), 54 A (at 780 W/cm2) normally-ON 4H-SiC JFET with 0.143 cm2 active area. IEEE Electron Dev. Lett. **29**(10), 1132–1134 (2008)
61. V. Veliadis, T. McNutt, M. McCoy, H. Hearne, P. Potyraj, C. Scozzie, Large area silicon carbide VJFETs for 1200 V cascode switch operation. Int. J. Power Manag. Electron. **2008**, 523721 (2008)
62. V. Veliadis, H. Hearne, T. McNutt, M. Snook, P. Potyraj, C. Scozzie, VJFET based all-SiC Normally-off cascode switch for high temperature power handling applications. Mater. Sci. Forum **615–617**, 711–714 (2009)
63. V. Veliadis, B. Steiner, K. Lawson, S.B. Bayne, D. Urciuoli, H.C. Ha, N. El-Hinnawy, S. Gupta, P. Borodulin, R.S. Howell, C. Scozzie, Reliable operation of SiC JFET subjected to over 2.4 million 1200-V/115-A hard switch stressing events at 150 °C. IEEE Electron Dev. Lett. **34**(3), 384–386 (2013)
64. P.G. Neudeck, D.J. Spry, L. Chen, N.F. Prokop, M.J. Krasowski, Demonstration of 4H-SiC digital integrated circuits above 800 °C. IEEE Electron Device Lett. **38**, 1082–1085 (2017). https://doi.org/10.1109/led.2017.2719280
65. P.G. Neudeck, D.J. Spry, C. Liang-Yu, G.M. Beheim, R.S. Okojie, C.W. Chang, R.D. Meredith, T.L. Ferrier, L.J. Evans, M.J. Krasowski, N.F. Prokop, Stable electrical operation of 6H-SiC JFETs and ICs for thousands of hours at 500C. IEEE Electron Device Lett. **29**(5), 456–459 (2008)
66. P.G. Neudeck, S.L. Garverick, D.J. Spry, L.-Y. Chen, G.M. Beheim, M.J. Krasowski, M. Mehregany, Extreme temperature 6H-SiC JFET integrated circuit technology. Phys. Status Solidi **206**(10), 2329–2345 (2009)
67. https://unitedsic.com/

Rudolf Elpelt received his PhD in physics in 2000 from the University of Erlangen-Nürnberg with a thesis on GaAs doping superlattices. In 2000, he joined the silicon carbide development at SiCED Electronics Development GmbH & Co. KG in Erlangen, where he was involved in TCAD simulation and technology development. Since 2011, Rudolf Elpelt is part of the silicon carbide technology development team of Infineon Technologies AG located in Erlangen.

Andrea Piccioni holds a M.Sc. degree in electrical engineering from Politecnico di Torino. He dedicated 3 years to R&D in the automotive industry, focusing on electric drive traction inverter technology. Building on this experience, he transitioned to Infineon Technologies, where he has spent over 3 years as a product application and definition engineer for SiC MOSFET.

Jake Choi , a seasoned engineer with a background in material and electrical engineering, earned his bachelor's degrees from Hanyang University and the Samsung Institute. With over two decades of experience in the semiconductor industry, he started at Samsung Electronics and later transitioned to Fairchild Semiconductor, where he spent 8 years as a BCD technology development engineer, contributing to cutting-edge solutions. Certified with Lean Six Sigma Black Belt, Jake led numerous training sessions. Currently, he is an intrinsic reliability engineer and manager at onsemi. In addition, he oversees SiC MOSFET and power discrete technologies, playing a pivotal role in driving innovation and product reliability.

Byeongyeop Lee is accomplishing and contributing high-voltage semiconductor reliability engineering with over 18 years of experience in the industry. He has a background in electrical engineering and has earned his bachelor's degree from Inha University, and he holds a Master's Degree in Management of Technology from Hanyang University. He started reliability engineering work at Fairchild Semiconductor and later transitioned to onsemi. He is specializing in reliability engineering for new product qualification of high-power discrete module and contributed to the development of reliability test hardware and software setup readiness. Currently, he is a package level reliability engineer and manager at onsemi.

Sandol Lee graduated from Chonnam National University and received a bachelor's degree in electrical engineering. He worked in a semiconductor industry for over 10 years. His first job was a process engineer and reliability engineer at Amkor Technology. He is a certified ISO 16949 internal auditor and Six Sigma Black Belt. Currently, he is an extrinsic reliability engineer in charge of SiC MOSFET and power module at onsemi.

Paul Salmen has been with Infineon Technologies since 2017 and is principal engineer SiC Module reliability. He studied physics at the Technical University of Dortmund, where he also completed his doctorate in the same subject.

Olaf Zschieschang received his degree (Dipl.-Ing.) in mechanical and electrical engineering in 1991 from the University of Applied Sciences Mittweida. He started his career with the characterization and test of power semiconductors, transitioning into the power module development few years later. In over 30 years in various companies in the power semiconductor industry, he and his teams have developed a large number of power semiconductor modules for a wide range of applications, including the manufacturing processes required for their production. His current activities are focused on the development of future packaging technologies for power semiconductors at onsemi.

Sebastian Fahlbusch brings more than a decade of experience in power electronics, specifically with silicon carbide (SiC) and wide bandgap technologies. He successfully completed his PhD thesis from the Helmut Schmidt University in Hamburg on a new multilevel power converter concept using SiC MOSFETs. Sebastian joined Nexperia's Product Application Engineer team in 2019, and his primary focus is to support the implementation of new power products that strengthen Nexperia's power portfolio. Today, he is heading the Power Product Application Competence Center at Nexperia.

Ryo Takeda is a solution architect at the Automotive and Energy Solutions of Keysight Technologies. He joined Hewlett-Packard in 1989 and initially served as an application development engineer and manager for semiconductor parametric test solutions. Transitioning into product planning, he has guided power device analyzer product definition. Currently, his focus is on projects that integrate power device measurement, modeling, and circuit simulation, as well as the development of power device dynamic testers. He holds both a BSEE and MSEE from Keio University.

Takamasa Arai joined Keysight Technologies Japan (formerly Agilent Technologies Japan) in 2007, initially serving as an application engineer supporting impedance measurement and semiconductor parametric tests. Since 2021, he has worked as a solution application engineer at Keysight Technologies, providing global technical support for power semiconductor dynamic testing, power device modeling, and power electronics circuit simulation solutions. He received his B.S. and M.S. degree in physics from Tokyo Institute of Technology, Tokyo, Japan, in 2005 and 2007, respectively.

Llew Vaughan-Edmunds is currently Senior Director, Product Management at Navitas. He has over 21 years of power semiconductor experience at Infineon, International Rectifier, ON Semiconductor, and Applied Materials. He has been responsible for strategy, engineering, and product line ownership for Gate Drivers, IGBTs, Diodes, and Wide Band Gap (SiC/GaN). He holds an MBA from Chapman University, BS in Applied Electronics from Liverpool University, and a patent in GaN device technology. He is currently Chairman for PowerAmerica and holds various board positions in power semiconductor companies and consortiums.

Victor Veliadis is the executive director and CTO of PowerAmerica, a member-driven consortium of industry, universities, and national labs accelerating the commercialization of energy-efficient silicon carbide and gallium nitride power semiconductor chips and electronics. At PowerAmerica, he has managed a budget of $156 million that he strategically allocated to over 210 industrial and university projects to catalyze SiC and GaN semiconductor and power electronics manufacturing, workforce development, and job creation. His PowerAmerica educational activities have trained 430 full-time students in collaborative industry/university WBG projects and engaged over 7000 attendees in tutorials, short courses, and webinars. Dr. Veliadis is currently negotiating a $64 M US Department of Energy PowerAmerica renewal to further catalyze WBG power technologies.

Victor Veliadis is an ECE professor at NC State and an IEEE fellow and EDS distinguished lecturer. He has 27 issued US patents, 6 book chapters, and over 150 peer-reviewed publications to his credit. He has given over 150 keynote/tutorial/invited presentations including keynotes at ICSCRM, APEC, ECCE, and WiPDA. Prior to entering academia and taking an executive position at PowerAmerica in 2016, Dr. Veliadis spent 21 years post-PhD in the semiconductor industry where his work included design, fabrication, and testing of SiC devices, GaN devices for military radar systems, and financial and operations management of a commercial semiconductor fab. Dr. Veliadis has a Ph.D. degree in electrical engineering from Johns Hopkins University (1995).

Chapter 5
Applications of SiC in Power Conversion

Sanbao Zheng, Jonathan Dodge, Andrea Colognese, and Levan Bidzishvili

This chapter of the book delves into diverse applications of silicon carbide (SiC) in power conversion technologies. Authored by industry experts Sanbao Zheng, Jonathan Dodge, Andrea Bolognese, and Levan Bidzishvili, the chapter explores SiC's transformative impact. Sanbao Zheng elaborates on SiC FETs' advantages in various converter designs, showcasing their speed and low losses in power conversion. Jonathan Dodge discusses SiC JFETs in solid-state circuit breakers, highlighting reduced losses and robustness. Andrea Colognese and Levan Bidzishvili elaborate on SiC metal-oxide-semiconductor field-effect transistors' (MOSFETs') role in enhancing electric vehicle traction inverters, emphasizing efficiency gains and semiconductor technology advancements for vehicle performance. This collaborative effort showcases SiC's pivotal role across power conversion applications.

5.1 SiC FETs and Power Conversion Topologies

Sanbao Zheng

Modern alternating current (AC)/direct current (DC) converter designs are converging to a few variants that have been proven to give best efficiency and power density. This article discusses some of today's popular converter topologies and how the SiC FET cascode is a good choice for a power switch in all of them.

S. Zheng · J. Dodge (✉)
Qorvo, Greensboro, NC, USA

A. Colognese · L. Bidzishvili
onsemi, Milan, Italy

M. Di Paolo Emilio (ed.), *SiC Technology*,
https://doi.org/10.1007/978-3-031-63418-5_5

5.1.1 'Faster the Better' Was an Early Mantra

In the early days of switched-mode converter design, faster edge rates were always deemed to be better, to avoid overlap of voltage and current producing spikes of high dissipation and loss of efficiency. The extreme example was the 'tail current' of IGBTs on switch-off, which could remain for a significant time after the collector voltage had risen to a supply rail, due to their bipolar technology. As a result, faster switching frequency just meant more dissipation spikes per second and excessive loss. However, operating at higher frequency was also a goal, as that could mean smaller passive components, particularly magnetics. MOSFETs solved the tail current problem, so switching frequencies could increase more than 10× with acceptable dynamic losses, until the next practical limit—increased copper and core losses in the magnetics at high frequency. MOSFETs of the silicon variety however had another application limit, their on-resistance, particularly at higher voltage ratings, producing excessive conduction losses at higher power levels, leaving IGBTs still in the game with their relatively fixed voltage drop.

5.1.2 SiC and GaN Are Now as Fast as You Could Want

Now, wide bandgap (WBG) technology switches in silicon carbide (SiC) and gallium nitride (GaN) varieties can demonstrate edge rates of 100 V/ns and 5 A/ns or higher, and modern magnetics materials can operate in the MHz region with tolerable core losses. GaN and SiC devices also have significantly lower conduction losses than silicon for the same voltage class. Despite all this, the market expectation for ever higher efficiency and high-power density of power converters has forced designers to try to actually minimize switching frequency for lower dynamic losses while keeping filter and transformer magnetics size acceptable. When safety creepages and clearances are required, these can be a practical limit to how small a converter transformer can be anyway.

Among WBG devices, a version of a SiC switch, the SiC FET, pioneered by Qorvo [1], is undoubtably the best performer in most metrics. A cascode combination of a SiC JFET and a Si MOSFET, the cascode SiC FET outperforms Si and SiC MOSFETs and GaN HEMT cells in the most important figures of merit for conduction and switching losses (Fig. 5.1).

In the cascode SiC FET, we now have a switch that is so fast it practically has to be slowed down to avoid EMI and voltage overshoots and which features an on-resistance down in the milliohms, even at high-voltage ratings.

Technology	SiC Cascode 650V – 45mΩ (UJC6505K)	Commercial SiC MOSFET	Commercial GaN HEMT	Commercial Si-Superjunction MOSFET
R_{DSA}	0.75 mΩ-cm^2	2-3 mΩ-cm^2	3-7 mΩ-cm^2	10 mΩ-cm^2
Normalized Die Area	1	2.6x	4x	13x
E_{OSS}	7.5 μJ	32 μJ	12 μJ	14 μJ
R_{DS}*E_{OSS}	255	960	350	480
V_{TH}	5	2.8	1.3	3.5
Avalanche Capability	YES	YES	NO	YES
Short Circuit	YES	YES	NO	YES

Fig. 5.1 Figures of merit compared between the cascode SiC FET and other current technologies

5.1.3 *There Is No 'One Size Fits All'*

One device can never suit absolutely every converter topology though, when all parameters are considered, including cost and packaging options. IGBTs, for example, are still being developed and still have their place at very high power, especially when magnetics cannot practically be miniaturized, such as in motor drives where switching frequency might still be sub-10 kHz. Si MOSFETs are rugged, easy to drive, and low cost, and at low voltages, their on-resistance can be measured in hundreds of micro-ohms. GaN HEMT cells switch fastest when that is the bottom line.

So where do the cascode SiC FETs excel? They are commonly available at 750, 1200 and 1700 V rating and can be stacked for higher voltages still, so they suit higher power levels than GaN or Si MOSFETs. Even in the 600/750 V class, cascode SiC FETs have about 4× lower on-resistance than the competing technologies in the same package (Fig. 5.2).

Comparing dynamic losses is more complex, as this depends closely on the converter topology chosen and its operating mode. In pursuit of higher overall efficiency, resonant or soft-switching converters are mostly chosen in modern designs, to achieve low losses through zero-voltage or zero-current switching, ZVS and ZCS. For conversion with isolation, the LLC and CLLC topologies are popular as is the phase-shift full bridge (PSFB) and variants such as the dual active bridge (DAB). It might be assumed though that dynamic losses are minimized with any of Si, SiC

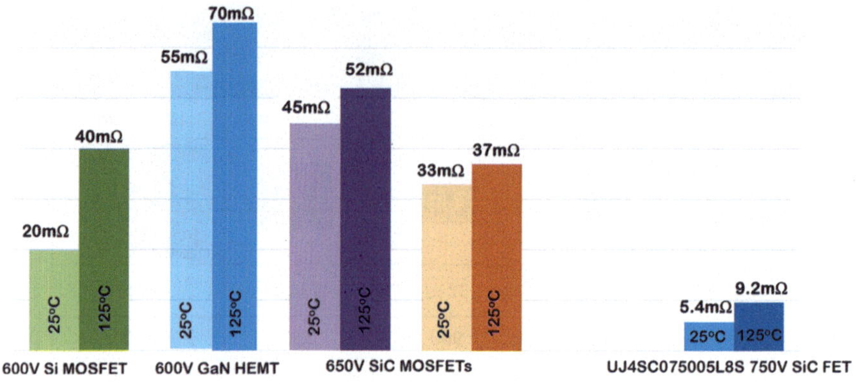

Fig. 5.2 Cascode SiC FETs' leading competing technologies in TOLL packages for conduction losses in the same voltage class

or GaN technologies in a soft-switched converter, but other parameters come into play, namely, device capacitances, particularly C_{OSS} which is charged and discharged each cycle, producing losses in the switch channel, windings and interconnections. Also, body diode characteristics can affect soft-switched converters, if only under transient conditions—the cascode SiC FET body diode has much lower reverse recovery energy than a Si MOSFET and much lower forward voltage drop than a SiC MOSFET.

Another standout characteristic of the cascode SiC FET is its simple gate drive. Similar to a low-voltage silicon MOSFET, the threshold voltage is stable at around 5 V, and it can be driven safely with a range of voltages, typically 0–12 V. Compare this with a SiC MOSFET, which must be driven to around 20 V for full enhancement, worryingly close to its absolute maximum value of sometimes just 23 V. GaN HEMT cells are also more difficult to drive, looking like a PN junction, the threshold can be lower than 1 V so is typically driven to only about 2.5 V for the on-state and noise immunity can therefore be an issue. If a marginally higher voltage is applied, hundreds of mA can flow, far over the absolute maximum of typically 20 mA continuous.

Gate drive power required by a cascode SiC FET is set by gate voltage swing, frequency and total gate charge Q_G, which is lower than a SiC MOSFET, due to the cascode SiC FET's lower Miller capacitance. With a lower gate voltage swing for the cascode SiC FET of 0–12 V compared with 0–20 V, the result is lower gate drive power required for otherwise similar parts.

We will now consider some examples of topologies of isolated converters fed from a DC rail derived from AC mains and how cascode SiC FETs benefit their operation.

5.1.4 Cascode SiC FETS in an LLC Converter

In this topology, the primary circuit has two resonances formed by a capacitor and the transformer magnetizing inductance and a separate 'uncoupled' inductor, either a discrete part or sometimes the transformer leakage inductance. The primary switches are driven at 50% duty cycle in anti-phase with output regulation achieved by varying the switching frequency up and down the upper slope of the resonance curve, effectively varying the impedance and gain characteristics of the primary, controlling energy transfer to the output. A basic outline and waveforms are shown in Fig. 5.3.

In the basic LLC circuit, ZVS is achieved at the turn-on transition with low loss. For turn-off, ZVS is not possible although ZCS can be, depending on operating conditions. Without ZCS, turn-off is therefore a 'hard' transition and the fast switching of the cascode SiC FET minimizes overlap and dissipation. This is particularly important at light load where the frequency can go very high. During 'normal' ZVS operation, the output capacitance $C_{OSS(er)}$ of the switches is discharged through transformer magnetizing current, and the resulting circulating energy E_{OSS} causes losses through the primary magnetics and switch channels. High C_{OSS} requires high magnetizing current to discharge it, restricting useable resonant and transformer inductances and ability to achieve ZVS in combination with a short dead time. In this case, the very low output capacitance of cascode SiC FETs is a distinct advantage. The switch body diode conducts to allow $C_{OSS(er)}$ discharge and sufficient dead time is needed to achieve this, and the body diode therefore should have fast recovery and low forward voltage drop. It is more problematic however under start-up and abnormal loading when the circuit can momentarily lose ZVS with 'hard' switching. Now, the body diode is transiently highly stressed and the parasitic BJT in the switch can turn on, so the fast, low loss characteristics of a cascode SiC FET body diode are vital to avoid failure. As the topology switches at higher frequency with light load, fast body diode recovery is also important to avoid loss of control as the dead time becomes more of a percentage of the cycle time.

If there is a downside with cascode SiC FETs, it is their inherent extreme speed and finding practical ways to control this for minimum dynamic losses. Because the

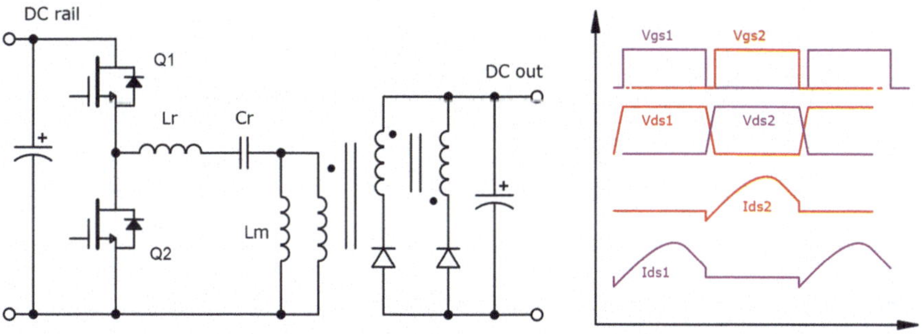

Fig. 5.3 LLC converter and basic waveforms

internal SiC JFET gate is not directly accessible, adjusting the cascoded Si MOSFET external series gate resistance is only an indirect way to control edge rates. Snubbers can help here with minimal loss, and variants of the cascode SiC FETs with an accessible SiC JFET gate have been developed, which does allow direct control of edge rates through an external chosen resistor.

5.1.5 Cascode SiC FETs in a Phase-Shift Full-Bridge Converter

The PSFB converter topology (Fig. 5.4) is often used at medium- to high-power levels and, unlike the LLC, operates at fixed frequency, which can be an advantage in EMI control. An 'H' bridge is switched at 50% duty cycle for the two cascode SiC FETs in each 'leg', and a change of phase difference between the two legs increases or decreases the duty cycle of voltage seen by the transformer primary, achieving output regulation. All switches can operate in ZVS with a resonant inductor Lr shaping current to reduce dynamic losses to a very low level. With the PSFB, maintaining ZVS at light load can be problematic though, unlike the LLC.

As with the LLC, resonant currents flow through the output capacitances of the switches and through body diodes in the dead time. Similarly, therefore, low values of $C_{OSS(er)}$ and body diode Q_{RR} and T_{rr} are important for low loss, and the class-leading performance of cascode SiC FETs is a major advantage here. In the PSFB topology, the body diodes recover with low voltage imposed, slowing charge removal, so low Q_{RR} is therefore important. Fast T_{rr} also enables dead time to be shorter for wider output control.

The cascode SiC FET excellent figures of merit for on-resistance with chip area, E_{OSS} and C_{OSS}, at high and low temperatures also of course contribute to lowest conduction losses.

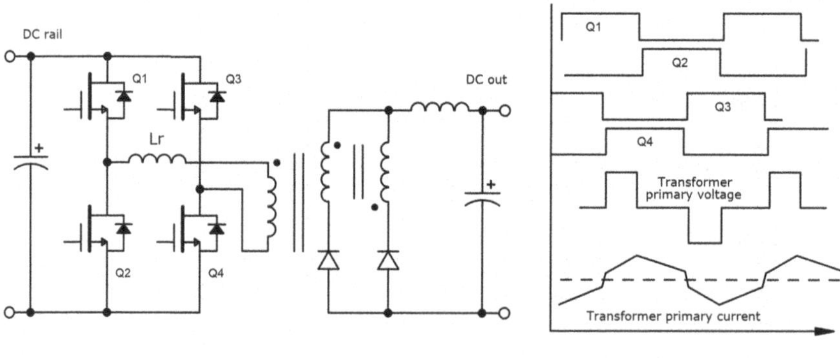

Fig. 5.4 The phase-shift full-bridge topology outline and waveforms

5.1.6 SiC FETs in a Totem Pole PFC Stage

Latest efficiency targets such as 80+ titanium for AC/DC converters practically cannot be met if a traditional diode bridge is used for AC rectification, due to the diode losses. A solution is to use a 'bridgeless' totem pole PFC stage that incorporates the AC line rectification function into the boost converter stage typically used to correct power factor. The circuit outline is shown in Fig. 5.5 where high-frequency switches Q1 and Q2 alternate between a boost switch and diode function, depending on the AC polarity, with Q3 and Q4 switched at line frequency to 'steer' the return current appropriately. Boost converters can operate in two modes—continuous or discontinuous (inductor) current conduction (CCM and DCM) or sometimes on the boundary between the two. Simply, in CCM, the inductor current ripple is low and continuous, and in DCM it is high, dropping to and remaining at zero for part of each high-frequency switching cycle. DCM inherently allows zero current switching but is not suitable for high power due to the high peak currents involved which stress the switches and make the magnetics unduly large. CCM reduces stress and allows a smaller inductor, but is 'hard' switched, with conduction of the body diode and consequent high recovery losses that leave the topology still unattractive using silicon MOSFETs. SiC MOSFETs have a much faster body diode and are an improvement, and SiC FETs are better still, with fast body diode recovery and a lower forward voltage drop than the SiC MOSFET. Again, the superior E_{OFF} and on-resistance figures of merit of a SiC FET also help to maintain efficiency.

Because Q3 and Q4 only switch at line frequency, dynamic losses can be ignored, but cascode SiC FETs in these positions still give a useful boost to efficiency when this is a design goal, due to their very low conduction loss. The waveforms in Fig. 5.10 are shown around the line AC zero crossing point with the circuit operating in CCM.

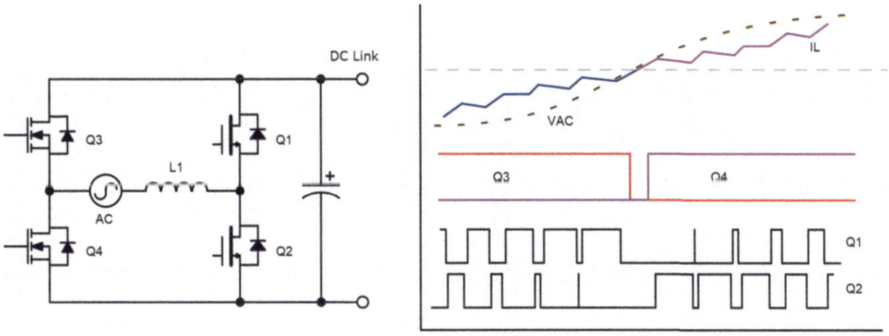

Fig. 5.5 The totem pole PFC topology in continuous conduction mode with SiC FETs

5.1.7 Flyback Converters and SiC FETs

While we have considered the use of SiC FETs in topologies that suit relatively high power, they can also be used in a traditional lower-power, offline flyback circuit as might be used in external power supplies for laptops, for example (Fig. 5.6). Silicon MOSFETs have been used successfully, but wide bandgap devices allow faster switching while maintaining low losses and promise the ever-smaller package that consumers demand. In a single-ended flyback, the switch sees the fully rectified line voltage, plus a 'flyback' voltage set by the designer, along with an overshoot and ring from the inevitable transformer leakage inductance. For single-phase 230 VAC with tolerances, the total peak voltage is often close to 600 V. This means that 650 V-rated devices such as GaN HEMT cells are uncomfortably close to their maximum and AC line transients must be securely limited. The GaN devices also have no avalanche rating, so they would fail immediately with excessive voltage. GaN is used in the application, but extreme care has to be taken to limit voltage and at the same time, the gate drive is also critical to avoid stress and failure.

With their ratings at 1200 V and 1700 V, SiC FETs can be used in single-ended flybacks at higher nominal AC voltages, for example, 690 V in industrial applications where a low-power auxiliary rail might be generated from the flyback circuit connected between phases of a three-phase system. SiC MOSFETs are another choice with ample voltage rating with an avalanche capability, but SiC FETs have these as well and yield higher efficiency with lower on-resistance and an easy, stable gate drive. If the flyback is operated in hard-switched, continuous conduction mode, the fast edge rate and low output switching energy of the SiC FET keep dynamic losses low. A snubber would typically be included to limit the drain voltage as shown with some dissipation, but with the cascode SiC FET robustness, this can be minimal size.

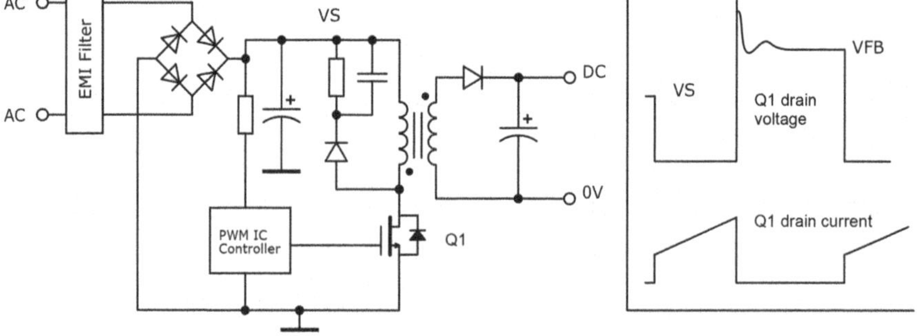

Fig. 5.6 The single-ended flyback topology using a SiC FET, shown in continuous conduction mode

5.1.8 Conclusion

There is little downside to using SiC FETs in most power conversion topologies, as they have objective figures of merit for efficiency across all operating conditions that are as good as if not better than other technologies. An obvious sweet spot is at higher voltages and higher power levels where their benefits are even stronger. As mentioned, slowing SiC FETs down with snubbers may be needed to avoid undue EMI or transient overshoots. The stacked die arrangement of an Si MOSFET and SiC JFET in the SiC FET cascode is necessarily more complex than a single MOSFET or HEMT cell resulting in extra unit cost. However, more die can be produced per wafer than for the other technologies for economy at scale, and when system benefits of easy gate drive and improved efficiency, leading to lower cooling overhead, are factored in, any cost difference for the SiC FETs becomes less significant or disappears.

5.2 Solid-State Circuit Breakers Using SiC JFETs

Jonathan Dodge

Ever since electricity was first harnessed, inventors and engineers have struggled to find the best ways to switch it off again in a reliable and safe way, without just relying on a manual switch or a fuse opening after a fault. Edison started off the effort in 1879 with a patent for a 'circuit breaker' to protect early lighting circuits, and it is now a global industry, with AC and DC applications from domestic to military. EVs and alternative energy sources are also driving demand. As a consequence, the market value for circuit breakers is projected to be $14B by 2031.

5.2.1 Mechanical Circuit Breakers Have Low Loss

The traditional circuit breaker solution has been electromechanical, with devices that can isolate a load manually or automatically after a detected overcurrent or short circuit. In normal operation, the breaker contact resistance can be very low, measured in micro-ohms, and when open, the isolation is very high and can be enough to meet the highest grade safety standard.

The problem is what happens in between these two states. As the contacts break with the fault current passing, an arc naturally occurs, and with AC, it can persist over a large contact separation distance until a voltage waveform zero crossing. At this point, it self-extinguishes, but at 50 Hz, this could be 10 ms at worst. During this time, 'let-through' energy is still being supplied to the faulty load, and the delay in isolating it is added to by the current sensing mechanism and reaction time of the circuit breaker coil, typically up to 5 ms. For DC, there is no self-extinguishing effect and extra-large separation is needed, often supplemented with complex techniques to divert the arc, either magnetically, or even with puffs of compressed air. With AC or DC, there is also a shorter-duration arc as the contacts make, due to the normal running current and any surge into the load. All this inevitably causes contact wear, resulting in a finite lifetime for the breaker. If it welded itself closed the last time the contacts made, you do not know about it until it fails to break after a subsequent fault, with potentially damaging consequences. To counter this, circuit breakers in critical applications might be changed out regularly as a precaution, at significant cost in time and hardware.

5.2.2 The Solid-State Option

A solid-state circuit breaker (SSCB) has been an option for decades, but there has been no semiconductor switch technology that could match the low-voltage drop of a mechanical breaker. However, there have been low to medium current applications where the drop and consequent power dissipation has been tolerable. IGBTs lose up to around a relatively fixed 2 V at higher currents, and silicon MOSFETs, with their characteristic on-resistance, can drop less at light loading but quickly overtake IGBTs in dissipation at higher currents, with the 'square' in I^2R doing the damage. For either technology, current above a few amps starts to produce significant dissipation and that is problematic, requiring cooling hardware. Another downside for SSCBs is the lack of galvanic isolation, but they score on speed, controllability and reliability, with the absence of arcing. For MOSFETs, unfortunately, voltage breakdown rating is traded for on-resistance, so the most robust parts will not be the lowest loss. Figure 5.7 summarizes the pros and cons of mechanical and solid-state breakers with higher values being better in each parameter.

5.2.3 Do Wide Bandgap Materials Make the Difference?

Wide bandgap switches are now available in silicon carbide (SiC) and gallium nitride (GaN) varieties and have characteristics that make them increasingly viable as SSCB switches. In particular, a SiC MOSFET switch has lower on-resistance than a similar size and voltage rating of silicon MOSFET and has much better temperature resistance and thermal conductivity as a material, easing cooling arrangements. Devices are also commonly available to 1700 V rating, compared with 1200 V for Si MOSFETs. SiC MOSFETs can be easily paralleled too, for example, nominally halve the total on-resistance and each sees a quarter the power of a single device, so temperature rise is much lower, contributing to yet lower on-resistance. Multiple devices are paralleled and then extend their viability in SSCBs to high current levels, limited by cost.

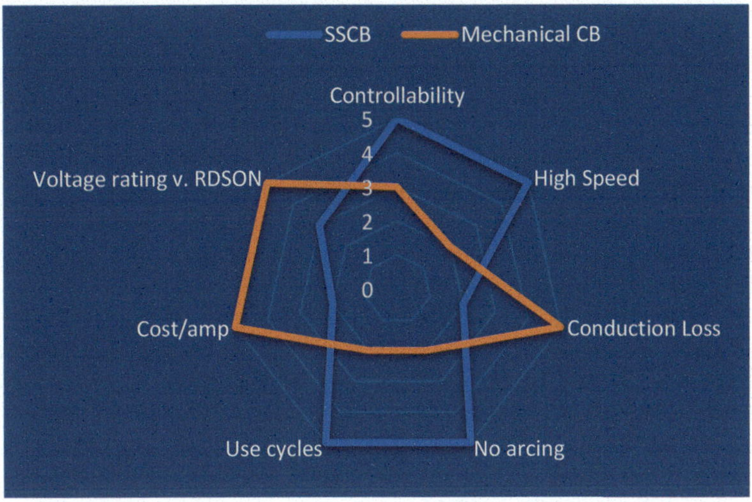

Fig. 5.7 Mechanical and solid-state circuit breaker comparisons

GaN devices available as HEMT cells are unlikely to be seen in the high-voltage SSCB application as they are best suited to lower operating voltages, have no protective avalanche capability and have a worse thermal conductivity than even silicon.

Problems remain however using SiC MOSFETs in SSCBs, and these mainly relate to residual resistive loss and confidence in the robustness of the devices—circuit breakers reacting to overloads or short circuits are there to prevent damage to personnel or property, so they must be highly reliable. Apart from having immunity to 'everyday' overvoltages on power rails, the switch must be robust against the high-energy voltage spikes generated from breaking high current through connection inductances. Snubbers and surge arrestors are therefore normally employed in SSCBs to limit transients and reduce the possibility of stress to the gate oxide of the SiC MOSFET. Unfortunately, compared with silicon, SiC has a higher inherent density of defects, acting as centres for onset of breakdown. Additionally, the oxide exhibits an ageing effect and is increasingly susceptible to total breakdown over time with repeated stress events. The defects also affect carrier mobility and consequently on-resistance achievable with a SiC MOSFET.

The gate threshold voltage in SiC MOSFETs is also subject to some variation with repeated stress, although not perhaps an issue in the 'static' SSCB application.

5.2.4 A Better Solution

Existing SiC MOSFETs are mostly targeted at switching applications, where the isolated gate construction makes them normally-off and easy to drive. In the SSCB application, a better solution, pioneered by Qorvo, is a SiC JFET, which does not feature a fragile gate oxide or a MOS channel resistance, making the die much smaller, more robust and with lower total on-resistance compared with a SiC MOSFET in the same voltage class. Figure 5.8 compares the construction of a SiC MOSFET and SiC JFET showing the latter's relative simplicity and size.

With the SiC JFET, there is no voltage- and temperature-dependent breakdown of a gate oxide and the gate threshold remains stable. The conduction channel is also robust, with no change in leakage current after repeated short-circuit events and the device has a defined and high-energy avalanche rating. A JFET is normally-off with zero-volt gate drive, but we shall see that this is not necessarily a disadvantage.

With their robustness and simplicity, SiC JFETs address some of the reservations when using SiC MOSFETs in SSCB applications, but their lower loss is also a major step forwards. In a TOLL package, for example, parts from Qorvo, when co-packaged with a Si MOSFET to form a cascode, achieve an on-resistance of just 9.2 mΩ at 125 °C compared with 4× this value for the best available silicon or SiC MOSFETs, or even GaN HEMT cells. At the same time, the device is rated at 750 V and the others are rated at 600–650 V (Fig. 5.9).

The standard JFET/MOSFET cascodes offered by Qorvo, dubbed 'SiC FETs' are not best suited for SSCB applications however, as they switch too fast, risking voltage overshoots from connection inductance. Unfortunately, edge rates are not

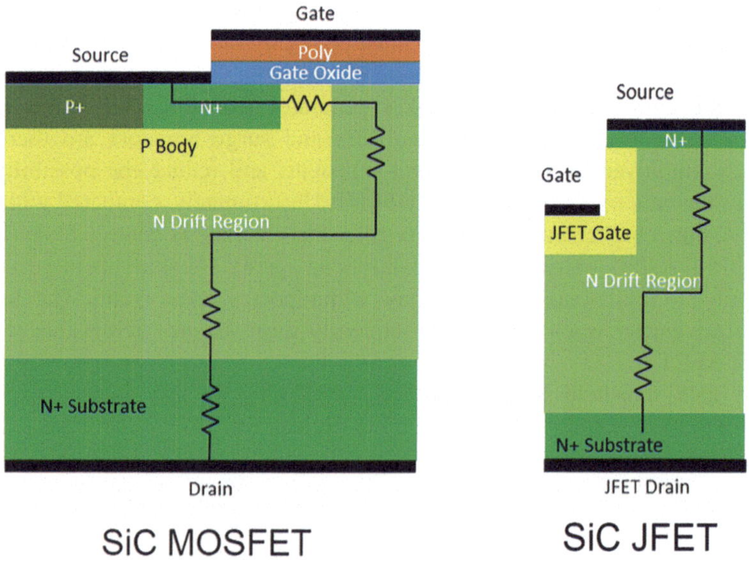

Fig. 5.8 SiC MOSFET and SiC JFET construction compared

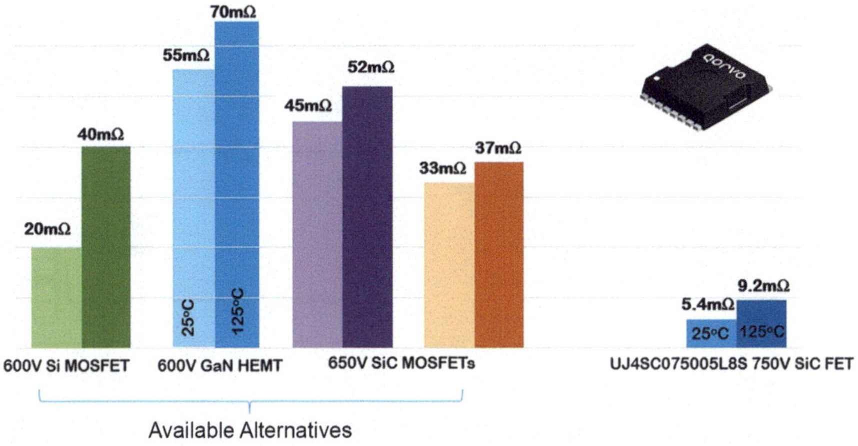

Fig. 5.9 Total on-resistance losses of SiC JFETs as cascodes are 4× less than competing technologies over a wide temperature range

easily controlled, because the gate of the internal JFET is not accessible to add series resistance. Additionally, the cascode's series Si MOSFET adds some unwanted on-resistance.

5.2.5 Dual-Gate MOSFET + JFET Suit SSCBs

The SiC FET cascode construction can be leveraged, however to form a part which is more suited to the SSCB application if the JFET gate is isolated and brought out to a separate pin (Fig. 5.10). In development at Qorvo, these 'dual-gate' products open up some new options and potential benefits:

The JFET gate could be connected to the Si MOSFET source through a resistor, to function as a classic cascode. Alternatively, in the SSCB application, the JFET gate could be directly driven, leaving the Si MOSFET continuously on, with its gate drive acting simply as an 'enable'. The advantage of this arrangement is that the JFET gate can be overdriven to about +2 V, for the on-state, to reduce on-resistance by a further 15%, more than compensating for the resistance introduced by the Si MOSFET channel. Drive to +2 V is still low enough to ensure negligible injection of minority carriers into the gate-source region. It may seem that the Si MOSFET could be eliminated and just a SiC JFET used but the device requires a negative drive of about −9 V to be driven off, and there is a concern that loss of auxiliary power would leave the SSCB in an unsafe state. However, the Si MOSFET, with the addition of the two Zener diodes shown in Fig. 6.4, ensures that the arrangement reverts to 'normally-off' with loss of JFET gate drive.

The Si MOSFET could be a separate device and could then act as an enable and safeguard for multiple paralleled SiC JFETs, giving higher current rating or lower

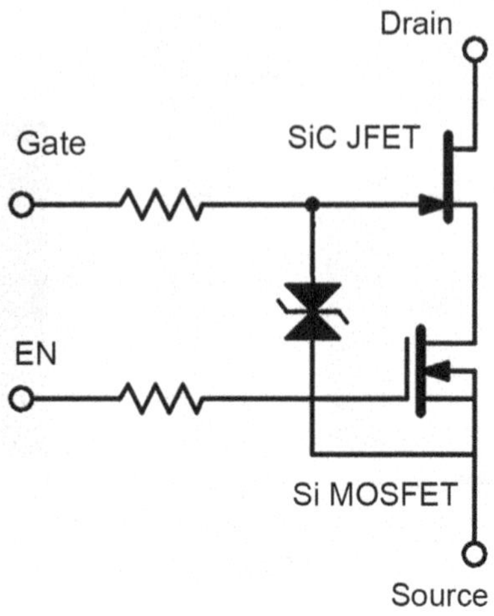

Fig. 5.10 The dual-gate SiC cascode

loss. In this case, the direct control of the JFET slew rate with a gate resistor enables close synchronization of paralleled devices as they switch to avoid transient stress. However, when one SiC JFET is adequate in the application, the co-packaging of the Si MOSFET and SiC JFET saves board space.

5.2.6 SiC JFETs Have an Integrated Temperature Sense Capability

As a bonus, because a JFET gate is similar to a forward biased diode for the on-state, the 'diode' curves for a particular device can be plotted and operating conditions directly interpreted as representing junction temperature to good accuracy, providing a valuable monitoring function. This is done in practice by applying the positive on-state bias of say 2 V, measuring the milliamp-range gate current and comparing this with calibration curves for the device, showing variation of gate current with temperature at a specified voltage (Figs. 5.10 and 5.11). In the static SSCB application, this is easy to do and provides a junction temperature measurement that has faster response and is more accurate than a thermistor.

5.2.7 Not so Fast!

A characteristic of a SiC MOSFET or JFET is its phenomenal switching speed, easily 25/50 ns, which can be a real asset in a power converter. In an SSCB application, however, this would be too fast on turn-off, risking voltage transients from connection inductance and ringing, increasing risk of breakdown. A desirable slew time for

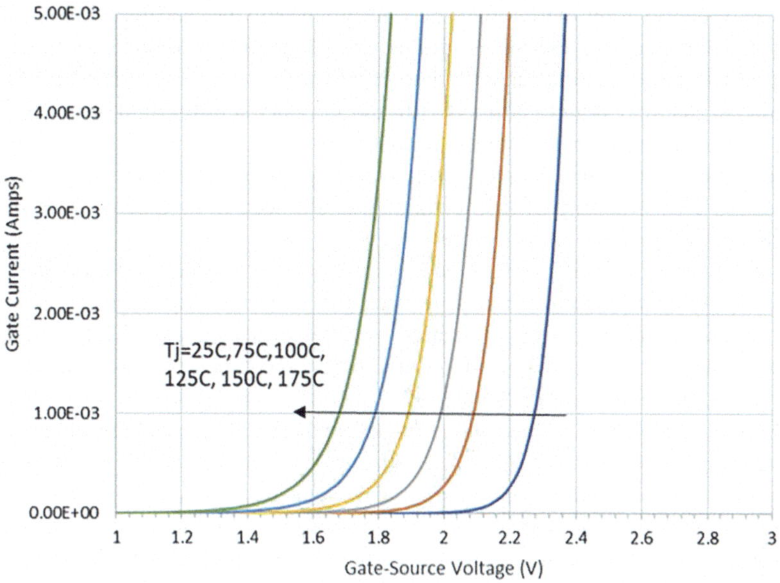

Fig. 5.11 JFET gate current under positive bias has an accurate relation to junction temperature

an SSCB would be about 500 ns minimum, and this is achieved with the dual-gate arrangement by 'tuning' of the JFET gate resistor.

5.2.8 Is 'Normally-Off' Really What We Want?

Designers of power converters generally prefer switches that are normally-off, defaulting to the blocking condition when there is no gate drive power. This is understandable, as it avoids possible continuous conduction under start-up conditions or when faults occur, for example, an auxiliary power rail dropping. Normally-off MOSFETs and SiC FET cascodes are therefore popular. In the SSCB application, however, the 'normal' condition is to be in conduction, so is a normally-on switch a better option? The device can certainly have lower losses—a simple SiC JFET without the series Si MOSFET would have about 10% cost reduction over a cascode SiC FET, with 10–15% lower on-resistance due to elimination of the Si MOSFET. If the now-accessible gate is overdriven, there is yet another 10–15% on-resistance reduction. Alternatively, for the same on-resistance, a JFET might be 30–45% lower cost than a SiC FET cascode. With the potential advantages of easy slew-rate control and integrated temperature sensing already discussed, the proposition seems attractive.

A downside is that power is now needed to turn the JFET off—a negative gate voltage with respect to the source which must be available under the fault condition that caused the demand for the SSCB to open in the first place. The gate voltage must also reliably stay there while the fault is investigated and rectified ensuring that the SCCB does not spuriously turn on with damaging or unsafe consequences. The problem is not insurmountable though, when the cascode Si MOSFET is not included to ensure default to normally-off; the auxiliary supply could be interlocked with/derived from the power source so that if the negative gate drive drops, it can only be if the power source is not there anyway to provide damaging current through the SCCB. The reliability of this technique would need to be assured with in-use monitoring and perhaps a test feature.

In practice, use of SSCBs is limited by their lack of galvanic isolation, and certainly any AC application with back-to-back JFETs at unsafe voltage levels would anyway need a separate series 'air gap', practically a mechanical relay, to meet safety regulations such as UL 489I and IEC 60947–10. This is not as contradictory as it seems, as the breaker or 'safety relay' can be arranged only to open after the SSCB element has itself opened and close before the SSCB closes. This ensures that the mechanical contacts do not break a high current or voltage and therefore no arcs occur and it can be smaller with no significant wear-out. The relay may anyway be included to guarantee no leakage and provide a means of implementing an enable/lock-out tag. In this case, the safety relay could be interlocked with the JFET gate drive so that it cannot close without the appropriate negative gate drive voltage being present.

In a DC application where safety isolation is not mandated, the extra relay would only be necessary as a precaution against the failure of control circuitry, in which case the benefits of the SSCB solution would be weighed against the risk of control failure and extra cost of the relay.

5.2.9 Conclusion

Solid-state circuit breakers have many advantages over mechanical types. Eliminating wear-out and limiting 'let-through' energy under fault conditions are the prime benefits, but they also open up possibilities for the smart control of loads with features such as active surge current limitation and remote access and reporting through wireless links.

SSCBs will be increasingly seen at higher and higher current levels as on-resistance of the semiconductor switches improves over time and dissipation goes down. Silicon carbide switches are still some way from their minimum theoretical resistance and JFETs in particular are leaders, with on-resistances that are now comparable with that of wired connections in a power distribution system. A large part of the market is for circuit breakers to fit into standard slots in compact distribution panels where cooling was not a major issue with mechanical breakers. As a result, SSCBs face the challenge of reliable operation while dissipating some heat in a less than ideal environment, but with on-resistances reducing and SiC devices anyway capable of high-temperature operation, their future in the application is bright.

5.3 Traction Inverter

Andrea Colognese

A traction inverter is a critical component in electric and hybrid electric vehicles (EVs and HEVs). Its primary function is to convert direct current (DC) power from the vehicle's battery into alternating current (AC) power, which is then used to drive the electric traction motor. The traction motor, in turn, propels the vehicle.

In details, here below are the main functions implemented in a traction inverter:

- *Power Conversion*: The traction inverter is responsible for converting the high-voltage DC power supplied by the vehicle's battery into AC power that can drive the electric motor. This process is crucial for controlling the speed and torque of the motor, enabling efficient and smooth vehicle operation. The efficiency of this function determinates the efficiency of the traction inverter and impacts directly the milage of the car and/or the capacity of the battery to achieve the target autonomy.
- *Motor Control*: Traction inverters play a crucial role in controlling the speed and torque of the electric motor. By modulating the frequency and amplitude of the AC power supplied to the motor, the inverter regulates the motor's performance, allowing for precise control of the vehicle's movement. The control algorithm is implemented in the traction inverter processor (ECU) and following strict requirements in terms of safety.
- *Efficiency and Regeneration*: Traction inverters contribute to the overall efficiency of electric vehicles. They are designed to minimize power losses during the conversion process. Additionally, many modern traction inverters support regenerative braking, allowing the motor to act as a generator during braking events, converting kinetic energy back into electrical energy and storing it in the battery for later use.
- *Thermal Management*: Traction inverters generate heat during operation, and efficient thermal management is crucial to ensure their proper functioning and longevity. Cooling systems, such as liquid or air cooling, are often integrated into the design to dissipate heat and maintain optimal operating temperatures.
- *Integration with Vehicle Control Systems*: Traction inverters are closely integrated with the overall vehicle control system. They receive commands from the vehicle's electronic control unit (ECU) to adjust the motor's performance based on factors such as accelerator input, driving conditions and energy management strategies.
- *Reliability and Safety*: Traction inverters must meet stringent reliability and safety standards. They are designed to withstand the harsh automotive environment, ensuring stable and safe operation over the lifespan of the vehicle.

Overall, the traction inverter is a critical component that enables the efficient and controlled use of electric power in modern electric and hybrid vehicles, contributing to their performance, efficiency and environmental benefits.

5.3.1 Inverter Block Diagram

The power stage of an inverter is a critical component responsible for converting direct current (DC) power from the input source into alternating current (AC) power that can drive the electric motor. The power stage typically consists of semiconductor devices, such as insulated gate bipolar transistors (IGBTs) or SiC metal-oxide-semiconductor field-effect transistors (MOSFETs), arranged in a bridge configuration. With reference to the picture below, here is an overview of the key elements in the inverter power stage:

- *Inverter Bridge*: The core of the power stage is the inverter bridge, which typically consists of three half-bridges/half-phases, each half-bridge incorporates two power semiconductor switches (high- and low-side switch). Switching points (AC pin) of each half-bridge is connected to windings of the AC motor. The switching devices inside a power module, often IGBTs or MOSFETs, are responsible for delivery of the battery current to the windings of the AC motor. The desired AC waveform, which is needed to bring the rotor of the AC motor in rotation, is generated through the special switching scheme of the power devices.
- *Intermediate DC Link*: The intermediate DC link is a capacitor that smoothens the DC voltage. DC link capacitor is typically located very close to the power stage (bridge) so that it can act as a buffer between the DC power supply (battery) and the inverter bridge; through immediate delivery of charge to the power bridge during switching process, it can stabilize the DC voltage during inverter operation.
- *Gate Drivers*: Gate drivers are circuits responsible for switching and protecting the power devices in the inverter bridge. Typically, gate drivers are isolated devices that receive on the low-voltage side control signals from the controller or processor and generate on the isolated high-power side the appropriate amplified gate voltage to turn the semiconductor devices fully on or off. Gate drivers play a crucial role in ensuring precise control over the switching frequency and timing, influencing the quality of the AC output.
- *Output Filter*: The output filter is often included to mitigate the impact of harmonics generated during the switching process. Harmonics can affect the quality of the AC waveform and can potentially lead to issues with the motor or other connected equipment. Common types of output filters include inductors and capacitors, which help smooth the output waveform.
- *Current and Voltage Sensors*: To make possible closed-loop control of the motor drive algorithm, it is needed to get feedback from the system. For this reason, sensors are integrated into the power stage to monitor current and voltage. Current sensors measure the current flowing through the power devices, while voltage sensors monitor the DC link voltage and output AC voltage.
- *Heat Sink or Cooling System*: Every switching cycle of the power device is accompanied by some switching losses, which leads to the heat up of the device. Heat sinks or cooling systems, such as fans or liquid cooling, are employed to

dissipate this heat and maintain the devices within their safe operating temperature range.

The inverter power stage (Fig. 5.12) is crucial for the overall performance and efficiency of the electric or hybrid vehicle. Its design and components are selected to meet the specific power requirements, thermal constraints and reliability standards of the application.

5.3.2 *Operating Conditions and Power Capability of a Traction Inverter*

At high level, a traction inverter is normally specified in terms of power or phase current capability. There is no simple way to convert from/to power to/from phase current as it depends on other system conditions and operation points. The procedure below can help on this task.

Starting from traction inverter operating condition, normally defined at the system level:

- VDC = 400 V.
- Motor current (IǾ_rms) = 400 Arms.
- Power factor (PF) = 0.9.
- Modulation index (m) = 0.9.

The total AC power the inverter can deliver is

$$P_{AC} = \sqrt{3} \bullet V_{ll\_rms} \bullet I_{\varnothing\_rms} \bullet PF$$

where V_{ll_rms} is

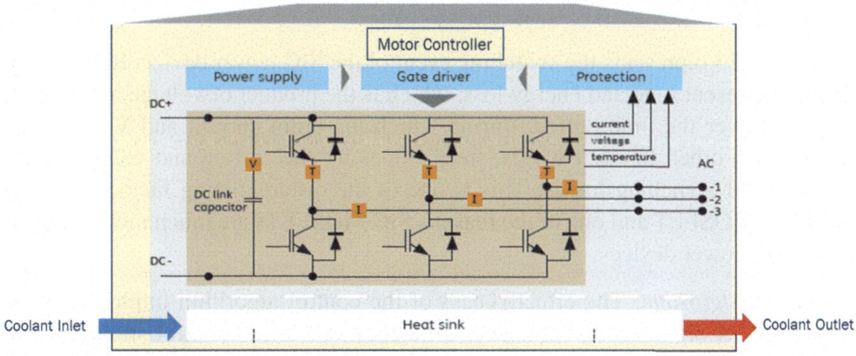

Fig. 5.12 Block diagram of a traction Inverter

$$V_{ll\_rms} = \frac{V_{DC}}{\sqrt{2}} \bullet m$$

With the data provided in the example,

$$V_{ll_{rms}} = \frac{400V}{\sqrt{2}} \bullet 0.9 = 255Vrms$$

$$P_{AC} = 1.73 \bullet 255Vrms \bullet 400Arms \bullet 0.9 = 159KW$$

In this specific case, a traction inverter with a phase current of 400 Arms under the provided conditions above, can be rated to deliver 159KW.

5.3.3 Inverter Efficiency

As mentioned before, the efficiency of a traction inverter is a crucial factor in determining the overall efficiency of an electric or hybrid electric vehicle (EV or HEV). It represents the ratio of the useful output power (AC power delivered to the electric motor) to the input power (DC power drawn from the vehicle's battery). In mathematical terms, efficiency (η) is expressed as

$$\text{Efficiency}(\eta) = \frac{P_{out}}{P_{in}} \bullet 100\%$$

Key factors that influence the efficiency of a traction inverter include the following:

- *Inverter Bridge Efficiency*: This is the efficiency of the power stage where the DC power is converted to the AC power. The switching losses in the semiconductor devices (IGBTs or MOSFETs) and conduction losses contribute to the overall efficiency.

Figure 5.13 represents the switching event of the SIC power device. Red lines in the plot represent time and energy loss, which is the product of voltage and current integrated over the time period, during which the drain current and Vds voltage overlap each other. In other words, steep slopes for the current and voltage could lead to lower switching losses. This is one of the differentiating factors between IGBT and MOSFET and one of the reasons SIC MOSFETs are much more efficient than IGBT power devices:

- *Control Algorithm*: The effectiveness of the control algorithm implemented in the inverter significantly influences efficiency. Advanced control strategies can optimize the modulation of the inverter to reduce losses and enhance overall system efficiency.

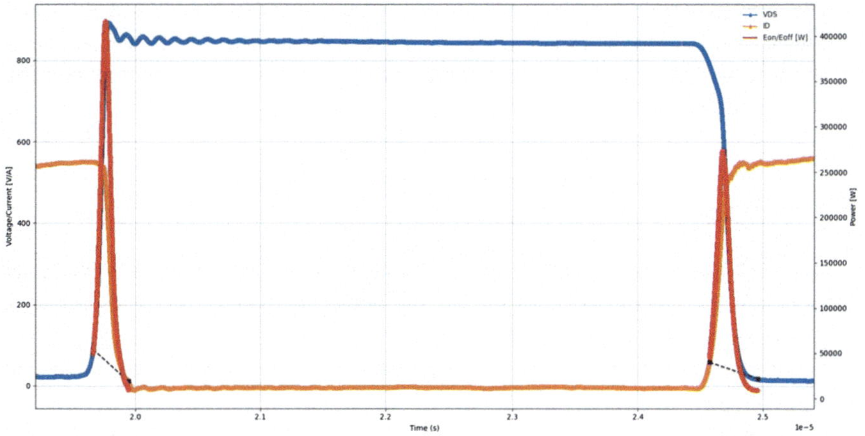

Fig. 5.13 Switching behaviour of a SiC device

- *Output Filter Efficiency*: If an output filter is used to reduce harmonics in the AC waveform, its efficiency in achieving this purpose without introducing additional losses is essential.
- *Cooling System Efficiency*: The effectiveness of the cooling system in dissipating heat from the power devices and other components plays a critical role. A well-designed cooling system helps maintain optimal operating temperatures, preventing overheating and ensuring efficiency.
- *Losses in Wiring and Components*: Resistive losses in wiring, connectors and other passive components contribute to overall losses. Minimizing these losses is crucial for improving efficiency.
- *Operating Conditions*: The efficiency of a traction inverter may vary under different operating conditions, such as load, temperature and voltage. Manufacturers often provide efficiency curves or maps to represent the performance across a range of conditions.
- *Regenerative Braking Efficiency*: Inverters that support regenerative braking can improve overall vehicle efficiency by recovering energy during deceleration. The efficiency of regenerative braking depends on the effectiveness of the energy recovery system.

Efficiency is a critical consideration in the design and selection of traction inverters for electric vehicles, as it directly impacts the range and energy consumption of the vehicle. Manufacturers strive to improve efficiency through advancements in semiconductor technology, control algorithms and thermal management systems. It is common for modern traction inverters to achieve high efficiencies, often exceeding 90% or even 95% under typical operating conditions.

The efficiency (Fig. 5.14) of the inverter is not the only factor to consider. We can introduce the losses in the motor as well and calculate the effective power available 'at the wheel'. Let us consider an example of system inverter and motor with the respective efficiencies $\eta_{inv} = 98.7\%$ and η_{motor}:

$$\eta_{\text{motor}} = \frac{P_{\text{mecchanical}}}{P_{\text{electrical}}} \bullet 100\% \eta_{\text{motor}} = 80\%$$

To have 100KW 'at the wheels', the motor needs to be fed with 125KW. With the efficiency of the inverter stage of 98.7%, the power the battery needs to deliver is 126.6KW and the losses are 1.6KW.

The losses in the inverter have a strong dependency on the performances of the power switches.

IGBT technology is quite a common, reliable and established technology adopted for traction inverter, but silicon carbide metal-oxide-semiconductor field-effect transistors (SiC MOSFETs) are gaining popularity especially in high-power and high-frequency systems. Due to the intrinsic characteristic of the material, SiC MOSFET offers key advantages in inverter applications and can improve the overall efficiency of the inverter by ~2–3%.

SiC MOSFETs have faster switching speeds compared to traditional silicon-based devices such as insulated gate bipolar transistors (IGBTs). This allows for higher switching frequencies, which can lead to smaller and lighter passive components (such as inductors and capacitors) in the inverter design. SiC MOSFETs exhibit lower switching losses, resulting in higher overall efficiency. The reduced losses contribute to lower heat generation, allowing for more compact and efficient cooling solutions. In addition, these devices generally have lower on-state conduction losses, particularly at high voltages and currents. This contributes to improved efficiency and power density. SiC MOSFETs maintain their performance at higher temperatures compared to silicon-based devices. This characteristic is advantageous in applications where elevated temperatures are common, such as in electric vehicles and industrial settings. The combination of high switching frequencies,

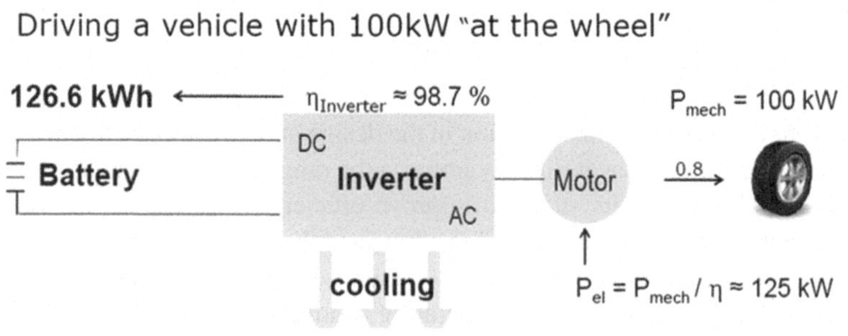

Fig. 5.14 Efficiency of a traction inverter

reduced switching losses, and lower conduction losses enables SiC MOSFET-based inverters to achieve higher power densities compared to traditional silicon-based inverters.

On the other side, SiC MOSFETs can be more expensive than traditional silicon devices. The gate drive requirements for SiC MOSFETs are different from those of silicon devices. Proper gate drive circuitry is crucial to maximize the benefits of SiC MOSFETs and ensure reliable operation. In addition, while SiC MOSFETs offer high-temperature performance, their long-term reliability in specific applications is an ongoing area of research and development. Manufacturers continue to improve the robustness of SiC devices. Integrating SiC MOSFETs into existing systems may require adjustments to control algorithms and gate drive circuits. System designers need to consider the compatibility of SiC MOSFETs with other components in the inverter system.

onsemi has a wide portfolio of modules for traction application targeting 400 V and 800 V battery architectures with IGBT and SiC technologies with different power ranges and different tab configurations. Two main package technologies are available:

- Single-side direct cooling, gel-filled, press fit for auxiliary pins and pin fin three-phase module (Fig. 5.15).
- Single or dual indirect cooling, transfer moulded half-bridge module (Fig. 5.16).

The two package platforms enable the customer to implement different mechanical and assembly strategies. To test the performances of the module without requiring the development of the full inverter, onsemi is offering practical and quick kit solutions to run evaluation measurement and preliminary software development. The kit is implementing a three-phase inverter with the cooling channel, the driver board and the DC capacitor. Figures 5.17 and 5.18 represent an example of two kits based on the above modules.

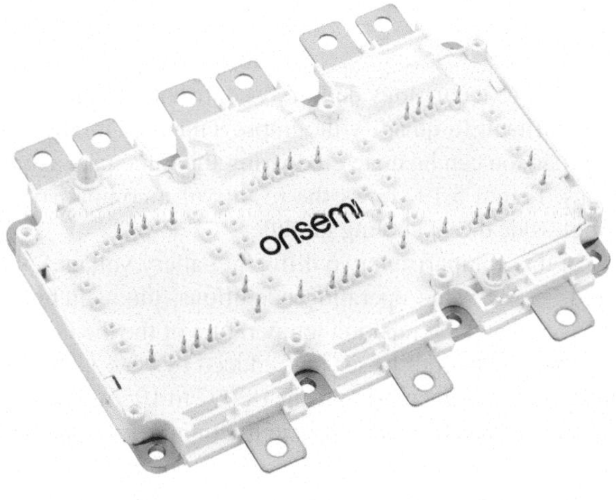

Fig. 5.15 Full-bridge module—gel-filled direct cooling

Fig. 5.16 Half-bridge module—transfer mould

Fig. 5.17 Single-side direct cooling inverter kit

The SiC technology definitely contributes to the overall efficiency of the inverter. It is difficult to quantify the improvement as it depends on many factors. A direct comparison can be done comparing the same package in the same operative conditions. Figure 5.19 shows the results of a comparison between SiC and IGBT on single-side direct cooling module.

In this comparison, two different battery voltages, 400 V and 650 V, are considered. In the same operating conditions, the switching and conduction losses are measured and the junction temperature of the die is monitored. Especially at higher voltage, the overall losses are reduced by 50% using SiC devices, and the junction temperature is decreasing from 189 °C to 105 °C. Based on this simple comparison, the benefit of SiC technology in traction inverter application is evident.

Fig. 5.18 Dual-side indirect cooling inverter kit

400VDC

Device Losses and Temperatures					
	Switching	Conduction	Combined Losses	Avg Junction Temperature	Max Junction Temperature
per switch	276.19 W	226.77 W	502.96 W	140.4 °C	150.0 °C
per diode	22.58 W	41.71 W	64.30 W	97.3 °C	99.8 °C
Converter Losses	1.793 kW	1.611 kW	3.404 kW		

-24% Pdiss

Device Losses and Temperatures					
	Switching	Conduction	Combined Losses	Avg Junction Temperature	Max Junction Temperature
per switch	191.73 W	197.54 W	389.27 W	101.8 °C	108.1 °C
per diode	1.52 W	40.56 W	42.08 W	101.8 °C	
Converter Losses	1.160 kW	1.429 kW	2.588 kW		

650VDC

Device Losses and Temperatures					
	Switching	Conduction	Combined Losses	Avg Junction Temperature	Max Junction Temperature
per switch	508.75 W	234.28 W	743.02 W	175.4 °C	189.8 °C
per diode	32.15 W	40.52 W	72.67 W	107.9 °C	110.6 °C
Converter Losses	3.245 kW	1.649 kW	4.894 kW		

-50% Pdiss

Device Losses and Temperatures					
	Switching	Conduction	Combined Losses	Avg Junction Temperature	Max Junction Temperature
per switch	164.00 W	200.40 W	364.41 W	99.8 °C	105.5 °C
per diode	5.65 W	40.48 W	46.12 W	99.8 °C	
Converter Losses	1.018 kW	1.445 kW	2.463 kW		

Fig. 5.19 Comparison of IGBT with SiC performances

Sanbao Zheng is currently a principal application engineer with Qorvo Inc. He holds a PhD degree with a focus on power electronics from Polytechnic Institute of New York University. He has been working in power electronics areas for over 20 years. His experiences span over a wide range of industries, including automotive, aerospace, LED lighting, laser, telecommunication, and data center. His expertise areas include power converters and their digital controls for high efficiency, high-power density, reliability, and EMC.

Jonathan Dodge received a Bachelor of Science Degree in Electrical and Electronics Engineering from Oregon State University in 1992, followed in 2000 by a Master of Science Degree in Electrical Engineering from the University of Idaho, with emphasis in power electronics and digital design. Jonathan is a licensed professional engineer in the state of Oregon. His experience includes automated manufacturing testing; design of analogue, digital and power electronics; working with power levels ranging from 1.5 to 500 kW in renewable energy and other applications.

Andrea Colognese earned his Master's Degree (1996) and PhD (2005) in Electronics at the University of Padua. Always focusing on the semiconductor area, he started his activity as an IC and SoC designer, becoming soon a project leader and team leader. In the second half of his career, he moved to more business-oriented positions in different semiconductor companies, covering worldwide business responsibility, marketing and product line manager roles in industrial and automotive power semiconductor markets. Now at onsemi, he is responsible for the automotive EV traction solutions business in EMEA and leading the team of product line mangers focusing on module and bare die product and development. The scope of his activity is to define and develop new SiC and Si bare die portfolio in response to market demand, and specific business engagements, considering system solutions as well on top of the dedicated semiconductors, in cooperation with all the OEMs and Tier1 investing in electric vehicles.

Levan Bidzishvili earned his Master's Degree (2010) in Electronics at the Karlsruhe University of Applied Sciences in (Germany). Always focusing on the automotive systems and semiconductor area, he started his activity as a HW designer in the automotive field (Tier1), becoming soon a systems designer/architect. In the second half of his career, he moved to semiconductor field keeping focus still on automotive applications. Now at onsemi, in the automotive EV traction solutions group in EMEA, he is a senior apps engineer responsible for power modules for EV traction applications. The scope of his activity is to define and develop new SiC devices (power modules) in response to market/customer demand, and specific customer engagements, considering system solutions as well on top of the dedicated semiconductors.

Chapter 6
Case Studies

Suresh Thangavel, Aparna Kumar, Mike Zhu, Christoph Ihm, Tobias Asshauer, Rémi Comyn, Zan Huang, Didier Balocco, James Victory, Stephen Oliver, and Maurizio Di Paolo Emilio

This chapter presents the widespread adoption and transformative impact of silicon carbide (SiC) technology across various industries. Suresh Thangavel (Sect. 6.1) from Infineon Technologies discusses trends in three-phase hybrid inverter designs for enhanced efficiency and cost reduction. Aparna Kumar and Mike Zhu (Sect. 6.2) of Qorvo emphasize SiC's role in improving efficiency and power density of data center alternating current (AC)/direct current (DC) power supplies. Christoph Ihm and Tobias Asshauer (Sect. 6.3) from Infineon Technologies highlight SiC's efficiency improvements in renewable energy systems such as photovoltaic (PV) inverters and energy storage. Rémi Comyn (Sect. 6.4) from KnowMade analyzes the

S. Thangavel · T. Asshauer
Infineon Technologies, Warstein, Germany

A. Kumar · M. Zhu
Qorvo, Greensboro, NC, USA

C. Ihm
Infineon Technologies, Neubiberg, Germany

R. Comyn
KnowMade, Valbonne, France

Z. Huang
Wolfspeed, Durham, NC, USA

D. Balocco
onsemi, Vélizy, France

J. Victory
Onsemi, Scottsdale, AZ, USA

S. Oliver
Navitas Semiconductor, Torrance, CA, USA

M. Di Paolo Emilio (✉)
Power Electronics News, Pescara, Italy

M. Di Paolo Emilio (ed.), *SiC Technology*,
https://doi.org/10.1007/978-3-031-63418-5_6

strategic importance of SiC patents in shaping the industry. Zan Huang (Sect. 6.5) from Wolfspeed explores SiC's potential in revolutionizing data center cooling systems. Didier Balocco and James Victory (Sect. 6.6) of onsemi discuss advancements in Simulation Program with IC Emphasis (SPICE) simulation for power semiconductor models. Stephen Oliver (Sect. 6.7) from Navitas Semiconductor explores high-voltage SiC technology in heavy-duty electric vehicles and aircraft power systems. Lastly, the Sect. 6.8 delves into advanced SiC solutions for DC fast charging, covering topologies such as full-bridge LLC, phase-shift full-bridge (PSFB) converters, and dual active bridge (DAB) converters, and their role in enhancing efficiency and reliability in electric vehicle (EV) charging infrastructure. This chapter collectively showcases how SiC is driving efficiency, performance, and sustainability across diverse applications.

6.1 Today's Megatrends for Photovoltaic Inverters in Big Residential and Commercial Applications

Suresh Thangavel

Some key design requirements will remain as they are for the foreseeable future. In residential and small commercial applications, the photovoltaic (PV) system voltage is expected to remain at 1000 V for three-phase systems. Such equipment will be useful for users whose primary concern is generating electricity for self-consumption. Hybrid inverters help achieve this using less materials and space while reducing the complexity needed to build PV systems.

Much like string inverters, hybrid inverters can connect multiple PV panels and convert DC current to AC. In addition, hybrid inverters can supply DC currents directly to an energy storage system (ESS) system, such as a battery, thereby eliminating unnecessary power conversions, which saves energy.

However, the core area for improvement is reducing the overall system cost. This has resulted in a push for integrating a DC-DC stage to charge and discharge batteries, leading to the search for high-efficiency DC-DC converter linking the battery and the string inverter (Fig. 6.1).

Another challenge is to reduce power conversion losses to minimize heat generation, and especially in residential applications, to create silent, fan-less designs. The overarching design goals, therefore, include improving overall system efficiency and power density, especially under partial load conditions. This is a highly challenging requirement that demands low conversion losses at a higher switching frequency operation. Here you can also see how a wide bandgap semiconductor could add more value to the system.

The energy requirement in areas where PV systems are installed is also changing. For example, in residential applications, there is a gradual move away from gas to electric heating pumps for heating purposes. Then, there is the rapid increase in

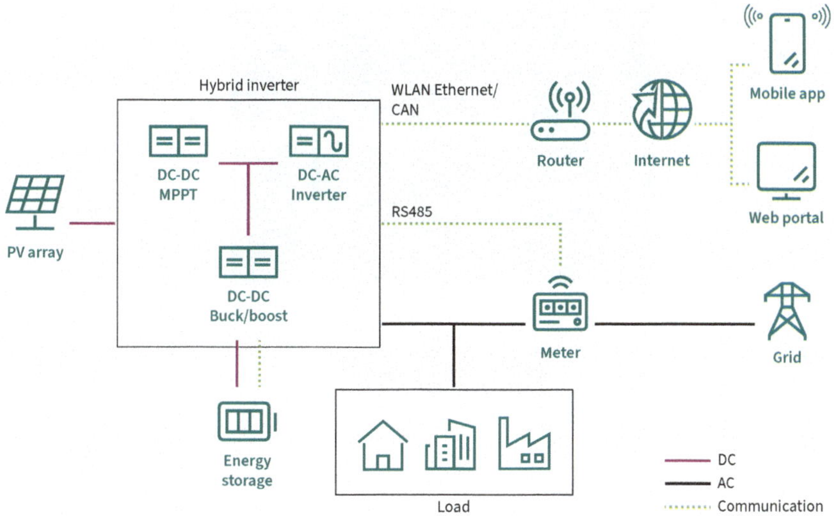

Fig. 6.1 Rising self-consumption is pushing designers toward a hybrid architecture with support for energy storage

electric vehicles (EVs) that require electricity for charging. Both residential and commercial applications are, therefore, keen to utilize self-generated electrical energy.

6.1.1 Improving the Three-Phase Hybrid Inverter Design

At the core of the PV inverter are the power conversion blocks. Inside the power converter functional block, a hybrid inverter implements two DC-DC stages, one for booster operation and another for battery charging and discharging operation, and finally a DC-AC stage to generate grid-compatible single- or three-phase AC power.

As shown in Fig. 6.2, the inverter is surrounded by a host of sensors that monitor everything from currents and voltages to operational temperature. These sensors feed into a control block that comprises a microcontroller that controls the power switches through pulse width modulation and runs real-time control algorithms. This system also supports advanced control algorithm and monitoring methods such as maximum power point tracking (MPPT) technique for the DC-DC booster. An additional microcontroller operates in the connectivity block, supported by appropriate security hardware, to provide remote monitoring and diagnostics.

Our focus in this section will be on power converter functional block within three-phase hybrid inverter. The first DC-DC stage converts variable DC voltage into a fixed DC voltage while simultaneously ensuring that the maximum power is extracted from the photovoltaic string via the maximum power point tracking

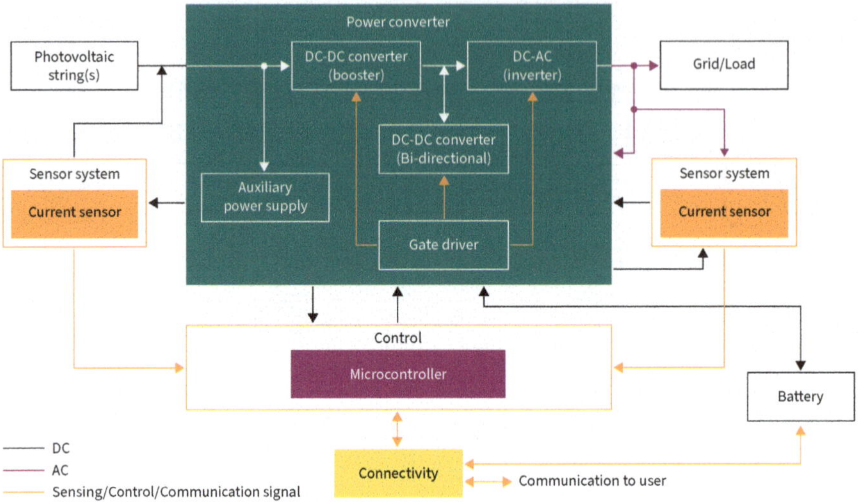

Fig. 6.2 PV inverters can be divided into four functional blocks: the power converter, sensor systems, control, and connectivity

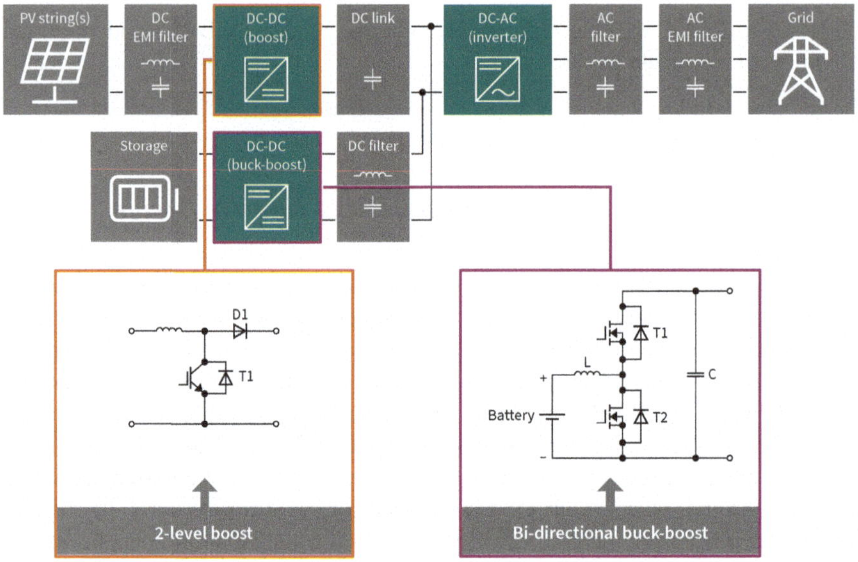

Fig. 6.3 Topologies commonly used in DC-DC functional block for three-phase hybrid inverter

(MPPT) technique. Usually, a simple two-level boost topology is preferred for this stage (Fig. 6.3).

And secondly, the DC-DC stage allows for bidirectional energy transfer between the battery energy storage system and the DC bus. In general, three-phase hybrid inverter comes with high-voltage batteries. High-voltage battery typically ranges

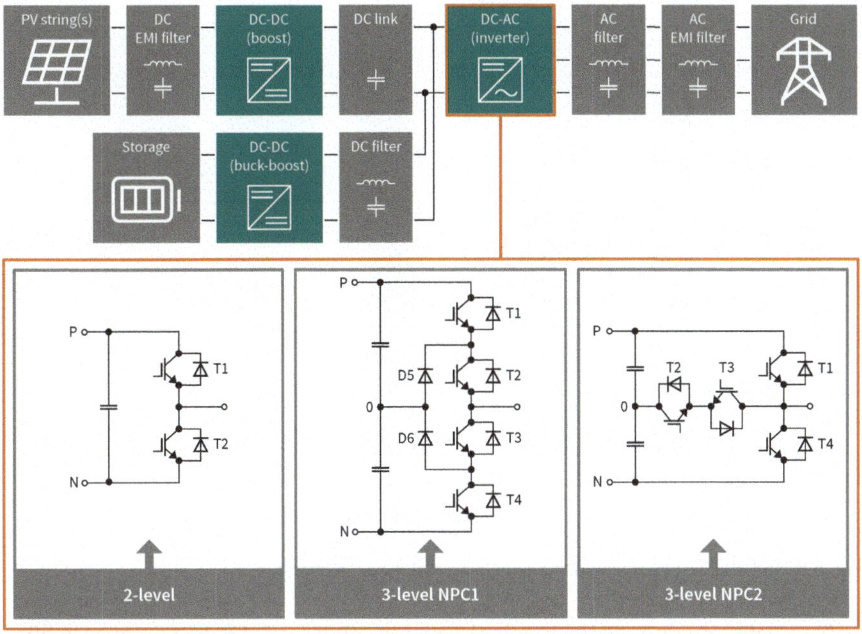

Fig. 6.4 Topologies commonly used in DC-AC functional block for three-phase hybrid inverter

between 150 and 600 V. Typically, non-isolated bidirectional DC-DC converters are preferred (Fig. 6.3).

The DC-AC stage converts DC power into grid-compatible AC voltage, current, and frequency. Either two-level or three-level topology is preferred in this stage (Fig. 6.4). However, for most simplified inverter designs, two-level or three-level NPC2 topology is recommended. In general, the choice of topology greatly depends on inverter design requirements such as efficiency, power density, and weight and size of the system. Further, the filter size, volume of the system, and cost greatly depend on the balance between the operating switching frequency and its cooling requirements.

Furthermore, to ensure the grid stability, different standard and grid codes recommend to fulfill the power quality requirements for grid-connected photovoltaic inverters. Generally, to meet the total harmonic distortion (THD) requirements, the inverter has to be switched at higher switching frequencies, typically higher than 10 kHz.

Inductance size and weight reduce with increase in switching frequency (Fig. 6.5). In this use case, a three-time increase in switching frequency (from 16 to 48 kHz) shall decrease the inductor size by 40%. In inverter design switching frequency plays a vital role in reducing the size, weight, and cost of the system. It is a highly challenging requirement that demands power semiconductors with low conversion losses at a higher switching frequency operation.

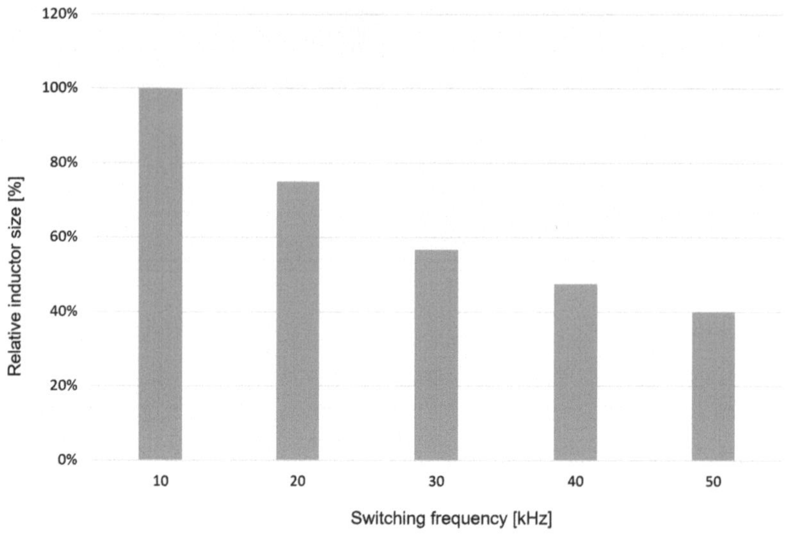

Fig. 6.5 Relative inductor size dependency on switching frequency

Now, let us go through the topologies and the choice of chip technologies within DC to DC and DC to AC functional blocks. This scheme shows the appropriate position of the choice of chip technologies within the choice of topology options (Fig. 6.6). We color-coded the corresponding switch chip technologies. In olive green, we have the silicon carbide chip and in black the silicon chip.

The silicon solution refers to the silicon chip set in all switch positions and the silicon carbide solution refers to the SiC chip in all switch positions, whereas the hybrid solution uses a combination of the silicon isolated gate bipolar transistor (IGBT) and SiC diode chip. However, for NPC2 topology, we consider a combination of the SiC MOSFET chip in outer switch positions T1 and T4 and the silicon chip in inner switch positions T2 and T3 as the SiC solution.

Before we deep dive, let me introduce the background for comparison. The target output power is 12 kW, the output voltage is 380 V, and the package under consideration is TO-247-3-pin. We used Plec's simulation tool for performance comparison, and in this simulation, we limit the junction temperature operation to 150 °C with an ambient temperature of 45 °C and it is a fan-less design.

DC-DC boost converters play an important role in PV inverter systems; they step up variable output voltage of a solar array to a suitable or fixed voltage required by the DC-AC inverter stage. Now, let us start with the choice of chip technology for a simple two-level boost topology. Note that each choice is evaluated based on its capability to operate efficiently, its size, and cost of implementation.

The very first full silicon solution reaches its upper limit at 16 kHz switching frequency operation in this design due to higher conversion losses and junction temperature limit (Table 6.1). Compared to a full silicon solution, the hybrid solution is more attractive if you want to achieve low conversion losses and higher

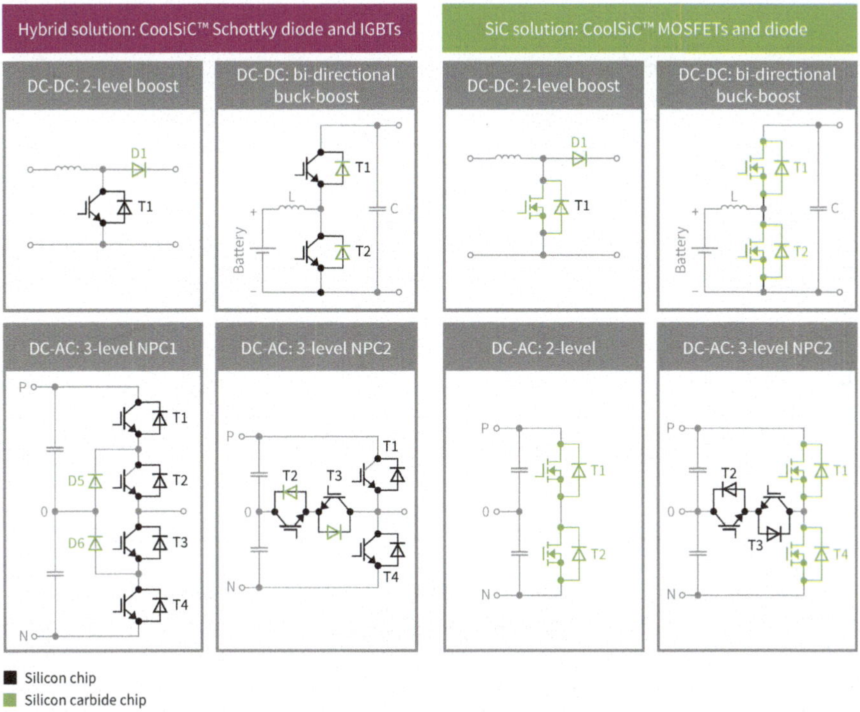

Fig. 6.6 The appropriate position of choice of chip technologies within the choice of topology options

Table 6.1 Comparison of chip technology for DC-DC booster with simple two-level boost topology

Solution	SiC	Hybrid	Si
No. of device per inverter*	1+1	1+1	1+1
Switching Frequency	32 kHz	16 kHz	16 kHz
Total converter losses (only device)	34 W	48 W	61 W
Efficiency	99.4 %	99.1 %	98.9 %
$T_{VJ(max)}$ IGBT/MOSFET	107°C	130°C	150°C
$T_{VJ(max)}$ Diode	97°C	118°C	148°C

efficiency at a good price performance compromise. Further marginal increase in switching frequency is feasible by allowing switch to operate at its temperature limits.

As for the silicon carbide solutions, it enables the designer to achieve compelling efficiency with higher power density, size, weight, and cost reduction, by switching

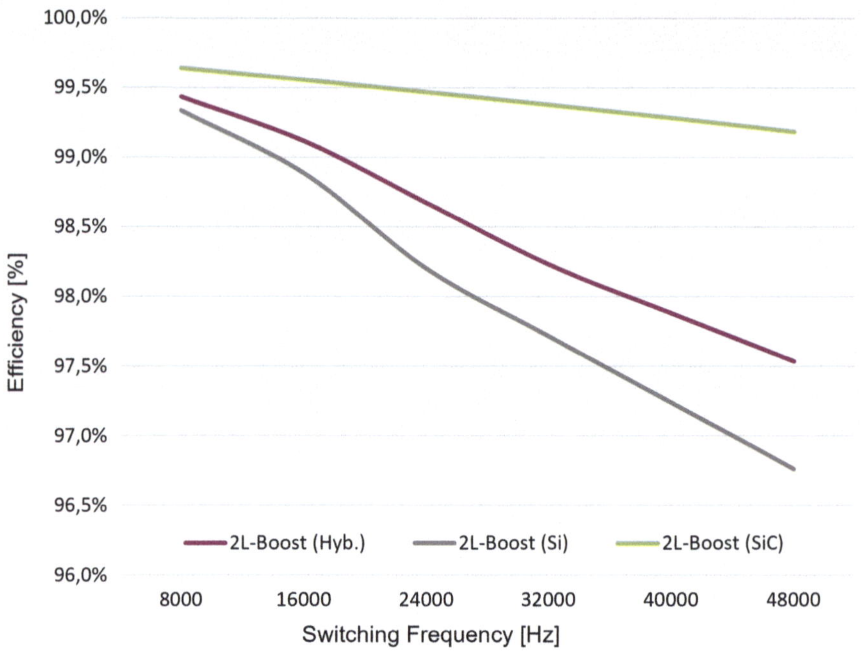

Fig. 6.7 Choice of chip technology for DC-DC functional block with simple two-level boost stage. Simulation condition: CCM, $V_{IN} = 400$ V, $V_{OUT} = 750$ V, duty = 0.47, $I_{IN} = 13.2$ A, Ta = 45 °C, $R_{th(c\text{-}h)} = 3$ °K/W (per switch) *No. of devices per DC-DC booster I Si: IKW40N120CH7 (x2), Hybrid: IKW40N120CH7 (x1) + IDW20G120C5B (x1) & SiC: IMW120R040M1H (x1) + IDW20G120C5B (x1)

two times faster when compared to a full silicon solution. Further it allows the designer to achieve compelling efficiency and lowest conversion losses (Fig. 6.7).

The above statements are valid for the second DC-DC converter stage while transferring energy from the battery energy storage system to the DC bus.

Now let us go through the choice of a topology and chip technology within DC-AC functional block. Here you can see a topology comparison, regarding inverter efficiency and switching frequency. The two-level topology suffers heavily on efficiency above 8 kHz switching frequency operation. It is evident that the three-level NPC-1 and NPC-2 topologies are widely used for their higher efficiency and lower power conversion losses (Fig. 6.8).

The three-level NPC-1 enables more than 24 kHz switching frequency operation, while the three-level NPC-2 is the most suitable topology for less than 24 kHz switching frequency operation.

Here you can see a topology and chip technology comparison between inverter efficiency and switching frequency. Best-in class hybrid inverters must be efficient, power dense, and compact.

It is evident that the three-level NPC2 topology with hybrid and SiC solutions enable the designer to further increase the switching speed two times faster than the silicon solution; however, the losses remain the same (Fig. 6.9 and Table 6.2). In

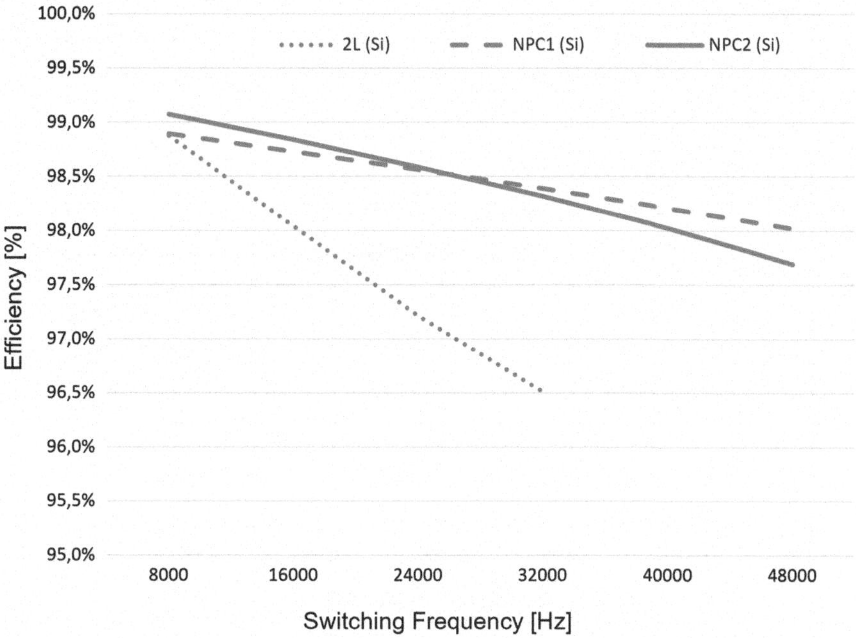

Fig. 6.8 Three-level topology is widely used for its higher efficiency. Simulation condition: SVPWM, V_{DC} = 750 V, VAC = 380 V, F_{OUT} = 50 Hz, MI = 0.83, I_{OUT} = 18.2A, pf = 1, Ta = 45 °C, $R_{th(c\text{-}h)}$ = 3 °K/W (per switch). No. of devices per inverter | 2 L (Si): IKW40N120CH7 (x6), NPC1 (Si): IKW40N65ES5 (x18) and NPC2 (Si): IKW40N120CH7 (x6) + IKW40N65ES5 (x6)

general, the increased power semiconductor cost shall be compensated by the reduction in filter inductance size, weight, and cost by the increased switching frequency operation. And lower operational temperature shall further reduce the size of the housing and further enable fan-less designs.

Such features make a big difference in residential and commercial installations, where quieter and smaller systems are required, and ultimately reduce the amount of materials needed and allow to build lighter and more efficient hybrid inverters.

6.1.2 *Today's Megatrends in Improving the Three-Phase High-Power String Inverter Design for Utility-Scale Applications*

SiC MOSFETs in solar and energy storage applications have clear benefits compared to silicon technologies, addressing the pressing need for energy and cost savings, particularly when bidirectional power conversion is required.

Ease of installation is one of the key features of high-power PV string inverters. It is beneficial if only two workers are needed to carry and install the system. Hence, size and weight are of special importance. Latest generation silicon carbide

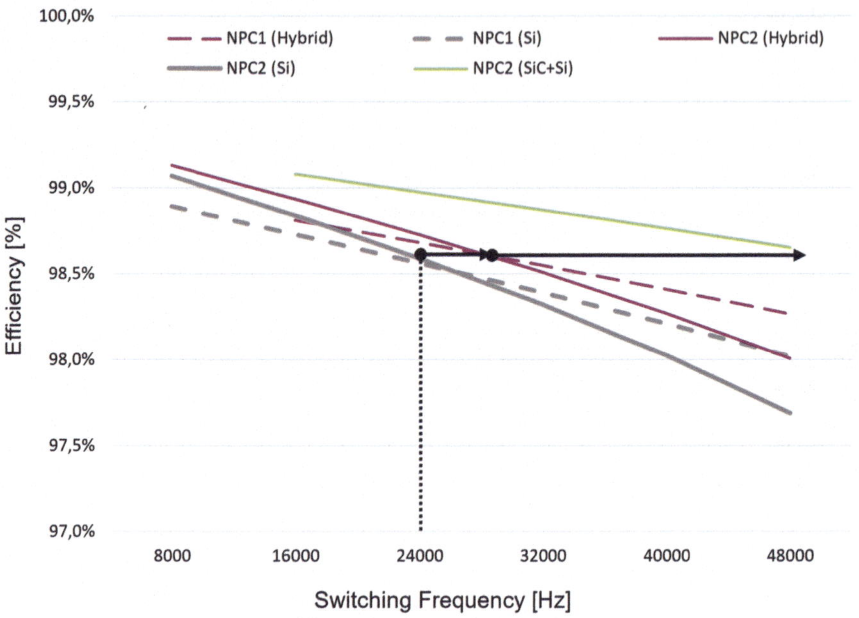

Fig. 6.9 Choice of topology and chip technology for DC-AC stage to build efficient and compact hybrid inverter. Simulation condition: SVPWM, V_{DC} = 750 V, VAC = 380 V, FOUT = 50 Hz, MI = 0.83, IOUT = 18.2A, pf = 1, Ta = 45 °C, Rth(c-h) = 3 °K/W (per switch). *No. of devices per inverter I NPC1 (Si): IKW40N65ES5 (x18), NPC1 (hybrid): IKW40N65ES5 (x12) + IKW50N65SS5 (x6), NPC2 (Si): IKW40N120CH7 (x6) + IKW40N65ES5 (x6) and NPC2 (hybrid): IKW40N120CH7 (x6) + IKW50N65SS5 (x6) and NPC2 (SiC): IMW120R040M1H (x6) + IKW40N65ES5 (x6)

Table 6.2 Comparison of topology and chip technology for DC-AC inverter

Solution	2L (SiC)	NPC2 (Si)	NPC2 (Hybrid)	NPC2 (SiC)	NPC1 (Hybrid)
No. of devices per inverter*	6	6 + 6	6 + 6	6 + 6	18
Switching frequency	32 kHz	24 kHz	32 kHz	48 kHz	32 kHz
Total inverter losses (device)	152 W	172 W	182 W	163 W	196 W
Efficiency	98.8%	98.6%	98.5%	98.7%	98.4%
TVJ (max) IGBT/MOSFET	150 °C	104°C	110°C	109°C	99 °C
TVJ (max) Diode	150 °C	92 °C	94 °C	109 °C	78 °C

semiconductors enable a significant increase in power density and conversion efficiency. This does not only save energy but also results in smaller and lighter equipment with lower associated capital, installation, and maintenance costs.

6.1.3 Key Application Requirements and Their Challenges

The high operation voltage of 1500 V in utility-scale installations requires a low cosmic radiation-induced failure rate combined with a higher system efficiency for the power devices. Because of these contradictory requirements, ANPC multilevel topologies are the preferred solutions today due to their highest efficiency over the full range of power factor operation (Fig. 6.10). Such inverters are perfectly designed for both solar and battery storage applications.

A cost-effective way of adapting the ANPC topology is to optimally combine Infineon's 1200 V CoolSiC™ MOSFET with TRENCHSTOP IGBT7 technology. Figure 6.11 shows a phase leg of the proposed implementation, whereby T1, T4, T5, and T6 are composed of silicon-based IGBTs with corresponding silicon freewheeling diodes (FWD). Transistors M2 and M3 consist of CoolSiC MOSFETs with internal body diodes. By using the modulation scheme presented in Fig. 6.4, the IGBTs are switched continuously *on* or *off* with grid frequency of 50/60 Hz. Hence, the IGBTs are optimized for lower conduction losses. In this way, switching losses only occur in the fast and highly efficient SiC MOSFETs. Thus, the number of SiC devices is reduced to a minimum, achieving an optimal cost-performance ratio.

6.1.4 Reference Design Proves the Benefits

In order to demonstrate the significant benefits of using SiC MOSFETs in PV string and energy storage inverters, Infineon has developed a modular reference design for 1500 VDC systems rated up to 300 kW. The design uses a novel bidirectional

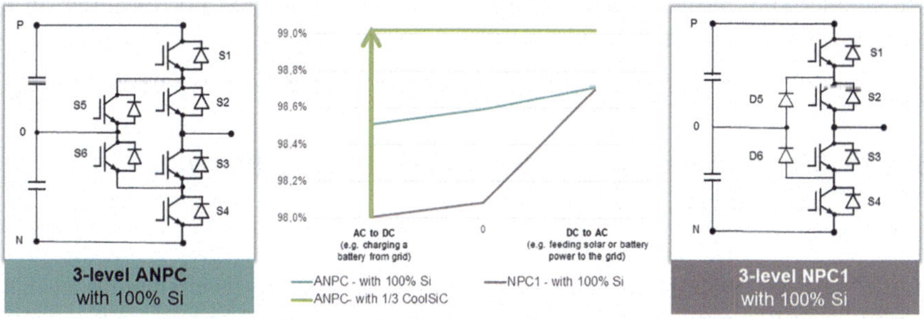

Fig. 6.10 Comparison of the two most common topologies for 1500 V PV inverters

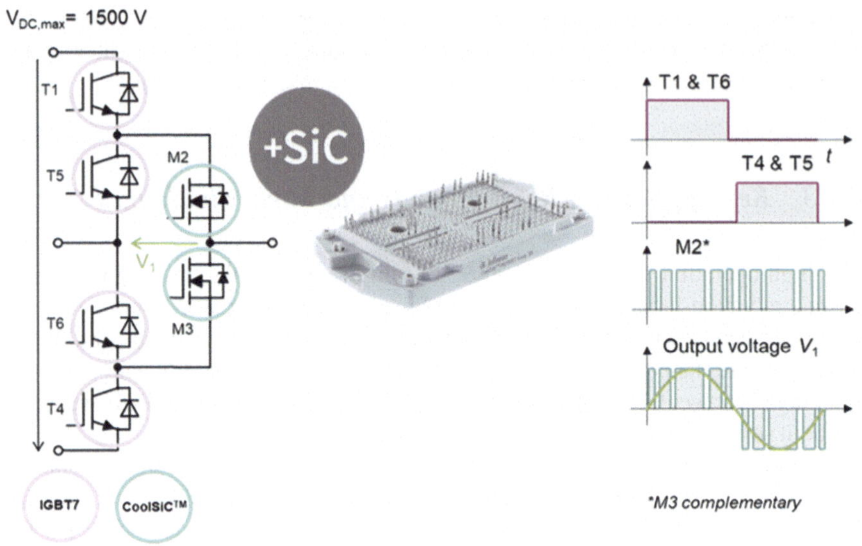

Fig. 6.11 ANPC topology in an Easy 3B power module and its modulation scheme

three-level ANPC topology that achieves an efficiency close to 99% in both directions and a switching frequency up to 96 kHz (interleaving configuration).

Further SiC semiconductors significantly reduce the size and weight of passive components such as capacitors and inductors, since they enable higher switching frequencies. This allows for smaller units, which in turn helps make inverters smaller, resulting in higher power density and lower costs (Fig. 6.12). Power density is greater than 5 kW/kg for a complete solution including heat sinking and all control, allowing 300 kW throughput in the ideal 100 kg maximum cabinet weight.

Energy usage and cost savings by using SiC can be easily calculated from overall efficiency improvements; for example, compared with a IGBT solution, 1200 V CoolSiC MOSFETs can halve the losses in an ESS installation and provide typically 2% extra energy and run-time. SiC MOSFET unit cost is not yet generally less than an IGBT for similar performance. But on the system level, hardware costs are reduced significantly, as high efficiency at higher switching frequencies allows to use smaller and cheaper magnetic components and heat sink. For example, in a 1500 V PV string inverter, at least 5–10% saving in cost per kW can be expected.

6.1.5 SiC Success Story in High-Power String Inverters for Utility-Scale Application

KACO decided to develop a new device in the large string inverter segment that would close the gap with central inverters and is now marketed under the name Blueplanet 125 TL3. The targets the engineers set seemed ambitious at that time. The capacity should be more than 100 kilowatts, but the unit should not weigh more

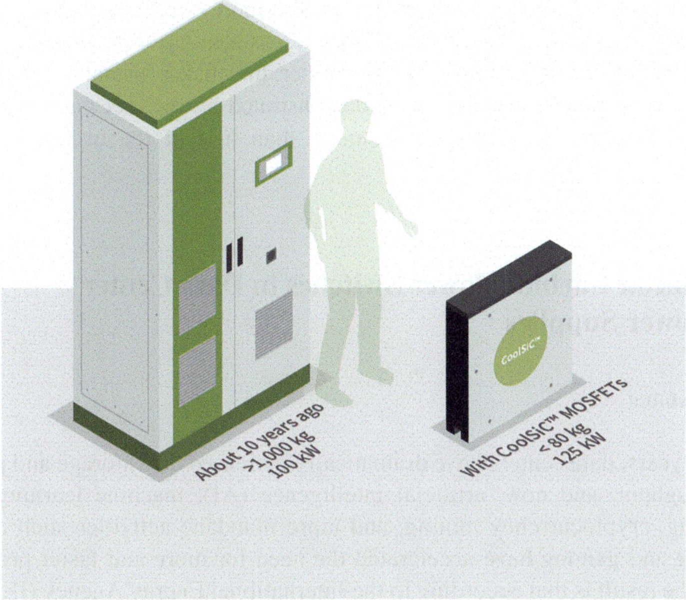

Fig. 6.12 The progress made over the past 10 years in semiconductor technology is also reflected in the size of the PV inverter. (Source: Matthias Hagg, CTO, Kaco new energy, "The time", PV magazine 12/2013)

than 80 kilograms. "Because when you install it in remote regions, it should be possible for two people to carry it and mount it," says Matthias Haag, CTO, KACO new energy. For the new development, Infineon was responsible for the module with the semiconductor components. KACO's job was to ensure optimal placement on the printed circuit board, interconnection with the choke coils, and measurement of the prototype. The real art is managing the high power over the printed circuit board, says Haag. Nevertheless, more than 130 amps are combined in two switching modules, each about half the size of an iPhone.

The complexity of the circuit design is also due to the fact that the strengths of silicon carbide only become apparent when the inverter's switching frequency is particularly high. The new device now chops the direct current at 48 kHz before using the coils to generate a smooth 50 hertz sinusoidal alternating voltage. Haag notes that this high clock frequency creates new phenomena.

If conductor paths are not arranged optimally, interference occurs because the conductor paths behave like tiny antennas at this frequency. The frequency of 48 kHz was intentionally selected so that the third harmonic would be below 150 kHz. Ultimately, this makes it easier to comply with electromagnetic compatibility guidelines.

The high switching frequency makes it possible to reduce the size of the choke coils that are found in lower frequency devices, cutting their weight by more than half and their cost by just under half of those in comparable inverters with the usual clock frequency for silicon components. Furthermore, since conduction losses in silicon carbide are lower than those in silicon, less heat has to be dissipated which also shrinks the size of the heat sink. Altogether, this makes a smaller housing possible. Just how relevant this is can be demonstrated by the fact that chokes, heat sinks, and housings still account for more than half of the overall weight of the device.

6.2 Silicon Carbide Power Switches in Data Center Power Supplies

Aparna Kumar

Over the years, data centers have dramatically increased data storage and processing throughput, and now, artificial intelligence (AI), machine learning, cloud computing, cryptocurrency mining, and more mundane activities such as video streaming and gaming have accelerated the need for more and faster processing power. The result is that according to the International Energy Agency (IEA), data centers accounted for 0.9% of global greenhouse gas emissions in 2020 and 1–1.3% of global electricity demand or 240–340 TWh in 2022 [1]. The effect depends strongly on geography, with Ireland, for example, spending 18% of its energy use powering data centers. New applications such as ChatGPT however are accelerating energy draw trends with just its "training" phase reportedly consuming 1.2 TWh and in operation drawing around 564 MWh per day, according to de Vries [2].

Hardware has evolved to keep up with demand, and it is a credit to the community of designers and engineers working to increase efficiency, that according to the IEA, data center energy consumption only increased 20–70% between 2015 and 2022, while workload increased 340%. However, limits are approaching for power density in data centers with cabinets slated to dissipate 100 kW in the near future, up from around 30–40 kW today. The AC/DC power supplies that provide the power, typically in racks at the bottom of the cabinet, add to the dissipation with any conversion losses, so these must be minimized to keep temperatures within reasonable bounds and accountants happy. The ACs/DCs must also be as small as possible to allow space for more server blades to increase the computing power per square foot of floor space—an important metric in data centers. Extreme power density has therefore become the goal in AC/DC design, requiring a combination of high conversion efficiency for minimal cooling hardware and small component size, particularly magnetics.

6.2.1 *Standards for AC/DC Efficiency and Size in Data Centers*

Standards have been introduced for efficiency and power density, for example, the familiar 80+ series for efficiency across different loading and input voltages. Today, the highest rating is expected, titanium, requiring minimum 96% efficiency end-to-end for single-output 230 VAC-rated equipment at 50% load. High efficiency must also be maintained across the load range—minimum of 90% at 10% load and 94% at full load. The European regulation 2019/424 or “Lot 9” now also requires the titanium standard, effective January 1, 2024. Physical format is also defined, for example, the Common Redundant Power Supply Specification (CRPSS) described in the “Open Compute” project by members including Google, Meta, Intel, Microsoft, and Dell. The standard is for replaceable redundant units that must fit in 1 U rack height (40 mm), 73.5 mm width, and 195 mm depth. The output is 12 V for further distribution, and the power available is expected to be in the multi-kW range, for a power density that might exceed 100 W/cubic inch.

Data center architecture designers continually look for ways to improve efficiency using the metrics of power usage effectiveness (PUE) and water use effectiveness (WUE) for the whole facility, including overheads such as cooling. An approach explored in the past by Meta, for example, as part of the Open Compute project was to distribute 275 VAC to each cabinet, with a conversion to 48 V with a local UPS attached [3]. The technology is available to down-convert 48 V directly to as low as sub-1 V for the board IC loads with or without isolation, but there will often be an intermediate bus at typically 12 V on the board to make the final “point of load” conversion more efficient. Distribution of high-voltage DC to the cabinets at perhaps 400 V has also been explored, but there is a large infrastructure built around AC distribution that would be expensive to change and ancillary equipment such as fans, air-conditioning, and cooling pumps may need AC anyway as standard parts. DC distribution allows a possible increase in overall efficiency and might be considered for newly built data centers.

6.2.2 *Traditional AC/DC Topologies Are Too Inefficient*

Whatever the system architecture, there is AC/DC conversion involved with power factor correction and typically further stages of DC/DC conversion. Now that “titanium” levels are mandated, this practically excludes the traditional conversion approach of bridge rectifier, boost power factor correction (PFC) stage, and perhaps a full-bridge PWM stage using silicon MOSFETs switching at say, 100 kHz with diode output rectification and then filtering. This scheme would result in an efficiency of around 85%, depending on the detail design, typically dropping off to a low level at light loads. To meet the titanium level of 96% would mean reducing

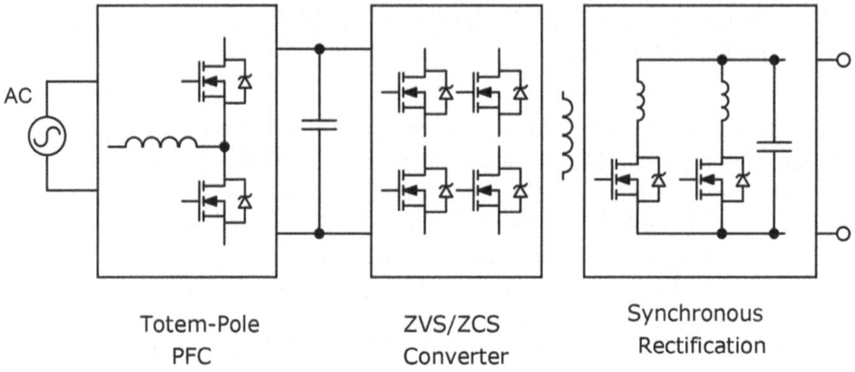

Fig. 6.13 A modern approach to high-efficiency AC/DC conversion

losses by a factor of more than 4×, involving a complete rethink of the approach taken. Dropping switching frequency is a way to reduce dynamic losses, but then transformer and filter magnetic size increase and power density becomes far too low.

A solution is to use a "bridgeless" PFC stage and a DC/DC conversion stage that utilizes zero-voltage and/or zero-current switching (ZCS), followed by synchronous rectifiers (Fig. 6.13). If wide bandgap semiconductors are used, such as silicon carbide types, there is an option to increase switching frequency substantially, to minimize magnetics size, although there are diminishing returns as core losses increase exponentially with frequency.

We will now look at each conversion stage in more detail.

6.2.3 The "Bridgeless" Totem Pole PFC Stage

It must have been quite a revelation when it was first realized that the diode bridge AC rectifier could be eliminated in an AC/DC converter and absorbed into the function of a boost PFC stage [3]. In its simplest form, the PFC boost diode is just changed to be a synchronous switch, the same as the main switch, and then the two are driven to form boost switch and diode alternately, depending on the AC main polarity, with diodes steering the return current back to the supply. For better efficiency, the diodes can be replaced by synchronous switches toggling at line frequency, or multiphase approach can be taken to share out losses (Fig. 6.14). For three-phase AC, it is relatively simpler still, with the "Vienna" rectifier a popular implementation.

At high power, however, the circuit is practically forced to operate in "hard-switched" continuous conduction mode (CCM) where current levels are manageable but dynamic losses are high, from any current/voltage overlap and from reverse recovery of the switch body diodes, which conduct during switching dead time. With silicon MOSFETs, target efficiency cannot be met without operating at a very low frequency and suffering large magnetics, but with silicon carbide switches,

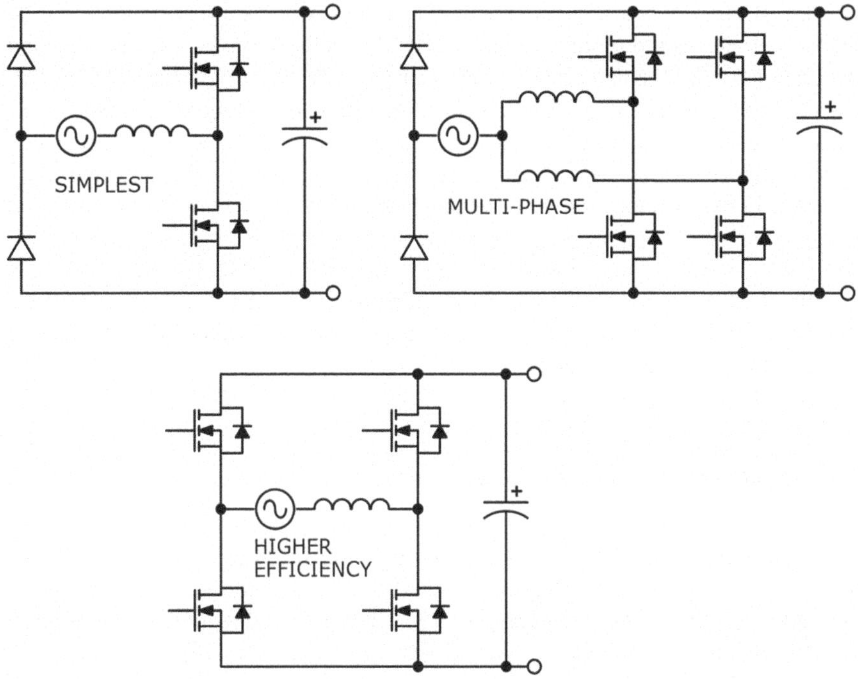

Fig. 6.14 Various implementations of a single-phase "bridgeless totem pole PFC" stage

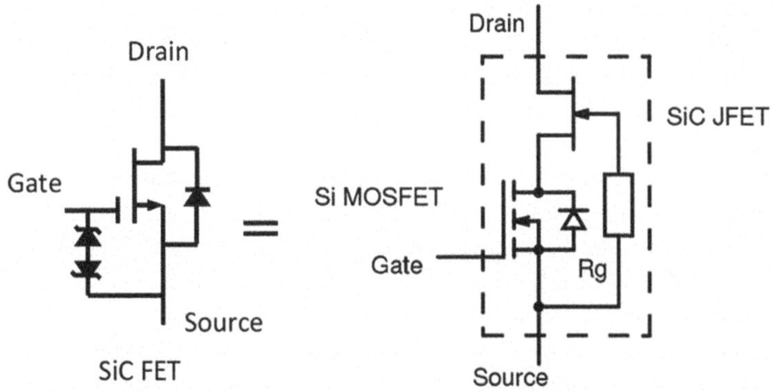

Fig. 6.15 The cascode SiC FET—a "stacked" silicon MOSFET and SiC JFET

edge rates can be higher for low overlap and the body diode is much faster with lower reverse recovery loss. A better solution still is a cascode SiC FET (Fig. 6.15), available from Qorvo [4], a cascode combination of a SiC JFET and a low-voltage silicon MOSFET, which has lower relative on-resistance and a body diode that is equally fast but with a much lower forward voltage drop than a SiC MOSFET. Cascode SiC FETs are available from 650 to 1700 V ratings and have the best figures of merit for on-resistance when compared with Si or SiC MOSFETs or gallium nitride (GaN) high-electron-mobility transistor (HEMT) cells. With its low dynamic and conduction losses, the cascode SiC FET can yield 99.28% efficiency for a bridgeless PFC stage, at 1.5 kW output from 240 VAC input, operating at 100 kHz, allowing a compact inductor size. This can be verified using the Qorvo FET-Jet online calculator [5] using their 18 milliohm parts UJ4C075018K4S for the high-frequency switches. In the example, the two switches toggled at line frequency are uncritical as dynamic losses are low, and at 1.5 kW, using UF3SC065007K4S cascode SiC FETs, the calculator puts their total losses at just 0.3 W each. The arrangement is capable of 3 kW output, and at 10% load or 300 W, the circuit still achieves 97.53% efficiency, allowing for losses further down the conversion chain while still meeting the titanium minimum efficiency.

6.2.4 DC/DC Conversion Stage

The main isolation and DC/DC conversion stage in a data center power supply can be one of several high-efficiency topologies. The *LLC* converter is popular for its simplicity, its zero-voltage switching (ZVS) capability, and its potential to require just one magnetic component. The primary resonant current is relatively high though, so low $R_{DS(on)}$ switches are important to keep conduction losses low. Dynamic losses can be very low down to very light loads, but the circuit can lose ZVS under abnormal loading or start-up, so precautions have to be taken to ensure stress levels are not dangerous under these conditions. The circuit switches with variable frequency to regulate the output and this can make EMI filtering less deterministic. Half-bridge and full-bridge versions of the LLC can be designed, no output inductor is required, and discrete or synchronous output rectification can be employed.

The *CLLC* converter operates similarly but with an additional resonant capacitor on the secondary side along with synchronous rectifiers, ensuring ZVS for all switches. Control can be by frequency modulation or phase shift. The symmetry of the topology makes it ideal for bidirectional power flow which is not a requirement in data center AC/DCs, so other topologies are typically chosen.

The phase-shift full-bridge (*PSFB*) topology is a favorite for many as it operates at fixed frequency with simple 50% duty cycle antiphase gate drives to each leg in an H-bridge of switches. Regulation is achieved by varying the relative phase of the gate drives to the two legs in the bridge. A resonant inductor is required in the primary allowing ZVS, but an output inductor is required and maintaining ZVS to light load can be problematic.

Another popular topology is the dual active bridge (*DAB*) (Fig. 6.16). First proposed in 1988, this is similar to a PSFB circuit with synchronous output rectifiers but does not need an output choke. A primary inductor is required but this might be the transformer leakage inductance. The converter acts as a current source with power regulated which, as a consequence, regulates the output voltage. In the

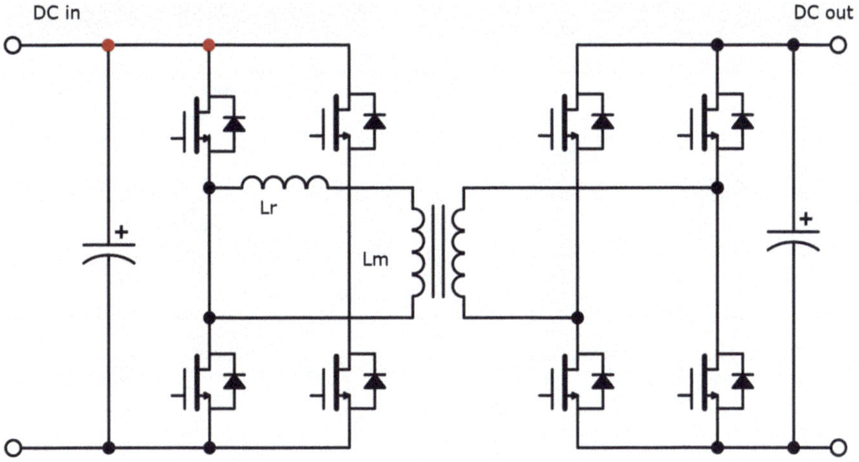

Fig. 6.16 The dual active bridge (DAB) topology

simplest implementation of the DAB, all the H-bridge legs are driven at 50% duty cycle with some dead time, and the legs are driven in antiphase to each other but the phase difference between primary and secondary gate drives is now controlled to regulate power flow. Control is complex and there are different modulation schemes with different control variables up to “triple phase shift” where primary and secondary duty cycles are varied as well as phase shift, with the advantage of potential reduction in RMS values of circulating currents. Output capacitor RMS current can be quite high, and there is a concern about flux imbalance in the transformer, which can cause magnetic runaway, saturation, and catastrophic failure. A DC blocking capacitor is a solution with a size and cost penalty, or other active control methods can be used to prevent the effect.

In the DAB, ZVS is naturally achieved in the primary and secondary switches at turn on; however, this relies on the output capacitance of the switches, C_{OSS}, being fully discharged during the dead time, implying that low C_{OSS} devices should be used and C_{OSS} variation with operating conditions carefully evaluated. With the simplest modulation schemes, turn-off occurs at the peak primary current, so this is a “hard” transition and fast edge rates are needed to avoid excessive dissipation. Body diodes of the switches also conduct during dead time, so like the other topologies, their forward voltage drop and recovery energy should be low.

Cascode SiC FETs are again a natural choice for best performance in the DAB topology with their low dynamic and static losses compared to other semiconductor types. The Qorvo FET-Jet calculator can be used again to explore what performance is possible. With 1500 W output selected as 50% load for a DC/DC converter operating with 400 VDC input and 48 VDC output with 100 kHz operation and a 10 μH resonant inductor, the semiconductor efficiency achieved is 98.83%. At 3 kW, the efficiency is about 97.7% and at 300 W, the efficiency is over 99%. Ohmic losses in the transformer and connections will add some losses, but it is clear that along with a cascode SiC FET-based bridgeless PFC stage, the titanium standard would be

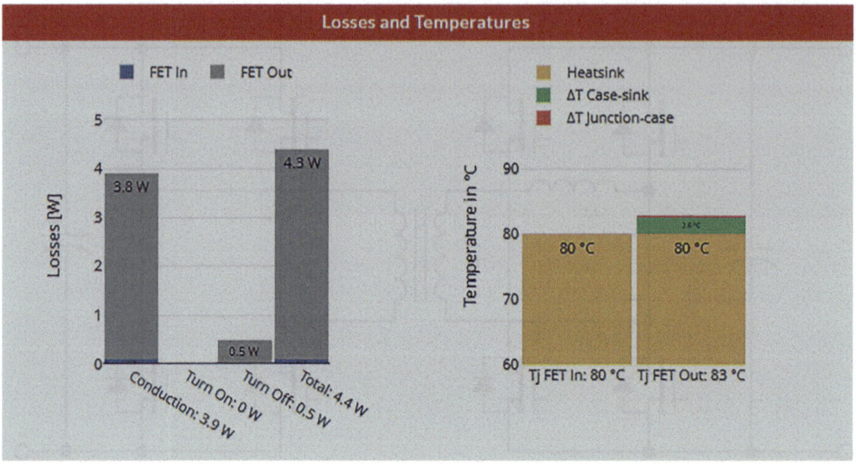

Fig. 6.17 Screenshot of the Qorvo FET-Jet calculator results for cascode SiC FET losses in a DAB converter at 1.5 kW output

easily met. A screenshot from the calculator, Fig. 6.17, shows that most losses actually come from the secondary side cascode SiC FET on-resistance. These were chosen as 6.7 milliohm UF3SC065007K4S types available from the standard Qorvo range. The 650 V device voltage rating is more than needed for a 48 V output, so SiC MOSFETs or GaN HEMT cells might be chosen here with perhaps 100 V rating, and a consequential lower $R_{DS(ON)}$ and yet lower conduction loss. Devices could also be paralleled for lower conduction loss.

6.2.5 Conclusion

The stringent efficiency and power density requirements of data center power supplies are met in theory and practice using the advantages of wide bandgap semiconductors, specifically silicon carbide types. Cascode SiC FETs also have the benefits of an easy, noncritical gate drive and robust short circuit and avalanche voltage ratings and are available in a range of through-hole and SMT packages.

6.3 Renewable Energy Systems

Christoph Ihm and Tobias Asshauer

The future belongs to renewable energies. The buildout of renewables and making them more efficient is a paramount task in global plans to decarbonize the world. In order to achieve net-zero CO_2 emissions by 2050, the share of renewables needs to

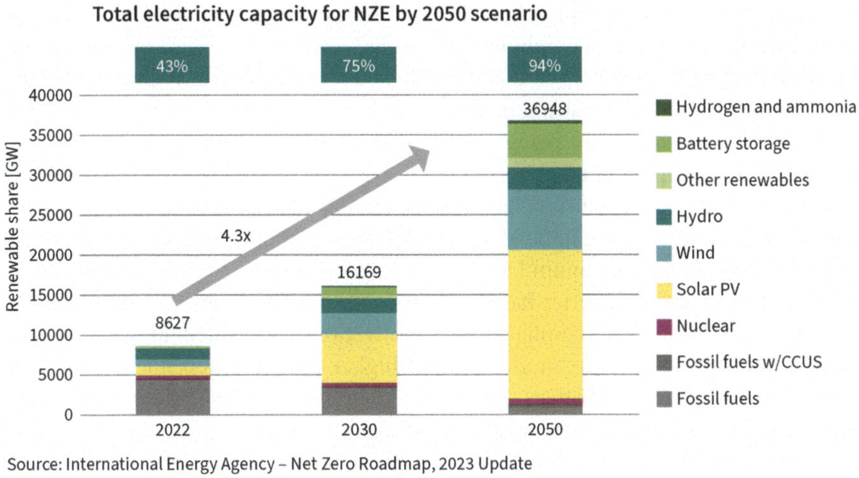

Fig. 6.18 Total electricity capacity for NZE by 2050 scenario

grow from 43% in 2022 to 94% in 2050, while at the same time, electricity capacity has to more than quadruple from 8.6 to 37 TW to serve future energy demand (source: IEA, 2023). This increase needs to be achieved by a significant buildout of multiple renewable energy systems. Solar/photovoltaic (PV), wind, and battery energy storage are expected to be the largest contributors comprising 51%, 21%, and 11%, respectively, of the total electricity capacity by 2050 (Fig. 6.18).

There has already been a huge rise in solar/PV, wind, and battery energy storage systems in recent years. Meanwhile, solar PV and onshore wind are the lowest cost options in most countries, granting them a key role in the ambition to make the world greener. Right from the start, semiconductors have been essential for the ramp-up of renewables and making them more and more efficient. And SiC provides a large potential to improve many applications along the green energy chain.

6.3.1 PV Inverter: SiC Providing Strong System Benefits

Photovoltaic systems are one of the most demanded applications to address carbon reduction and increase the share of renewable energy in the grid. In such a system, solar modules supply power in the form of direct current (DC), which has to be converted in a PV inverter to alternating current (AC) before it can be fed into the grid and be used or transmitted to the point of use. Therefore, PV inverters can be called power conversion systems, and depending on their implementation, they fall into the categories of microinverters, string inverters, hybrid inverters, or central inverters. Microinverters are directly attached to each solar module, while string, hybrid, and central inverters are connected to one or several strings of solar modules. Hybrid inverters are a strongly growing variant of string inverters that integrate a power conversion stage for linking a battery energy storage system. One variant of string inverter systems is a combination with DC optimizers. These optimizers are directly attached to each solar module and in such a setup do not require a DC/DC conversion stage in the PV string inverter.

Photovoltaic systems are being used in residential, commercial and industrial, and utility-scale deployments.

In residential deployments, solar modules are installed on a rooftop with a typically installed DC power of up to 10 kW. Installation variants include microinverter-based systems with a maximum DC input voltage of up to 60 V per microinverter/solar module or string inverter-based solutions with a DC input voltage of 600 V (one-phase) or 1000 V (three-phase). Key motivations for residential use are lower energy costs, the use of green energy, and higher energy independence. A clear trend is the integration of battery storage, EV charging, and heat pump systems, all managed by a smart energy management system.

In commercial and industrial (C&I) use cases, solar modules are also installed mostly on rooftops in small to medium deployments up to 1 MW connected to string inverters with a 1000 V DC input voltage. By contrast, large-scale C&I systems are mainly ground based with a typical power range of 1–5 MW and connected to 1500 V DC string inverters. Similar to residential deployments, the motivation includes a reduction of energy cost and higher energy independence but also a reduction in the CO_2 footprint. To support a lower CO_2 footprint, C&I companies are also enforcing the buildout of EV charging infrastructure. In addition, energy storage systems are also strongly adopted for higher self-consumption and lower energy cost. The operation of the whole installation is optimized by an energy management system that monitors and coordinates the energy flow.

In the utility scale, solar modules are ground based or in some cases floating on lakes and generate several megawatts up to even hundreds of megawatts of power. There, central inverters with a capacity of several megawatts or a cluster of high-power string inverters are being utilized. Key trends include enabling lower levelized cost of electricity (LCOE), and maximum flexibility for grid stability requirements.

At the core of the PV inverter are the power conversion blocks. In a hybrid inverter design, for example, three power conversion blocks are being used: a DC-DC converter to stabilize the variable DC voltage from a PV panel to a fixed DC link voltage, a bidirectional DC-DC converter that feeds the battery during charging, and a DC-AC converter supplied from the PV panels or the battery to deliver grid-compatible, single- or three-phase AC power.

Efficient and reliable power semiconductors and inverter technologies are needed for the power conversion and transmission with minimal losses. This is enabled by a broad portfolio of components to build inverters ranging from just a few hundred watts to several kilowatts for residential use up to several megawatts for the commercial- and utility-scale markets. It includes discrete IGBTs, MOSFETs, CoolGaN™, CoolSiC™ MOSFETs, and CoolSiC Schottky diodes in various package options, be it TO247 discrete switches or highly integrated Easy modules.

Silicon carbide solutions provide superior power density and efficiency for PV inverters. In the DC/DC booster stage, SiC drives lower system cost, system weight, and size through a higher switching frequency. In the DC/DC battery stage, SiC enables reduced cooling efforts for fan-less design, as well as an improved partial load efficiency. In the inverter stage, the increased switching frequency in 3-level designs drives lower inductance size and weight and reduces the AC filter size. CoolSiC™ MOSFETs from Infineon make photovoltaic inverters smaller, lighter, and more powerful. They allow for higher currents and reduced heat loss, enabling higher power density and smaller form factors for PV inverters, thereby strongly reducing cost in USD per watt.

6.3.2 *Wind Power Generation: IGBT-Driven Today but an Opportunity for SiC*

Power generation through wind turbines—onshore and offshore—is another critical contributor to the green energy transition. The vast majority of installed wind capacity today is onshore. However, while the offshore share is rather small, it is expected to grow strongly as large offshore wind farms are planned to be installed. Onshore turbines have the advantage of lower construction cost and are closer to the electricity grid but have an impact on the environment. Offshore turbines are larger power-rated turbines and have lower payback time but face higher construction cost. Wind turbines are operating under extremely harsh conditions and therefore require high reliability and robustness to ensure a long life cycle. Key trends include a shift to higher power ratings, higher system voltages, further improvements in efficiency and reliability, and system cost optimizations.

The power generated in the wind turbine generator is flowing through a converter before being transmitted into the grid. Commonly used converter topologies are doubly fed induction generator (DFIG) and full converter-based wind turbines. In a DFIG wind turbine, the stator of the generator is directly connected to the grid, while the rotor is connected to the grid through a back-to-back power converter. In this topology, only up to one-third of the wind turbine's power flows through the power converter. DFIG converters are typically deployed onshore in a power range up to 6 MW.

In a full converter wind turbine, the generator is fully decoupled from the grid through a back-to-back power converter, and the entire wind turbine power flows through the converter. Full converters are installed onshore and offshore and are available in a power range up to 15 MW. The AC-DC converter on the generator side is typically referred to as the machine-side converter (MSC), while the DC-AC converter on the grid side is typically named the line-side converter (LSC).

Wind power converters need to enable efficient energy generation, meet quality standards to withstand extremely harsh conditions, achieve highest availability to contribute to grid stability, and provide an extended lifetime. At Infineon, this is

	 Residential	 Commercial & industrial			 Utility scale	
Sub-segment		Small	Medium	Large	Small	Large
Characteristic: Place of installation	Rooftop or building integrated in the future (like walls made by panels)	Rooftop or building integrated in the future (like walls made by panels)		Ground based	Ground based, sometime floating (on lakes)	
Characteristic: Typical installed DC power	< 10 kW	10 kW to 100 kW	100 kW to 1 MW	1 MW to 5 MW	5–20 MW	> 20 MW
Characteristic: Maximum DC input voltage	60 V / 600 V / 1000 V	1000 V	1000 V / 1100 V	1500 V	1500 V	
Characteristic: Typical total system cost in 2022 [USD/W]	1.51	0.89			0.76	
Today's business model	Saving of electricity bill through own consumption of electricity generated from PV installation			Revenue generation through selling of electricity from PV installation		

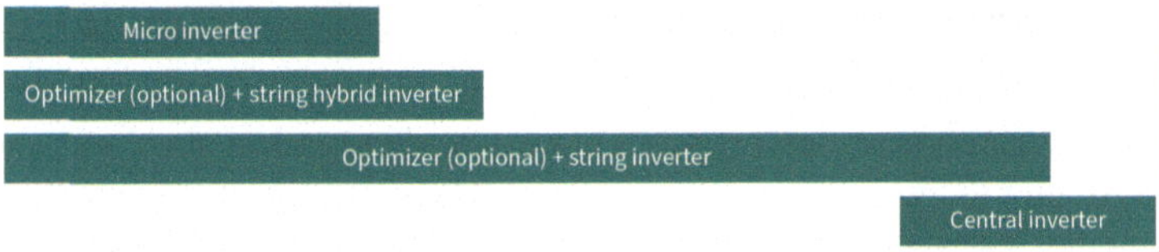

Fig. 6.19 Photovoltaic inverter—market segmentation

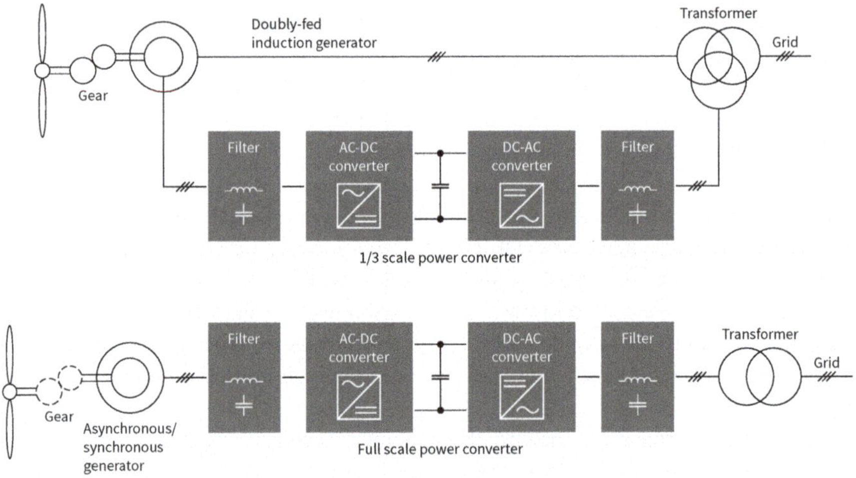

Fig. 6.20 Partial converter (DFIG) and full converter. (Used for IG and PMSG)

achieved today through a broad and leading portfolio of high-power semiconductor modules combing the latest IGBT chip technology, advanced .XT joining technology, and state-of-the-art package technology such as Prime PACK and XHP 2 in a cost optimized way (Figs. 6.19 and 6.20).

Going forward, SiC is expected to provide additional benefits in terms of higher efficiency and power density improvements. Efficiency benefits will be highest in offshore and high wind onshore projects especially under partial load conditions. Power density benefits are driven through a higher switching frequency which may result in a smaller converter size and system cost savings. SiC adoption may also lead to less cooling effort and lower system weight.

6.3.3 SiC Solutions Boost Efficiency of Energy Storage Systems

Renewables are the energy of the future, and their efficient use together with battery energy storage systems (BESS) is key to the global energy transition.

Globally, power systems are undergoing a pivotal phase of development. The exponential growth in renewable energy installations in the past decade, especially from wind and solar, that has exposed the grid infrastructure makes delivering power reliability, where and when it is needed, a bigger challenge than ever before.

Energy storage systems provide a wide array of technological approaches to manage our supply-demand situation and to create a more resilient energy infrastructure and bring cost savings to utilities and consumers. Energy storage is crucial for managing the irregularity of renewable energy sources. It stores surplus energy produced during periods of high production and releases it during times of high demand or when renewable sources are not generating electricity.

Battery-based ESS technology can respond to power dropouts in under a second, making use of clean energy sourced from collocated solar or wind plants. In such commercial and utility before-the-meter cases, ESS functions as energy backup, grid frequency stabilization, grid voltage stabilization, and black start capability. In behind-the-meter residential application, use cases are the renewable integration from photovoltaic, smart home integration, microgrids, and energy buffer for EV charging infrastructure.

Product competitiveness is affected by energy storage systems' power density and efficiency, which in turn affect product cost. Energy conversion efficiency is directly affected by the size, weight, and cost of energy storage systems.

Centralized and Smart String ESS Power Conversion System architectures are dominating the commercial and utility ESS market. Especially for string power conversion systems up to 300 kW, AC/DC, such as three-level NPC1, NPC2, and ANPC, and DC/DC topologies, such as DAB and LLC, request high switching frequencies to increase the power density besides high efficiency.

SiC MOSFETs up to 1200 V rating can be used directly in the DC-DC boost stage up to 1000 V and in the downstream single-phase or three-phase DC-AC inverter. In large commercial/utility installation with up to 1500 VDC output, SiC MOSFETs can still be used in a DC-DC three-level boost arrangement, keeping the MOSFET voltage stress below the 1200 V rating. Subsequent three-phase inverters can be multilevel types, where the voltage is shared across series switches, again

Topology		Properties	Products
DAB	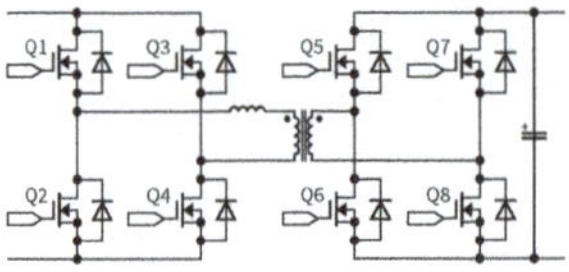	- High switching frequency possible to increase power density - Capable of operating in wide input and output voltage variation condition (with reduced efficiency) - Efficiency ~97% possible - Easy to parallel multiple modules - Preferable to use SiC or GaN - Suitable for HV batteries	**Discretes IGBT** - IKWH50N65EH7 (20 kW) - IKWH100N65EH7 (50 kW) - IKZA75N65EH7 (50 kW) - IKZA100N65EH7 (75 kW) - IKY120N120CH7 (100–150 kW) - IKY140N120CH7 (100–150 kW) **Discretes SIC** - IMW65R027M1H (20–200 kW) **Modules IGBT** - FF450R12ME4 (200 kW) - FF450R12KE7 (200 kW) - FF450R12ME7_B11 (200 kW) **Modules SIC** - FF08MR12W1MA1_B11 (200 kW) - FF6MR12W2M1H_B11 (200 kW)
LLC / PS LLC	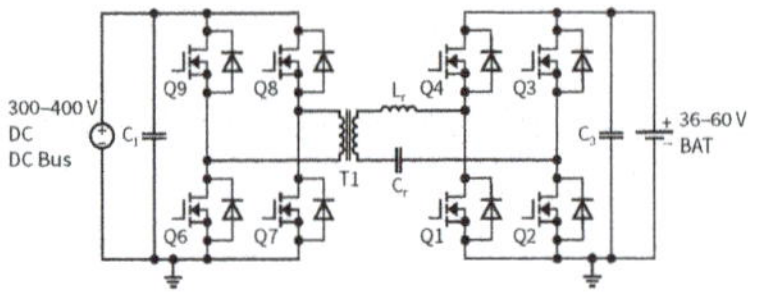	- ~ 95% efficiency - High switching frequency possible to increase power density - Slightly complex topology for control - In cost sensitive applications more suitable for narrow voltage range operation - For wide input/output voltage range operation, need to use GaN or SiC switches - Suitable for LV and HV batteries	
Active clamp current fed full bridge	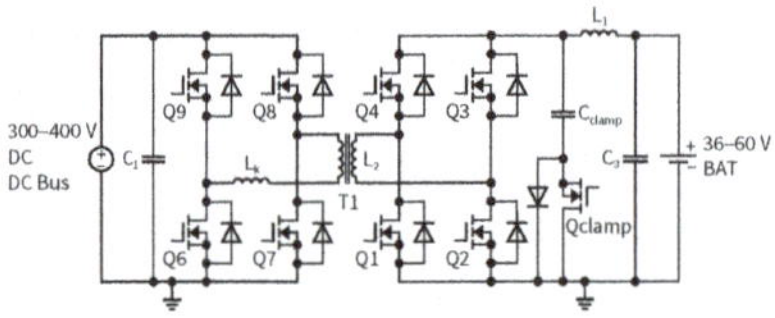	- ~ 96% efficiency - Easy system paralleling - Si MOSFETs can be used - Low power winding at startup needed - Switching frequency limited to ~150 kHz - Suitable for LV batteries	

Fig. 6.21 DC-DC topologies (< ~300 kW)

allowing 1200 V SiC MOSFETs to be used. DC-AC inverter switching frequency is not usually pushed very high; even though SiC can switch at MHz rates efficiently, inverters only have magnetic components for filtering rather than energy storage and coupling, so magnetics do not scale down as dramatically as in AC-DC or DC-DC converters, with their large transformers and storage chokes. SiC inverters switching around 100 kHz are therefore a good choice for very low dynamic and static loss, along with reasonable size filter components (Figs. 6.21 and 6.22).

Discrete CoolSiC MOSFET switches can cover applications as far as 30 kW, whereas power modules such as Easy CoolSiC should start being considered for systems beyond 30 kW.

By utilizing Infineon's silicon carbide devices, energy storage products achieve enhanced compactness and flexibility, along with notably improved efficiency and reduced losses.

This results in decreased costs associated with heat dissipation in systems, ultimately aiding end users in enhancing operational stability and expediting their return on investment.

Infineon, with over two decades of expertise in SiC product development and application, has been tirelessly striving to create increasingly advanced SiC products. Infineon's 1200 V CoolSiC MOSFETs have a high-power density, allowing them to decrease losses by 50% and give around 2% more energy without needing

<table>
<tr><th></th><th>Topology</th><th>Properties</th></tr>
<tr><td>DAB</td><td>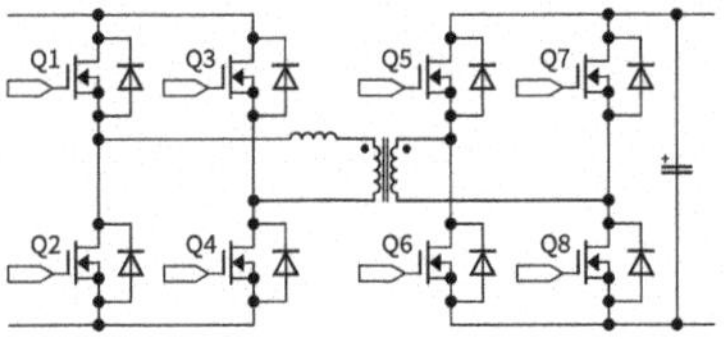
</td><td>- High switching frequency possible to increase power density
- Capable of operating in wide input and output voltage variation condition (with reduced efficiency)
- Efficiency ~97% possible
- Easy to parallel multiple modules
- Preferable to use SiC or GaN
- Suitable for HV batteries</td></tr>
<tr><td>LLC / PS LLC</td><td>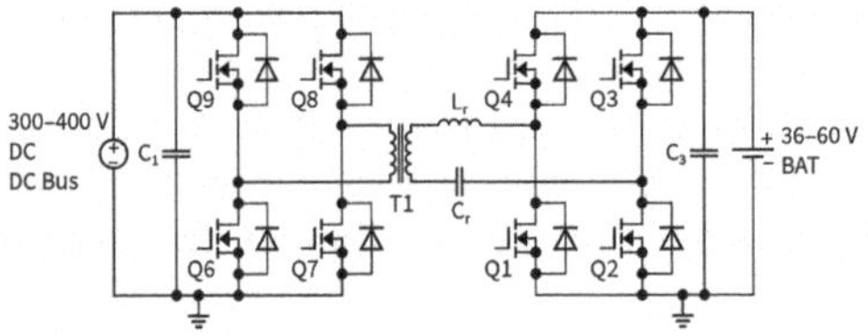
</td><td>- ~ 95% efficiency
- High switching frequency possible to increase power density
- Slightly complex topology for control
- In cost sensitive applications more suitable for narrow voltage range operation
- For wide input/output voltage range operation, need to use GaN or SiC switches
- Suitable for LV and HV batteries</td></tr>
<tr><td>Active clamp current fed full bridge</td><td>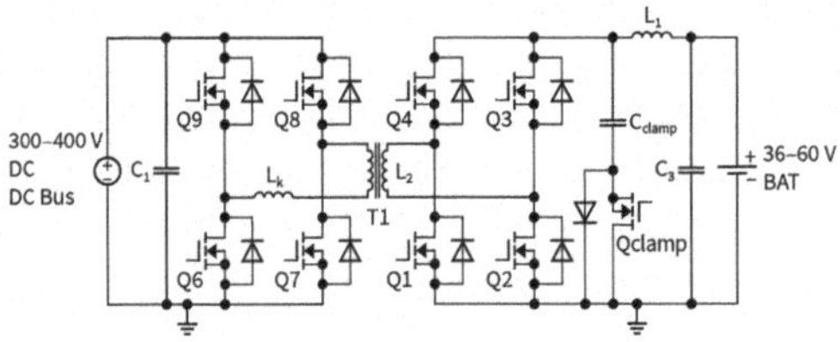
</td><td>- ~ 96% efficiency
- Easy system paralleling
- Si MOSFETs can be used
- Low power winding at startup needed
- Switching frequency limited to ~150 kHz
- Suitable for LV batteries</td></tr>
</table>

Fig. 6.22 AC-DC topologies

a larger battery. This is particularly advantageous for energy storage solutions that require high performance, lightweight, and compactness.

6.4 Power SiC Patent Landscape

Rémi Comyn

Patents include technical, market, and legal information. Accordingly, patent analysis can provide meaningful information about the competitors, the level of their R&D investment, their technological roadmap, the future products and targeted markets, and the risks and opportunities existing in the field.

KnowMade has been researching and monitoring the intellectual property (IP) landscape across the whole power silicon carbide (SiC) value chain (>19,000 inventions): from bulk SiC crystal to SiC epiwafers, devices, packaging, modules, circuits, and applications. This chapter provides insights from this research. There is no mention of patent numbers that could link a patent to its owner.

6.4.1 *The Rise of Chinese Players in the Power SiC IP Landscape*

Until the mid-2010s, Japanese players maintained a strong inventive leadership in power SiC (Fig. 6.23). Since 2015, Chinese research organizations and companies have progressively ramped up their inventive activity in the field. In 2023, Chinese players have produced more than 70% of all power SiC inventions. The rise of China's IP has been observed in most sectors of the semiconductor industry monitored by KnowMade. It stems from China's ambition to leapfrog the global competition, leveraging tax policies, subsidies, and loans to support the emergence of a domestic supply chain for various semiconductor technologies, including compound semiconductors. As part of this strategy, the Chinese government has set up incentives encouraging patenting activities. In the nascent power GaN and power SiC industries, Chinese patenting activities successfully covered the complete supply chain with large amounts of patents. In 2022, KnowMade released an analysis of the emerging Chinese SiC ecosystem based on the patenting activities of Chinese players across the SiC supply chain [6]. Even though the quality of such patents may be questioned, this situation brings about new challenges for global competitors in the semiconductor market looking to develop their manufacturing and business activities in China.

Chinese SiC patents are equally distributed between the materials (including bulk SiC bare and epitaxial wafers) and the power devices (SiC MOSFET, SiC diodes, etc.). Since more than 80% of the SiC substrate market has been controlled by three main companies outside China (Wolfspeed, Coherent, ROHM/SiC Crystal),

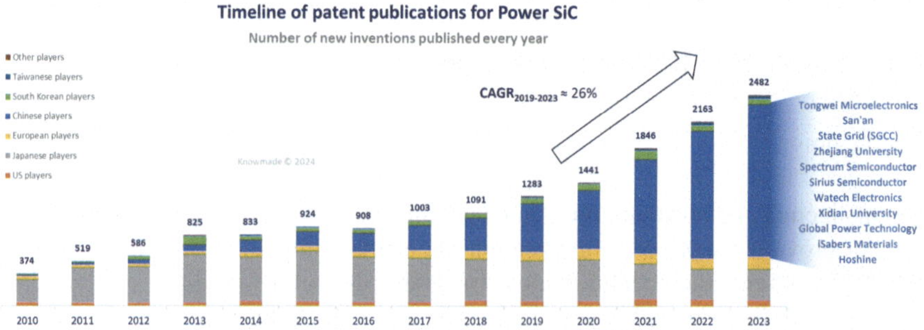

Fig. 6.23 Inventive activity in the power SiC IP landscape since 2010

the development of a domestic SiC substrate industry has become increasingly critical for Beijing to avoid supply issues and reduce its dependency on foreign countries. However, there was no standard growth furnace available for SiC crystal production, and key players in this field have maintained relatively confidential IP strategies (e.g., using trade secrets instead of patents) to protect their technological lead. Accordingly, about half of the Chinese inventions related to SiC materials have focused on raw materials, apparatus, and tools for SiC crystal growth. In order to accelerate the development of a domestic SiC substrate technology, Chinese players have been using multiple levers [6]:

- The transfer of technology, often with the transfer of patent rights, from Chinese research organizations pioneering SiC crystal growth technology since the 2000s (e.g., CETC, Shandong University, SICCAS) to industrial players. It led to the emergence of several domestic industrial players since the mid-2000s (Tankeblue, Synlight Crystal, SICC, etc.). Furthermore, certain Chinese integrated device manufacturers (IDM) have been backed by Chinese research organizations to develop SiC substrate technology (e.g., CENGOL by SICCAS).
- The acquisition of foreign startup companies pioneering SiC crystal growth technology (e.g., Sweden's Norstel acquired in 2017 by An Xin Capital, before its acquisition by STMicroelectronics in 2019).
- The establishment of joint ventures (JV) with foreign companies (e.g., Ascatron and Bronze Technologies' established Basic Semiconductors in 2017).
- The attempt of Chinese players mastering the sapphire substrate technology (e.g., Bright Semiconductor, JSG) to leverage their know-how and expertise for the growth and processing of SiC substrates, just like the US company GT Advanced Technologies (GTAT).

The SiC substrate patent landscape has remained a very dynamic area in China so far. In 2023, additional domestic players have emerged in this space (Fig. 6.23), such as Tongwei Microelectronics, Hoshine, and iSabers materials, focusing on SiC engineered substrates. The continuing entrance of newcomers despite the small market size attests to the growing interest in SiC technology and confirms that Chinese players integrate IP as a fundamental part of their strategy to enter the power SiC market.

Historically dominated by research organizations (Xidian University, UESTC, IMECAS) or state-owned companies (CRRC, SGCC, CETC), the Chinese SiC device patent landscape has also become more dynamic lately [6]. Just like in the SiC substrate patent landscape, Chinese research organizations have been driving the emergence of new companies since 2015, joining early SiC players such as CENGOL, BASIC Semiconductors, Global Power Technology, and PN Junction Semiconductor. In recent years, the number of Chinese industrial players from other industries (optoelectronics, RF, GaN) entering the SiC device patent landscape has grown fast. A notable example of such player is San'an Optoelectronics (San'an) [7]. San'an implemented most strategies described earlier for the SiC substrate patent landscape. In addition to leveraging its know-how in compound semiconductor manufacturing, San'an acquired a SiC substrate startup company in 2020, Nortel

New Materials (Nortel), and then announced a JV for advancing high-volume 200 mm SiC device manufacturing with STMicroelectronics in 2023.

Chinese patenting activity related to modules and circuits took off in 2015. The patenting activity related to SiC circuits is strongly supported by Chinese academic players such as Nanjing University of Aeronautics and Astronautics, South China University of Technology, and Xi'an Jiaotong University. Power module market players Starpower and Macmic entered the SiC patent landscape in 2020 and 2021, respectively, but have filed relatively few patent applications for SiC specifically so far. Interestingly, compared with the SiC substrate IP space, there is a much denser collaboration network between Chinese IP players in the downstream supply chain (from devices to circuits), especially between research organizations and industrial players.

Accordingly, IP dynamics suggest that China has made an important progress in establishing a SiC ecosystem that will support the development of a complete domestic supply chain and that will secure their power SiC product lines. Interestingly, few Chinese players have notably extended their IP activities beyond the Chinese borders, demonstrating a limited interest from Chinese companies in expanding their business into other markets so far. Indeed, foreign patent applications account for less than 5% of all patent applications filed by Chinese patent applicants. This situation is expected to last, especially in a context of low domestic production capacity with respect to the potential size of the SiC market in China.

6.4.2 *The Role of IP in the Strategies and Development of Established SiC Companies*

As patents provide an exclusive right to control the use of an invention, a patent assignee may leverage them to stop or limit the activity of its competitors (manufacturing, using, selling products resulting from the invention). Importantly, this right is limited to a given geographic area. Therefore, players must file several patent applications to protect an invention in the main markets targeted by the resulting products. Eventually, the budget related to a patent portfolio may become significant for the company with increasing the number of patents, due to application fees (for pending patent applications) and maintenance fees (every year once the patent is granted). Therefore, patents relate not only to R&D investments but also to the investments supporting a business strategy.

A straightforward way to describe and start analyzing the current IP competition consists of comparing the number of granted patents and the number of pending patent applications assigned to each player, as shown in Fig. 6.24. Depending on their position in these graphs, IP players may be split into different categories:

- IP leaders (*in the top right corner*): They own the highest numbers of granted patents and are assigned to numerous pending patent applications, thus combin-

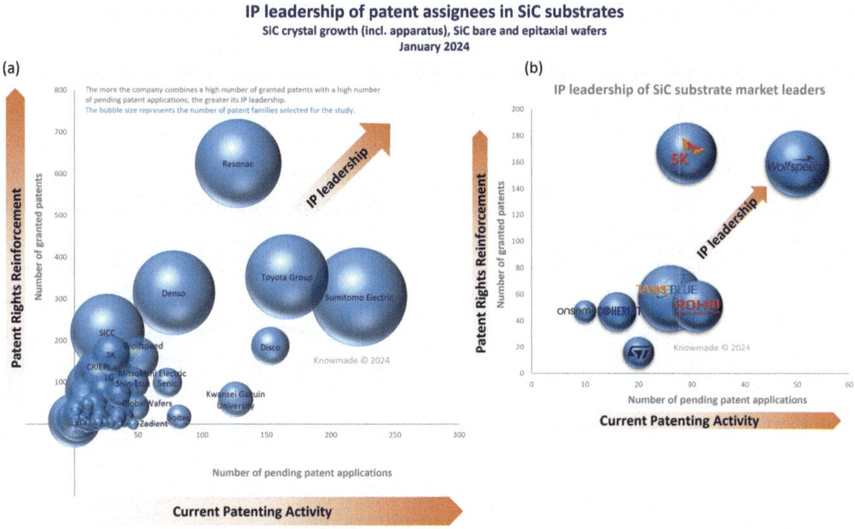

Fig. 6.24 (**a**) IP leadership of main patent assignees in the SiC substrate patent landscape and (**b**) IP leadership of the main SiC substrate market players in 2022. (According to McKinsey & Company and Yole Group [8])

ing a strong IP position and a significant potential for further strengthening this position (depending on how many pending patent applications will be granted).

- IP challengers (*in the bottom right corner*): They own limited numbers of granted patents compared to IP leaders, but they are assigned to the largest numbers of pending patent applications, which provide them with the potential to close the gap with leading IP players in terms of granted patents.
- Well-established IP players (*in the top left corner*): They own high numbers of granted patents but are currently assigned to limited numbers of pending patent applications. They maintain their IP rights alive but are less active in terms of inventive activity (number of new inventions).
- Noncompetitors/newcomers (*in the bottom left corner*): They own limited numbers of granted patents and pending patent applications.

This approach was applied to the main segments of the power SiC value chain: SiC substrates, devices, packaging and modules, circuits, and applications. It highlights different ecosystems emerging to support the fast-growing power SiC market as well as contrasted levels of IP competition between the segments.

In general, few Chinese players have emerged as well-established IP players, IP challengers, or IP leaders in the power SiC patent landscape, because of their limited patenting activity outside China: in most cases, a Chinese invention has been protected by only one domestic patent, in contrast with IP leaders filing multiple patent applications to protect their inventions in different regions (e.g., in the USA, Japan, China, South Korea, Taiwan, and Europe).

Figure 6.24a includes the various technologies covered in the SiC substrate patent landscape, to grow SiC crystal (PVT, LPE, CVD), to manufacture SiC bare and engineered wafers (slicing, finishing, bonding, layer transfer), and to provide high-quality SiC epitaxial layers thereon. The graph reflects the IP leadership of Japanese players in most aspects of SiC substrate technology (Toyota/Denso, Resonac, Sumitomo Electric, Disco) and the emergence of relatively new players with global IP strategies, in Europe (Soitec, Zadient Technologies), South Korea (Senic), and Taiwan (GlobalWafers).

Resonac—the merger of Showa Denko K.K. and Showa Denko Materials Co.—is far ahead of its competitors in terms of enforceable patents (granted patents). This IP leadership results from an intense patenting activity in the last decade, split between bulk SiC and SiC epitaxial wafers (accounting for 60% and 40% of its SiC substrate inventions, respectively). What is more, many of the inventions have been protected in most major geographic patenting areas, leading to a global IP leadership. In this strategy, Resonac successfully acquired key patents from historical Japanese players in the SiC substrate patent landscape, such as Bridgestone, Nippon Steel, and Hitachi Chemical, and engaged into many IP collaborations, especially with Toyota and Denso (adopting Resonac's SiC epitaxial wafers for its SiC power device technology in 2020) [9].

The notable IP position of Toyota group (including the patenting activities of Toyota Motor, Toyota CRDL, Toyota Industries, and Toyota Tsusho) and Denso in the SiC substrate patent landscape is analyzed as part of a broader strategy aiming to secure Toyota's future in the global electric vehicle (EV) market [10]. However, this strategy is also expected to benefit other OEMs (Denso's customers) looking for solid know-how and expertise all along the value chain of EV, from raw materials to advanced electronic systems. Lately, Toyota accelerated its patent filings in the SiC substrate IP space, in the context of a new development with Kwansei Gakuin University. The patent co-filings describe an approach to remove subsurface damage induced by mechanical processing, such as scratches, from SiC wafers, and create a defect-free buffer layer suitable for CVD growth of SiC epitaxial structures.

Figure 6.24b shows the discrepancy that exists between market leadership and IP leadership for SiC substrates. It suggests that early players in this space have mainly relied on confidential IP strategies (e.g., trade secrets) at a time when the market adoption of SiC technology was seen as a long-term possibility. As a result, they kept a strong technological lead in this field, which is well reflected by the current market situation, as newcomers keep struggling to catch up in terms of crystal quality and size of SiC wafers. So far, well-established market players have demonstrated limited IP activities to strengthen their patent portfolio, except for Wolfspeed and SK (including SK siltron, SKC, and SK Innovation). Interestingly, while the South Korean patenting activity has remained limited in the SiC patent landscape (accounting for $\approx$5% of all inventions worldwide), South Korean players have strongly focused their efforts on SiC substrate technology under the leadership of SK Group, LG Corporation (including LG Chem and LG Innotek), and POSCO [11].

In the South Korean SiC device patent landscape, Hyundai Motor and KERI have been leading the IP activity so far. Since January 2022, Hyundai Motor has increased by more than 60% its number of granted patents related to SiC power devices while maintaining a similar number of pending patent applications. However, despite this significant patenting activity, strongly focused on trench SiC MOSFET, Hyundai Motor has not emerged yet in the global IP competition for SiC power devices (Fig. 6.25). This situation reflects the much stronger IP competition occurring in the SiC power device IP space, in comparison with the current competition in the SiC substrate IP space.

In a previous IP report released in 2022, KnowMade pointed out that the global IP competition for SiC power devices was on the rise [12]. On the one hand, many well-established IP players and IP leaders, which used to focus on certain regions to protect their SiC inventions, were now patenting their inventions in additional geographical areas, especially in Europe and China. As electric vehicles were driving the emerging power SiC market, companies had been adapting their IP strategies accordingly. On the other hand, notable SiC device market players that had not been significantly involved in SiC-related patenting activities so far started protecting new inventions describing SiC power devices (e.g., STMicroelectronics, onsemi). Considering the level of investments that have been made to support the establishment of a robust power SiC supply chain, patents may be an important tool for early SiC companies to secure their market share as new competitors enter the field.

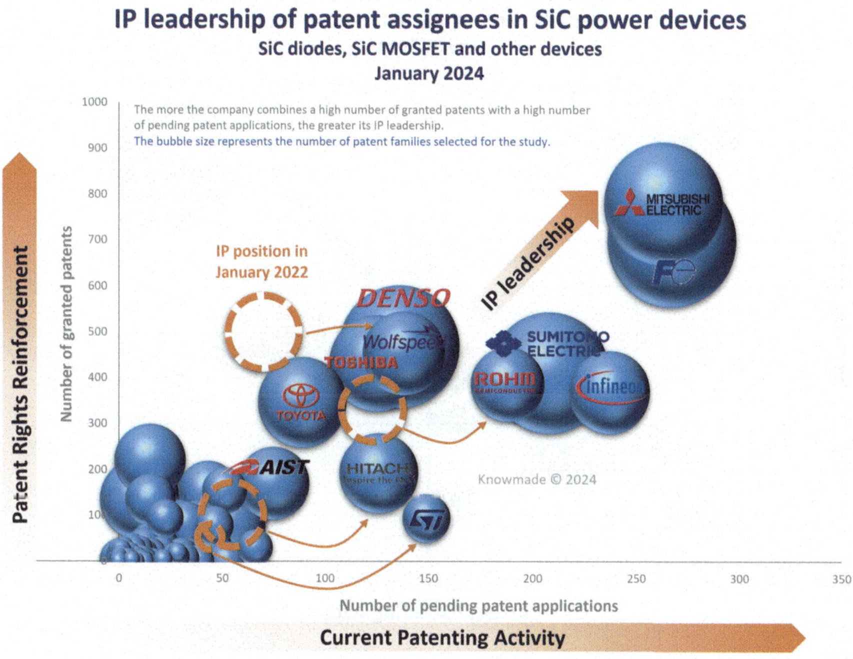

Fig. 6.25 IP leadership of main patent assignees in the SiC device patent landscape, highlighting the companies that have notably changed their IP position since January 2022. Note that generic patents, applicable to any of Si, GaN, or SiC semiconductors are not included in the graph

While most of the main IP players identified in Fig. 6.25 have significantly improved their IP position since 2022, in terms of granted patents, several companies have made notable moves in this IP space during the same period, according to the number of their patent applications. Importantly, most of them are leading market players (Wolfspeed, STMicroelectronics, ROHM) that had very contrasting IP positions as of 2022.

For instance, Wolfspeed has been one of the main well-established IP players in the SiC device patent landscape. However, many of its fundamental patents will expire in the next few years, so Wolfspeed will face the challenge of maintaining its dominant IP position [13]. What is more, the US company has strongly focused its marketing strategy on planar SiC MOSFET so far. Accordingly, Wolfspeed holds limited IP related to trench SiC MOSFET compared to the leading IP players in this field (Fuji Electric, Toyota/Denso, Sumitomo Electric, Infineon, ROHM, Mitsubishi Electric, etc.). Yet, making the parallel with Si power MOSFET and IGBTs, the SiC market may reorient to trench SiC MOSFET technology in the next few years, looking to further improvements in efficiency, yield, and cost [14]. Although Wolfspeed did not make any announcement to adopt this gate design in its next generations of SiC MOSFET, the company has already resumed its patenting activity in this space since 2022 (Fig. 6.26). Indeed, trench SiC MOSFET technology has become a very

Fig. 6.26 IP leadership of main SiC device market players in 2022. (According to McKinsey & Company and Yole Group [8]), for trench and planar SiC MOSFET. Note that SiC MOSFET patents applicable to any of planar or trench gate architectures are not included in this graph

competitive area in the SiC power device patent landscape, as illustrated by ROHM's IP trajectory in this space. Even though ROHM was already a well-established IP player for trench SiC MOSFET, the company has remarkably accelerated its patent filings since 2022 (Fig. 6.26). In general, patent applicants have focused on designing device structures that efficiently protect the gate oxide from high electric fields occurring at the bottom of the trench. However, it has become more and more difficult for IP challengers in this field to protect new gate trench designs.

Unlike Wolfspeed, STMicroelectronics was not well established in the SiC power device patent landscape in 2022. Therefore, the company has been facing the challenge to turn 30 years of SiC device R&D into valuable patents that address critical aspects of its current technology and that can be leveraged to maintain its early leadership in the power SiC market. Actually, STMicroelectronics was the first company targeted by a patent litigation in the SiC device patent landscape: in 2021, Purdue University sued the company for infringement of one of its fundamental patents filed in 2006 and describing a planar SiC MOSFET structure with a current spreading layer. Since 2022, STMicroelectronics have filed many patent applications in a view to improve its IP position in the SiC MOSFET patent landscape (Fig. 6.26). Most of these patent applications are still pending as of January 2024. Just like Wolfspeed, STMicroelectronics has focused its strategy on planar SiC MOSFET. Yet, unlike Wolfspeed, STMicroelectronics stated that it expects to maintain this strategy for at least the next two generations of devices. This statement is consistent with the absence of a significant patenting activity from STMicroelectronics in this space, contrasting with the high number of its patent filings related to the planar SiC MOSFET (Fig. 6.26).

In conclusion, the patent analysis of local ecosystems (China, South Korea, etc.) and the patent analysis of the global competition across the complete supply chain (from materials to devices, modules, and circuits) show that IP may impact the development of the SiC industry in many ways. The critical role of IP in developing the SiC industry is further confirmed by the strategy of companies pioneering the power SiC market. Today, most early SiC market players are IDM companies with a vertically integrated IP strategy: they build a patent portfolio covering most parts of the SiC supply chain. What is more, most early SiC substrate market players are IDM companies or are looking to adopt an IDM model for power SiC, acquiring early SiC device players with significant know-how and IP (e.g., SK/YesPowerTechnix) or establishing licensing agreements with key IP players in the SiC device patent landscape (e.g., Coherent/GE). Yet many companies, especially in the USA, apparently lack an IP strategy for SiC technology, despite unambiguous ambitions in the power SiC market. Accordingly, additional strategic moves such as IP licensing agreements, manufacturing partnerships, and M&A are still expected in the coming years, further reshaping the SiC IP landscape.

6.5 Design Next-Generation Data Center Cooling Systems with Silicon Carbide

Zan Huang

Data centers form the backbone of the digital world, housing the massive servers that power the Internet, cloud computing, and other data-driven services. As the demand for these services grows, so does the amount of energy they consume.

In the USA, data centers consumed an estimated 90 terawatt hours (TWh) of electricity in 2020, representing about 1.8% of total US electricity consumption. This is expected to grow to 140 TWh by 2030, an increase of 56%. A global increase in artificial intelligence (AI) applications is placing a further strain on data center power consumption, necessitating additional servers and, in turn, more power to run them. The global growth in data centers makes it imperative that system designers find new and novel ways to reduce power consumption while improving computing capacity.

Data centers are complicated systems and there are many ways to reduce overall power consumption. Among the various subsystems, data center cooling systems account for a significant portion of its total energy consumption, with estimates ranging from 20% to 45%. The exact percentage can vary based on the type of cooling system used, the climate of the location, and the IT equipment being cooled. The efficiency of cooling systems within data centers can be dramatically improved with silicon carbide leading to up to 50% reduction in system losses, achieving global efficiency standards, and enhancing system performance with reduced audible noise, faster rotation speeds, and more precise control.

6.5.1 Up to 60% Lower Losses System-Wide with 1200 V SiC MOSFETs

Data center cooling systems consist of an AC-DC (active front end (AFE)) stage followed by a DC-AC inverter stage. By using six 1200 V SiC MOSFETs (C3M0075120K) in place of IGBTs in the AFE of a three-phase-supplied, 11 kW cooling system, designers can achieve a 0.9% efficiency improvement. Additional improvement can be achieved when the inverter is also upgraded with drop-in 1200 V C3M0075120K SiC MOSFET.

Figs. 6.27 and 6.28 show the comparison of conduction loss and switching loss of a typical 1200 V SiC MOSFET and a 1200 V IGBT with the same rated current. Under lower loads, the SiC MOSFET conduction losses are as much as one-half that of the IGBT and offers up to 90% less switching losses than the IGBT by eliminating the turn-off tail current.

With lower switching losses, Wolfspeed SiC MOSFETs can operate at much higher switching frequencies than IGBTs, allowing system designers to use smaller

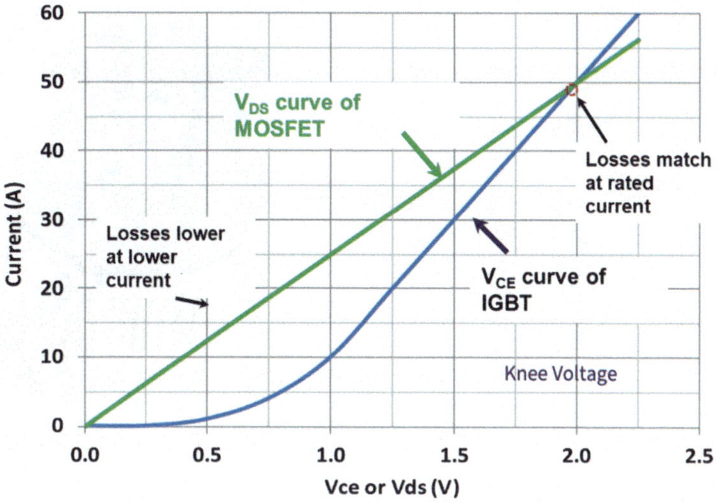

Fig. 6.27 Conduction loss, 1200 V SiC MOSFET versus IGBT

Fig. 6.28 Switching loss, 1200 V SiC MOSFET versus IGBT

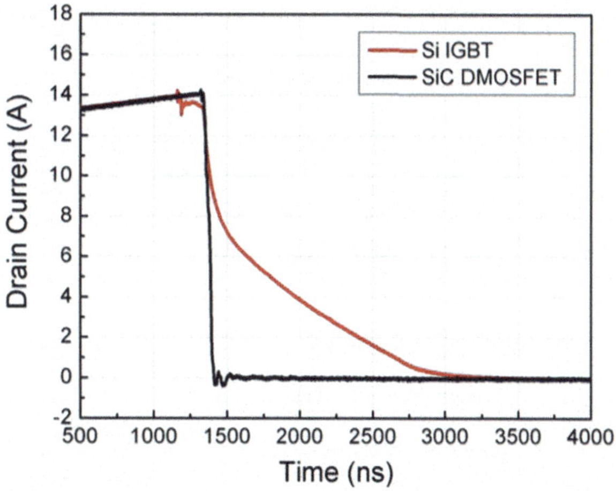

and lighter transformers in variable speed fan drives. Smaller transformers mean reduced copper losses, further contributing to energy efficiency improvements. Additionally, higher switching frequencies enable more precise speed control of the fan, ensuring optimal cooling performance at varying load conditions. With the improved light load conduction loss, SiC MOSFETs enable a much flatter, high-efficiency curve over all load conditions, further benefiting variable speed applications.

Shown in Fig. 6.29, Wolfspeed's newly released 11 kW High-Efficiency Three-Phase Motor Drive Inverter (CRD-11DA12N-K) was designed to test Wolfspeed's

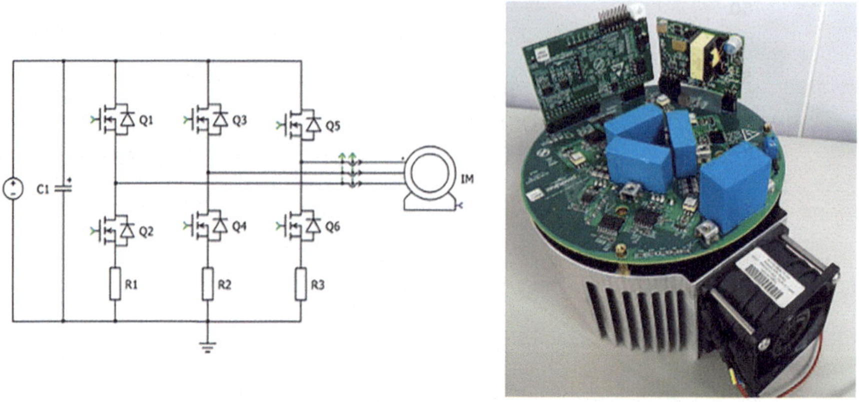

Fig. 6.29 11 kW Three-Phase Motor Drive Inverter with1200 V SiC MOSFET (C3M0075120K)

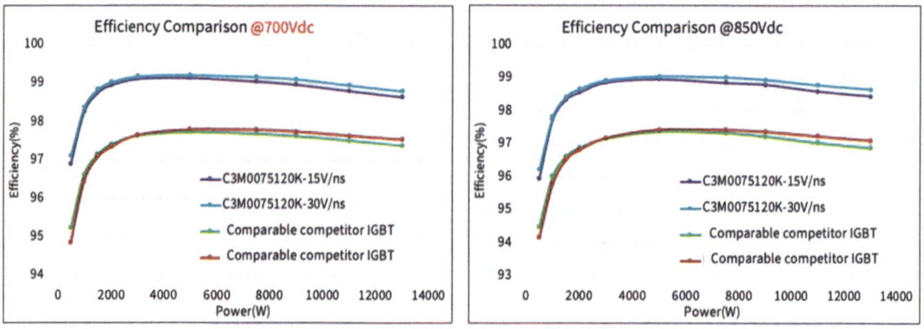

Fig. 6.30 Efficiency comparison @16 kHz

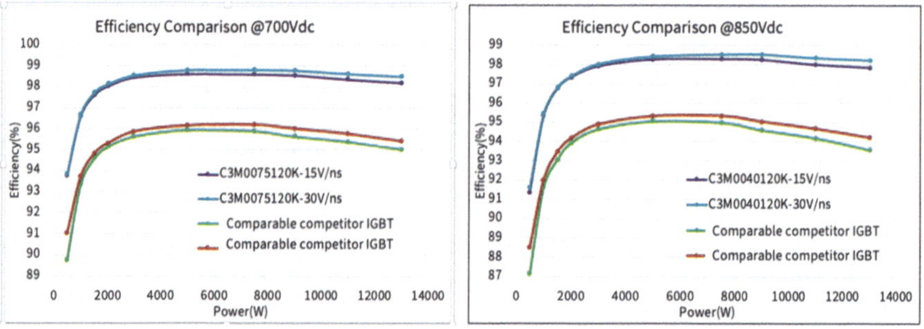

Fig. 6.31 Efficiency comparison @32 kHz

1200 V SiC MOSFETS in industrial heating and cooling applications. This design accomplishes peak efficiencies of 99% and a full load efficiency of 98.6%, with a wide operating voltage range of 550–850 VDC and closed-loop sensorless field-oriented control. The test results are shown in Figs. 6.30 and 6.31.

As shown above, by replacing silicon IGBTs with silicon carbide, running at 16 kHz, motor drive efficiency can be improved by >1.2%, a > 50% reduction in power losses. As IGBT device efficiency drops under high operating frequency, silicon carbide devices can easily double the operating frequency from 16 to 32 kHz to help shrink the sizes of passive components while maintaining >98% efficiency.

6.5.2 Achieve Higher Efficiency with Smaller Heat Sinks with SiC

Silicon carbides present a tremendous value at the system level in data center cooling systems, delivering high efficiency and power density with improved thermal resistance, and enabling overall smaller, less expensive cooling systems. For example, in a 25 kW compressor with a six-switch two-level active front end (AFE) switching at 45 kHz, designers can realize a 1.3% efficiency improvement with SiC in the front-end stage when benchmarking against a silicon switching at 20 kHz when discrete IGBTs are replaced with Wolfspeed's C3M0032120K discrete MOSFETs (1200 V/32 mOhm). A 1.1% efficiency improvement can be achieved in the inverter when Wolfspeed's 30 A rated full-bridge power module CCB032M12FM3 (Wolfpack) is conservatively benchmarked against a 50A and 100 A rated Si-IGBT module, both switching at 8 kHz. Together, these two changes yield an impressive 2.4% in efficiency improvement, with 50% reduction in losses system-wide.

One of the most noteworthy improvements that can be made in the inverter with silicon carbide is a significant reduction in system-generated heat, enabling designers to use smaller heat sinks and design overall smaller, lighter compressors for data center cooling systems, or enable more cooling using a similarly dimensioned compressor as before.

Results in the graphs above have been achieved with a 77% smaller heat sink for the silicon carbide inverter as compared to IGBT. Despite a much bigger heat sink for Si modules, the 50 A IGBT module's junction temperature is still significantly above the 150 °C temperature limit, but the 32 A SiC module and the 100 A IGBT module end up at the same junction temperature of around 129 °C. Also noteworthy is that efficiency in the silicon carbide inverter increases by 1.1%. Additionally, the efficiency improvement above is not only at peak loads but also at partial loads. At some partial loads, the efficiency improvement is higher which is ideally suited to the typical load profiles of data center cooling systems.

In summary, using a reduced and more optimized heat sink with silicon carbide in a three-phase-supplied, 25 kW system results in an overall 2.4% efficiency improvement with a 600 W reduction in losses while enabling smaller, lighter overall cooling systems (Fig. 6.32).

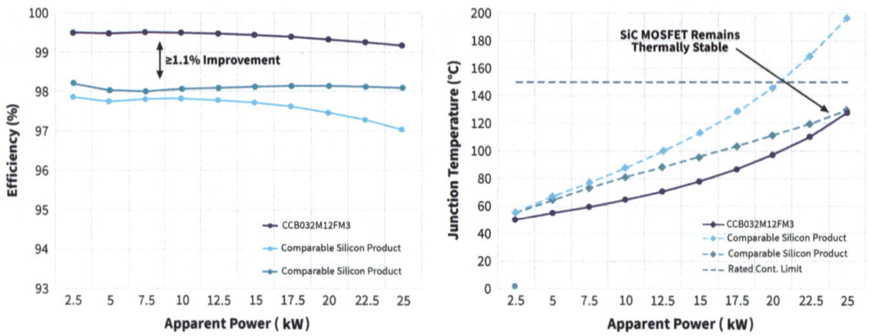

Fig. 6.32 25 kW inverter, Fsw = 8 kHz, larger Si IGBT heat sink: 1.37 L (0.7 °C/W), smaller SiC heat sink 0.8 L (0.99 °C/W)

6.5.3 In Conclusion

In conclusion, replacing traditional silicon carbide IGBTs with Wolfspeed silicon carbide 1200 V MOSFETS and power modules in 11 kW and 25 kW data center cooling systems can dramatically improve the overall system efficiency, delivering up to 2.4% increases in the overall system efficiency. High-efficiency improvement at higher power levels is possible throughout the load profile, resulting in massive energy savings. Silicon carbide also offers improved power density due to smaller passive components and heat sinks and leads to overall system cost and size optimization. Additionally, high junction temperature possibilities and improved thermal dissipation of SiC devices together with lower losses allow designers to build more compact systems, enabling an easy integration of drives and motors and ushering in the next generation of efficient data center cooling systems.

6.6 Spice Simulation and Models

Didier Balocco and James Victory

Circuit simulation was invented by the University of California, Berkeley, in the 1970s through a program called SPICE [15], acronym of "Simulation Program with IC Emphasis." The goal was to facilitate integrated circuit design by being able to predict the performance before having the first silicon fabricated.

SPICE includes library of predetermined devices to simulate electronic components such as resistors, capacitors, inductors, diodes, and bipolar or field-effect transistors. Each device has a model with a set of pre-defined equations and parameters to be able to customize and reproduce the components' performances.

To create a dedicated component model, the standard approach is to deploy curve fitting to extract the model parameters from measurement.

The last release made by the University of California, Berkeley, was in 1993. Since then, many public and private organizations have used the Berkeley SPICE code as a basis to create more advanced EDA platforms. As the SPICE simulators have evolved over decades, so have the component model libraries included within the simulators.

6.6.1 Power Devices and SPICE Models

The original models in Berkeley SPICE for MOSFETs were made for lateral devices in IC technologies. They are mainly valid for silicon material.

Even though, as previously mentioned, SPICE and its models have evolved over time, there has been a large negligence in power semiconductor models. Modern power components are built with complex structures (Fig. 6.33) to improve the component's performance in power applications. Such structures are not covered by the traditional models made for design in IC technologies.

For example, most power devices are vertical structures due to the high-voltage blocking requirements. For low- and medium-voltage (20–150 V) silicon MOSFETs, the gate structure moved from planar (Fig. 6.34) to trench to help reduce Miller capacitances and increase cell density for lower $R_{DS(on)}$. Further charge balanced, shielded gate transistors enable improved $R_{DS(on)}$ versus BV trade-offs while building in damping networks to improve circuit performance.

For silicon high-voltage MOSFETs, the superjunction MOSFET with deep charge balanced *P* and *N* pillar regions significantly improved the $R_{DS(on)}$ versus BV trade-off. With *P* and *N* pillar, during off state, the electric filed is horizontal instead of vertical. This helps to reduce the wafer thickness and the $R_{DS(on)}$.

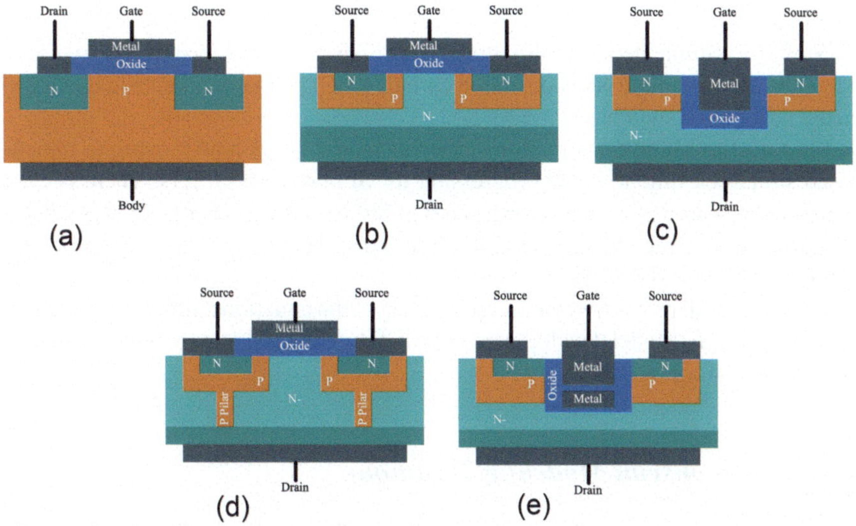

Fig. 6.33 Different types of MOSFET structures: (**a**) planar horizontal, (**b**) planar vertical, (**c**) trench, (**d**) superjunction, and (**e**) shielded trench

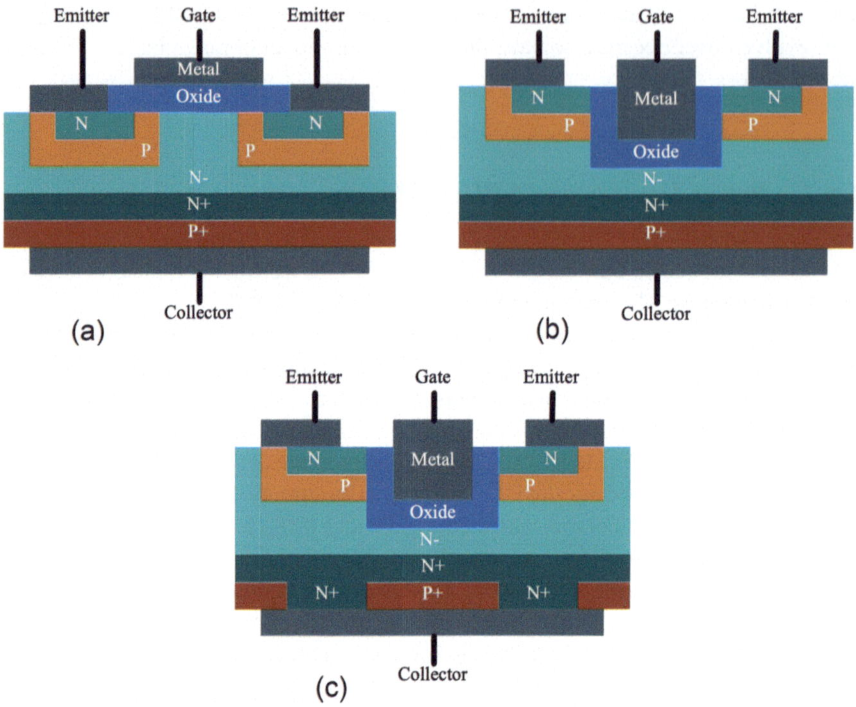

Fig. 6.34 Different types of IGBT structures: (**a**) planar, (**b**) trench, and (**c**) reverse conducting

The isolated gate bipolar transistor (IGBT) was invented in the 1990s. The IGBT structure (Fig. 6.34) has some advantages in power density for low-frequency operation.

Now, for simultaneous high-voltage and high-frequency operation, two new materials have been introduced, silicon carbide (SiC) and gallium nitride (GaN), to create new power switches.

For SiC, the MOSFET structure started as planar but is now moving to trench-based structures similar to the progression for Si power MOSFETs. There is even significant research into trench superjunction SiC MOSFETs. For GaN, it is a field-effect transistor with a heterogenic structure. This structure is called high-electron-mobility transistor (HEMT).

All these modern improvements are not part of the equations' library included by default in SPICE. Model development in SPICE has focused mainly on IC technologies chasing Moore's law.

6.6.2 New onsemi Modeling Technique

onsemi has introduced the unique physical and scalable SPICE modeling method for power devices in ISPSD 2016 [16, 17]. This technique associates process and layout parameters and dimensions to physics-based device equations by introducing new set of functions to capture modern device constructs such as trench gate, shielded gate, and superjunction (Fig. 6.35).

Such equations capture nonlinear current and charge behavior over voltage and temperature in all the various zones of modern power devices. Advanced effects such as reverse recovery are treated in detail.

The onsemi modeling approach is also applicable to wide bandgap devices. The direct link between layout and process parameter and the SPICE models enables design optimization. The models are used in application performance simulation during technology development and for the proliferation of new products.

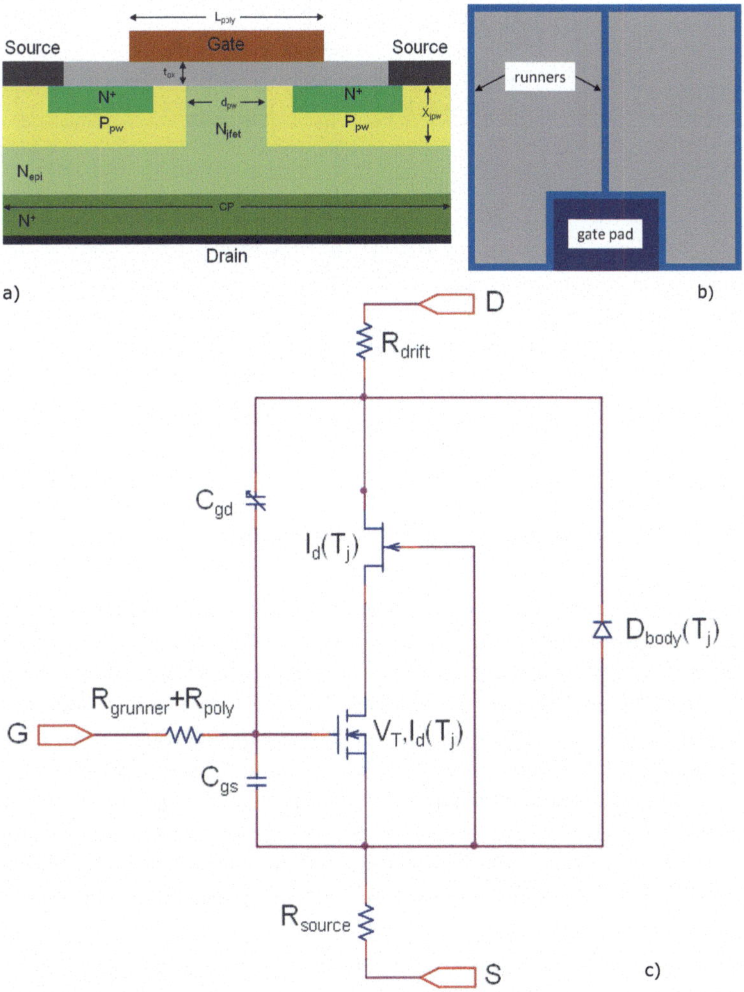

Fig. 6.35 (**a**) SiC MOSFET cross section, (**b**) SiC MOSFET typical layout, and (**c**) SiC MOSFET subcircuit model

SPICE JFET equation:

$$I_D = beta\left[2\left(V_{gs} - vto\right) - V_{ds}\right]V_{ds}\left(1 + \lambda V_{ds}\right)$$

Physical and scalable equations:

$$vto = \phi - \left(\frac{d_{pw}}{2\alpha}\right)^2$$

$$beta = \frac{2H_{bayeff}}{X_{jpw}\rho\left(-vto\right)}\left(\frac{d_{pw}}{2} - \alpha\sqrt{\phi}\right)$$

$$\phi = \phi_t \log\left(\frac{N_{jfet}P_{pw}}{n_i^2}\right)$$

$$\alpha = \sqrt{\frac{2\varepsilon_{Si}P_{pw}}{qN_{jfet}\left(N_{jfet} + P_{pw}\right)}}$$

$$\rho = \frac{1}{qN_{jfet}\mu}$$

where new technology and geometry parameters are introduced.

6.6.3 Device Simulation and Loss Extraction

It is possible to simulate any device characteristic with circuit schematics that replicate the real-world operation. The two major device characteristics that affect losses in power application are on-region conduction and capacitance. In power electronic applications, the two main loss mechanisms stem from conduction losses and switching losses that are induced by the device capacitance.

Understanding how losses vary with temperature is very important in power electronics. Relying on time-consuming lab measurements is very inefficient. Given reliable and accurate models, simulation is much faster. For example, simulation does not need temperature stabilization time/duration between two successive temperature points. In simulation, the junction temperature is a simple set in a virtual, instantaneous environment.

Figures 6.36 and 6.37 show some examples of circuits to extract those characteristics.

To compare devices between various manufacturers, a standard method has been defined to measure switching losses. This method is called "double pulse tester." This setup can also be simulated to obtain the losses.

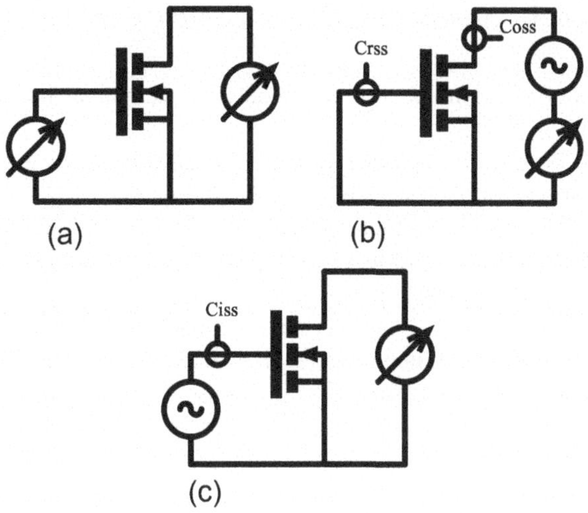

Fig. 6.36 (**a**) On-region extraction, (**b**) Coss and Crss extraction, and (**c**) Ciss extraction

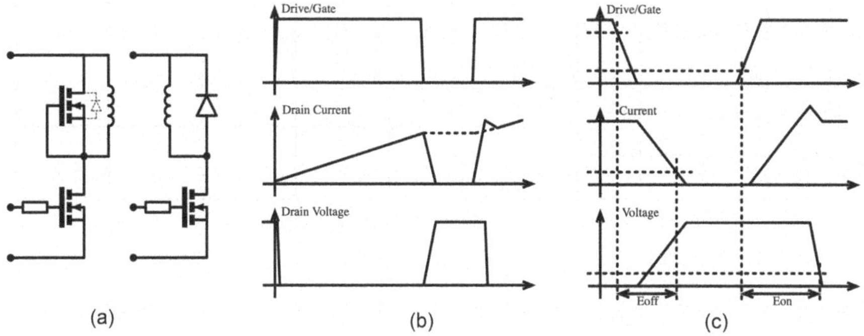

Fig. 6.37 (**a**) Double pulse tester schematic half-bridge versus quarter-bridge, (**b**) double pulse principle, and (**c**) measurement definitions

6.6.4 Limits of Double Pulse Tester Method

First, if we analyze modern power topology, there are various operating mode. A transistor is not only used as a main switch but can be also a synchronous rectifier. In this case, the switching events (turn-on and turn-off) appear not in the same quadrant as the main switch operation.

Second, modern topologies use soft-switching technique to reduce switching losses and increase switching frequency. Soft-switching techniques can be used to achieve several other goals on top of better efficiency such as size and weight reduction, application cost reduction, and longer lifetime.

6.6.5 *Simulation Result Comparison with Data Sheet Values*

For on-region, most of the models are accurate as it can be seen on the comparison of Fig. 6.38.

Now, when it comes to parasitic capacitors, there are big differences between various supplier models, as shown in Fig. 6.39.

If we look at a more critical parameter for switching application, i.e., the reverse recovery current or losses, it can be seen the superior results of onsemi physical and scalable models (Fig. 6.40).

The accuracy of power device models will affect the applications' simulations. Inaccurate and incomplete models will produce waveforms that do not match measurement (e.g., performed with an oscilloscope). Application important parameters (such as junction temperature, losses, and others) will not be matched as well.

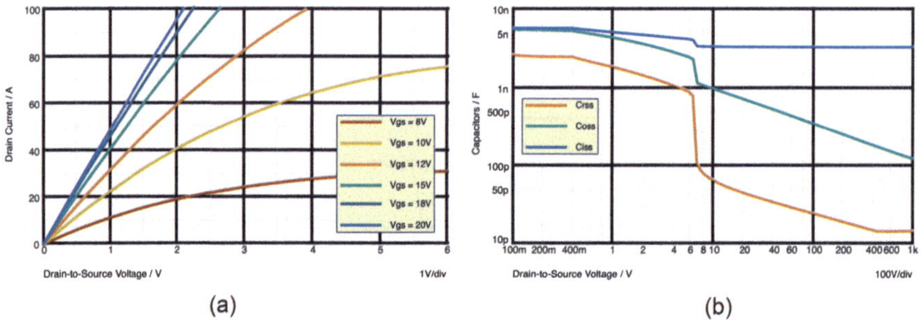

Fig. 6.38 (**a**) RDS(on) graph and (**b**) Cxss graph

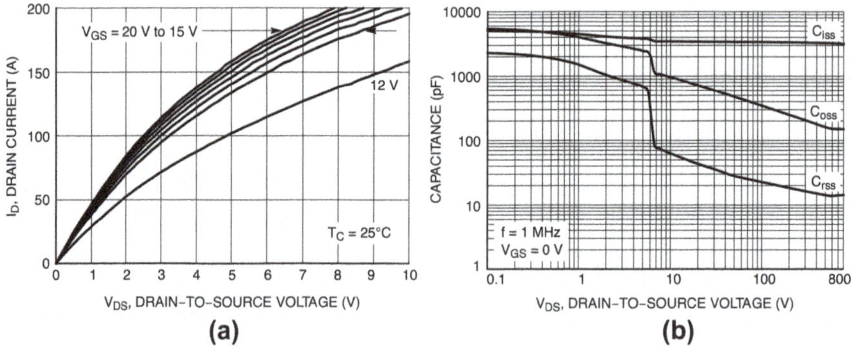

Fig. 6.39 (**left**) On-region data sheet graph and (**right**) capacitor sheet graph

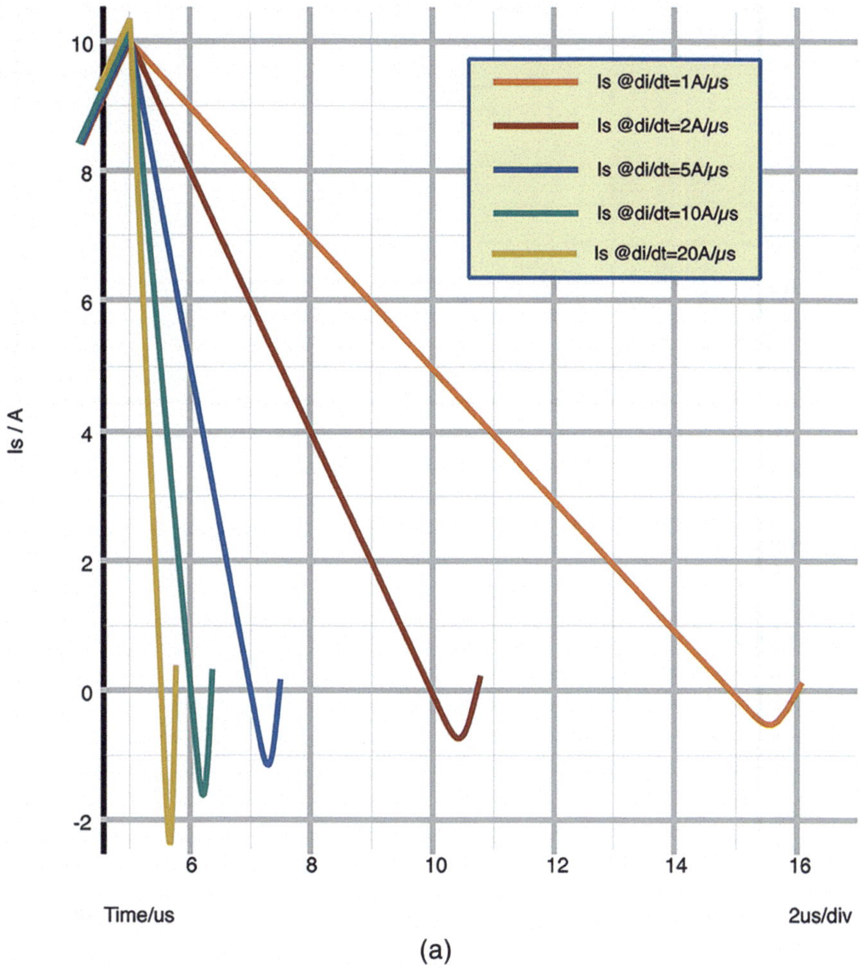

Fig. 6.40a Physical and scalable SPICE model simulation results

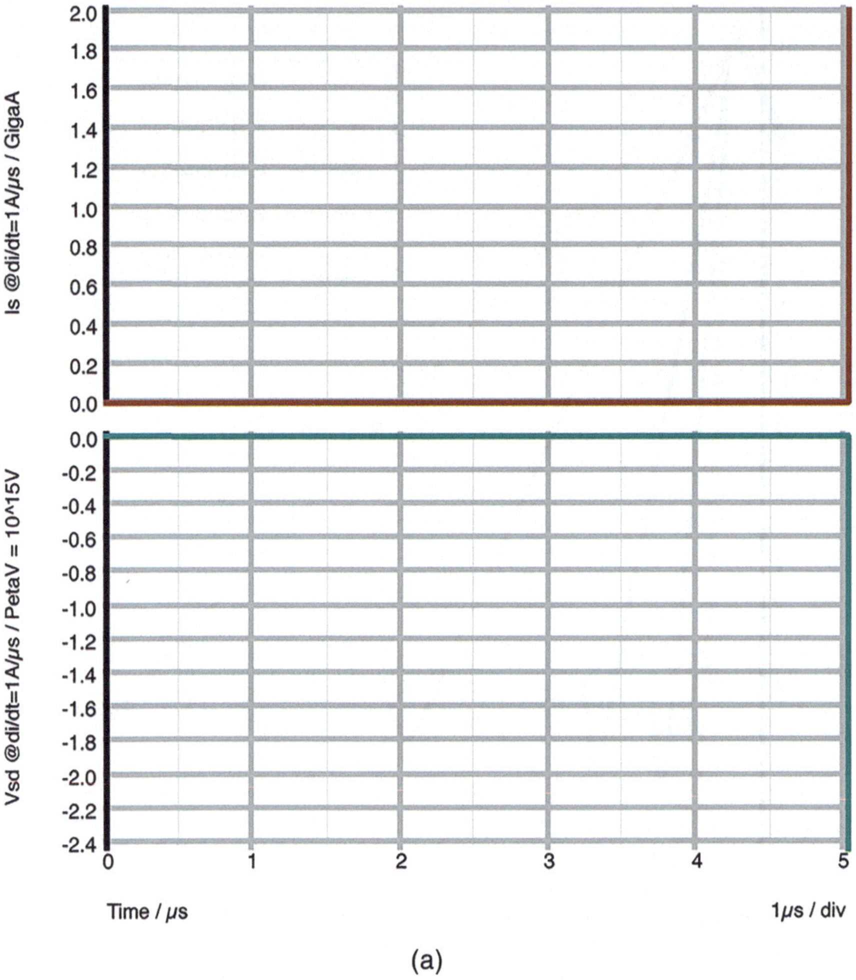

Fig. 6.40b Model not able to simulate reverse recovery and breakdown

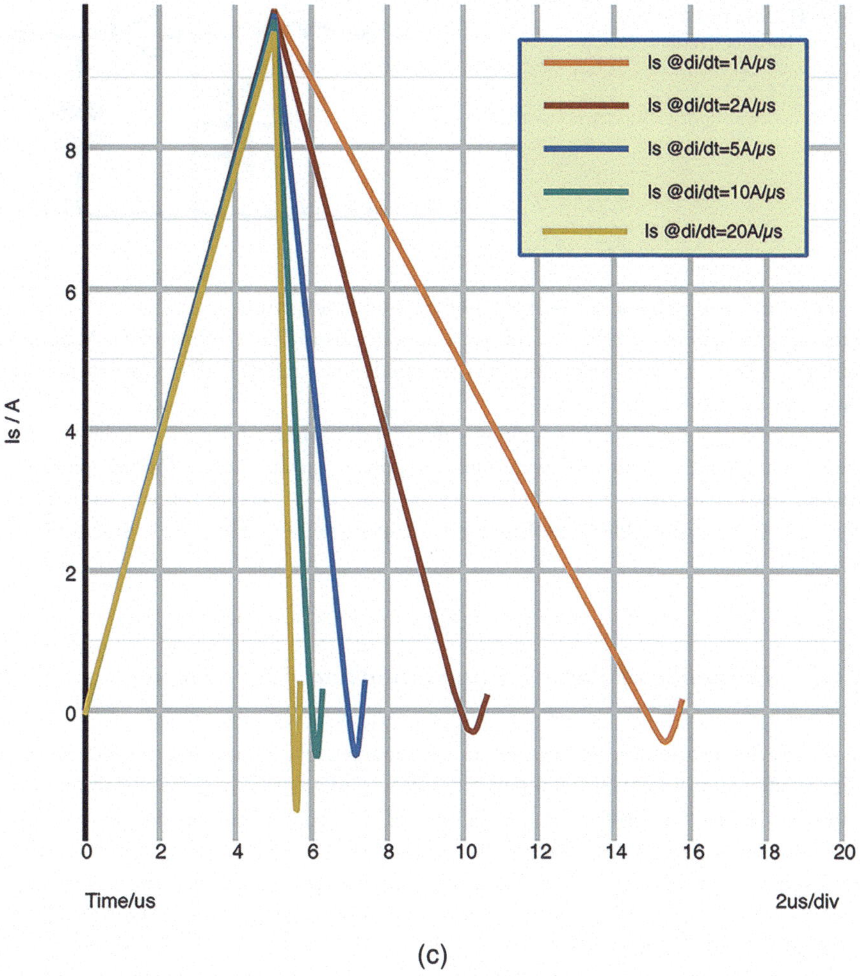

Fig. 6.40c Inconsistent simulation results

6.6.6 Power Application Simulations

Deploying SPICE and available component models, it is possible to make a complete application schematic. As an example, let us simulate a simple half-bridge stage. PWM and amplifier models are readily available as a library directly downloaded from the manufacturer's Internet site.

Let us now consider the circuit shown in Fig. 6.41, comparing then the results obtained with silicon IGBT and silicon carbide MOSFETs.

Fig. 6.41 Schematic of a DC-DC boost converter

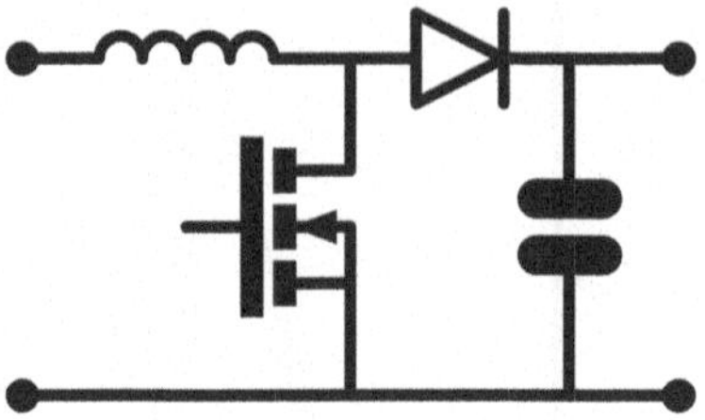

For IGBT, we will use a 20 kHz switching frequency, while we will use a 100 kHz with silicon carbide MOSFET. The boost operates from 400 V input to 800 V output with a 2 kW load. This load is relatively small for an IGBT. The waveforms are shown in Figs. 6.42a, b.

We can extract waveforms such as voltage and current on the power device. By multiplying those waveforms, we obtain the power, and by integrating this product, we get the losses over one switching cycle (Fig. 6.43). Using the electrothermal equivalence, it is possible to plot junction temperature too. The case temperature is set to 90 °C.

6.6.7 *Influence of Passive Components on Active Devices*

In the previous example, we used an ideal inductor and output capacitor. What happens if such components are not perfect? Let us add a parasitic capacitor in parallel with the inductor to simulate an inductor with a variable resonant frequency. The capacitor values are 10 pF, 100 pF, and 1 nF. The resonant frequencies associated are 1.6 MHz, 500 kHz, and 160 kHz. The SiC boost operates at 100 kHz. Let us then analyze the following waveforms (Fig. 6.44): drain voltage and current, junction temperature, and average losses.

We can see the inductor parasitic capacitor (inductor resonant frequency) affecting the switch losses. Thus, it is clear that for a complete and precise application simulation, accurate models of passive and active components are needed.

6.6.8 *System-Level Models (PLECS/PSIM) and Application Simulations*

As explained in Sect. 6.6.1, PSIM and PLECS are simulators that consider switches as ideal to simulate faster. In SPICE, given highly nonlinear component models (this is the case of modern power switches), simulations of switching transitions take time. This is because SPICE reduces its time steps to accurately capture the nonlinear regions. Clearly, smaller time steps reduce simulation speed with the trade-off being increased fidelity to the actual waveforms.

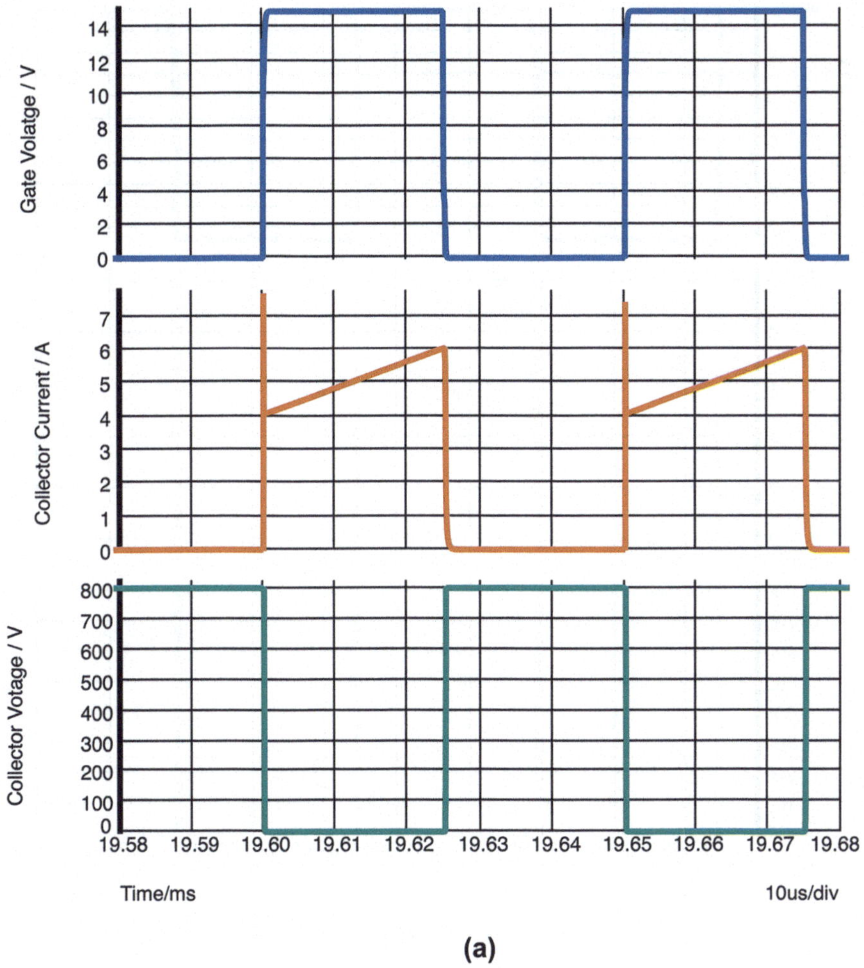

Fig. 6.42a onsemi Field Stop 7 IGBT waveforms

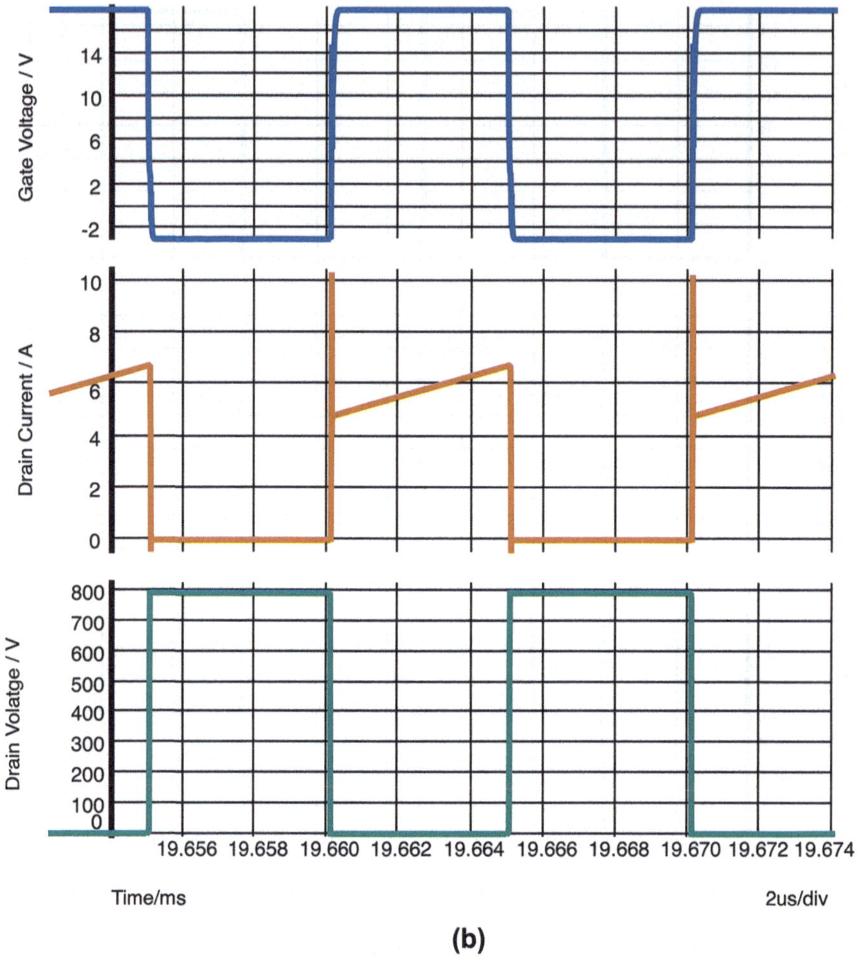

Fig. 6.42b onsemi M3 SiC MOSFET waveforms

By considering instantaneous ideal switching, PLECS and PSIM simulate much faster. However in an ideal switch, when multiplying drain voltage and current, there are no switching losses. We get only the conduction losses. To overcome this, both simulators use lookup table models with discrete values of losses at turn-on or turn-off over various operating and temperature conditions. When they need a value in between two points, they interpolate. When the value is outside the boundary, they extrapolate. As losses are non-linear curves, interpolation or extrapolation accuracy depends on data density.

There are three major ways to generate the loss tables:

1. Use values from the datasheet. However, even modern day datasheets may have limited datasets.

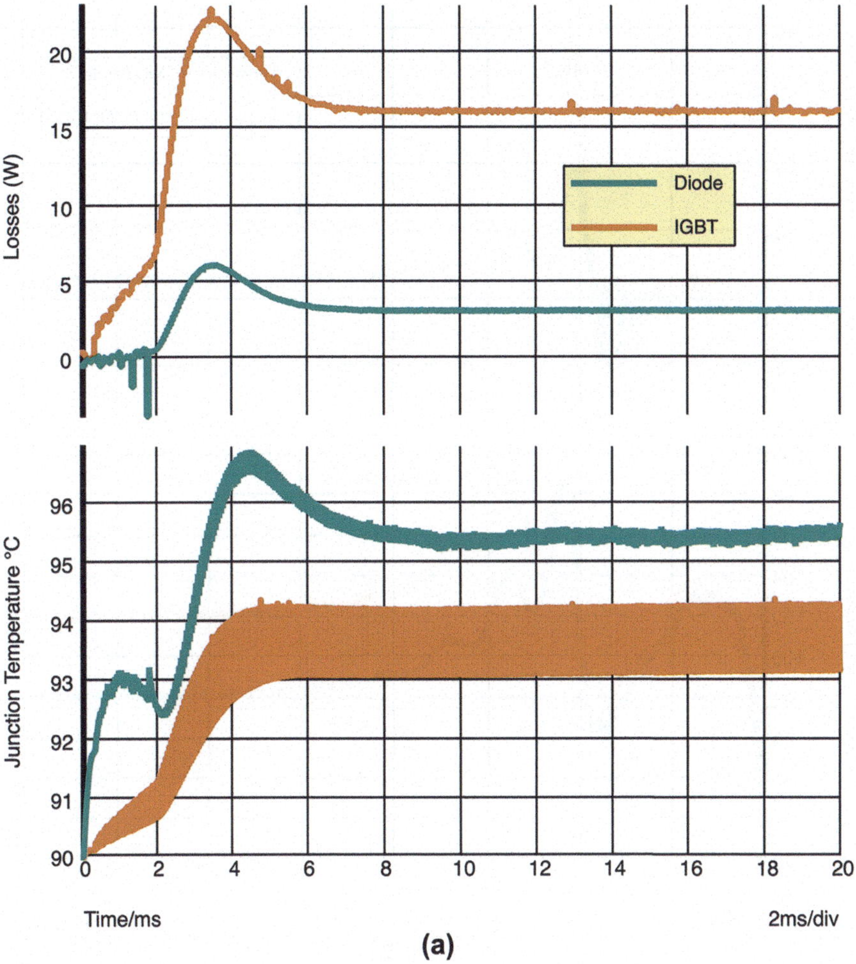

Fig. 6.43a onsemi Field Stop 7 IGBT losses

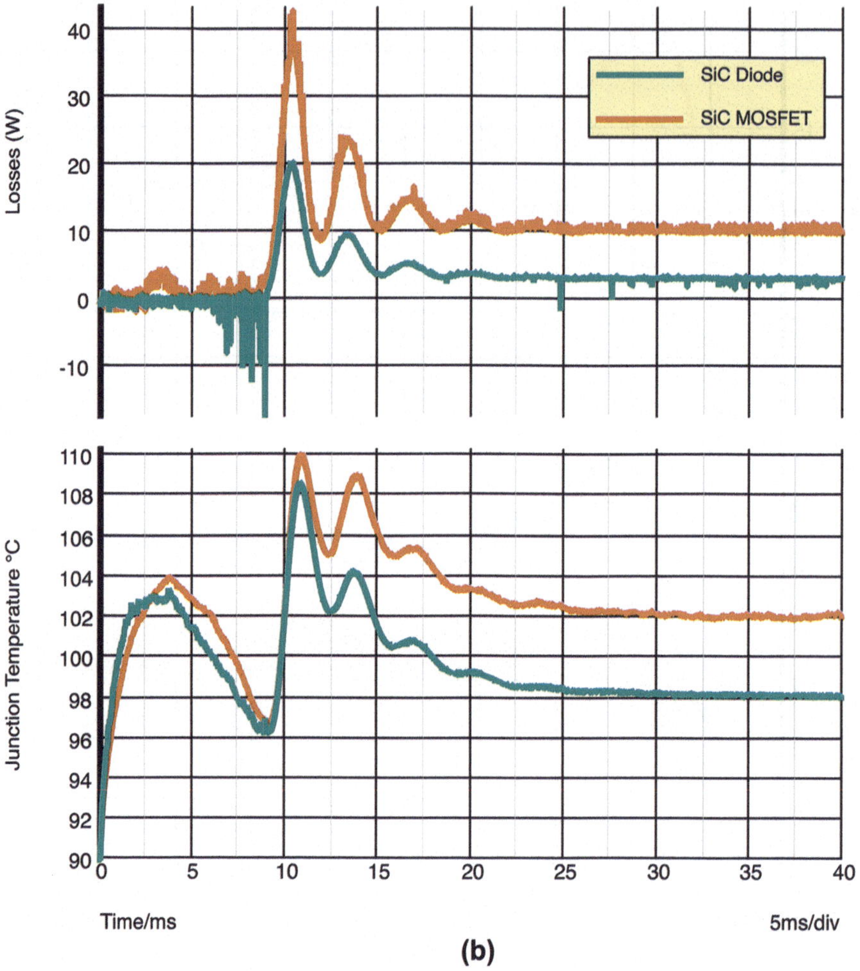

Fig. 6.43b onsemi M3 SiC MOSFET losses

2. Measure switching losses with a double pulse tester. In between each measurement, it is necessary to wait some time to have a stable temperature. This is a time-consuming method but more accurate than the previous one.
3. Use simulation. This method uses a virtual double pulse tester to generate the loss data. Obviously, this method is much faster than the measurement method. However, the method is futile without very accurate models. As we have seen in Sects. 6.6.5 and 6.6.8, the double pulse test covers only the main switch operation. Synchronous rectifier operation or soft-switching mode or transitions are not covered. On top of that, we also need to consider the influences of the passives and their parasitic elements.

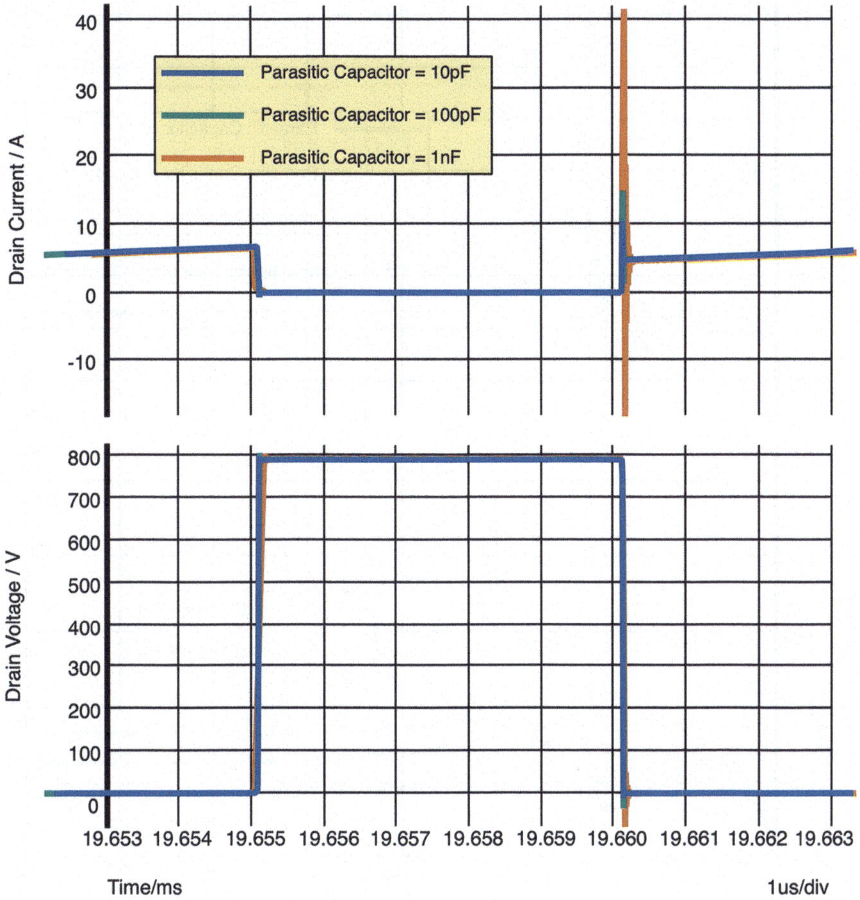

(a)

Fig. 6.44a onsemi M3 SiC MOSFET waveforms with inductors having various resonant frequencies

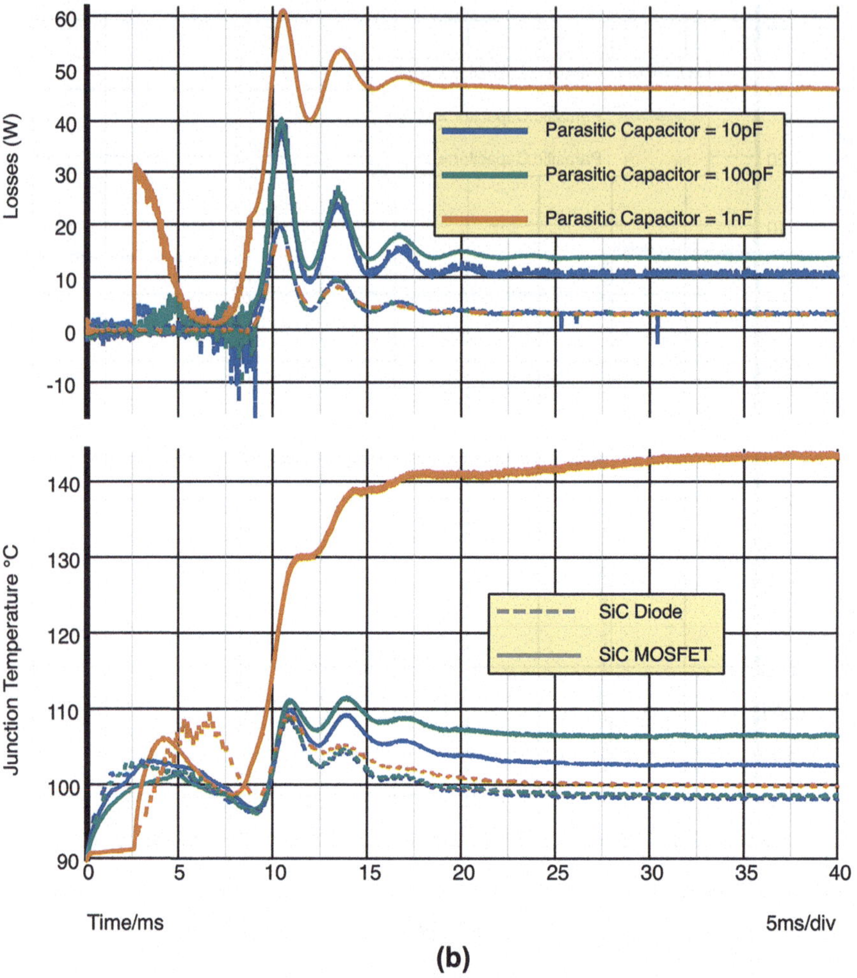

Fig. 6.44b onsemi M3 SiC MOSFET losses with inductors having various resonant frequencies

onsemi's innovative platform called the Self-Service PLECS Model Generator (SSMPG) solves all such issues. SSPMG uses onsemi's accurate physical and scalable SPICE models as the engine. The new virtual transition tester introduced by onsemi covers hard-switching, soft-switching, and partial soft-switching operations, AND/OR main switch (first quadrant), and synchronous rectifier (third quadrant) operations (Fig. 6.45).

Furthermore, the new transition loss tester includes more than 30 parameters to model the influence of passive parasitic elements. A system-level discrete device model can be obtained in 2 min with a very large dataset for accurate system simulations in PLECS. The basic steps required for operating the PLECS Model Generator are shown in Figs. 6.46, 6.47, and 6.48.

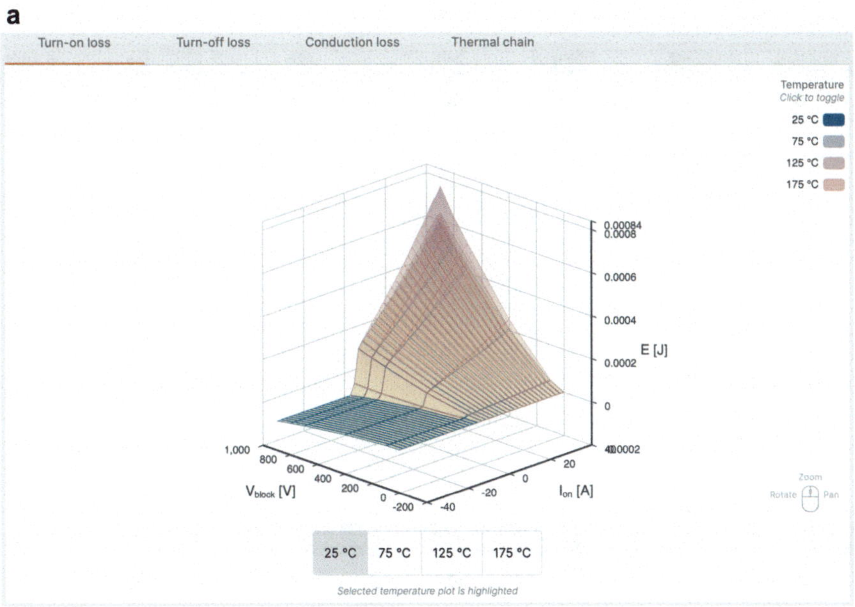

Fig. 6.45a Turn-on losses for the first and third quadrant operations. In the third quadrant, losses are negative because the energy stored in the output capacitor is recycled in the resonant network and gave back to the system

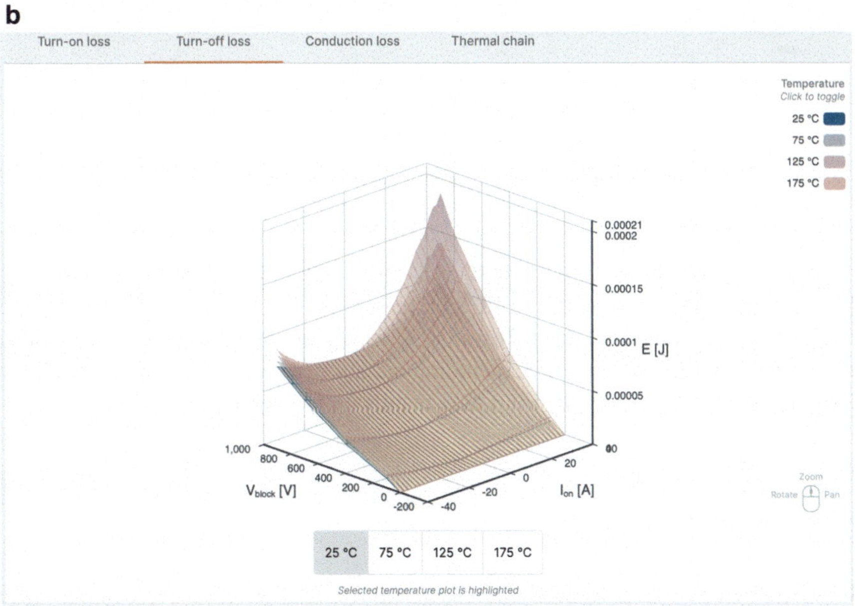

Fig. 6.45b Turn-off losses for the first and third quadrant operations. In the third quadrant, the energy is mainly the energy stored in the output capacitor, and there is no reverse recovery losses. In the first quadrant, the reverse recovery losses are on top of the output capacitor stored energy

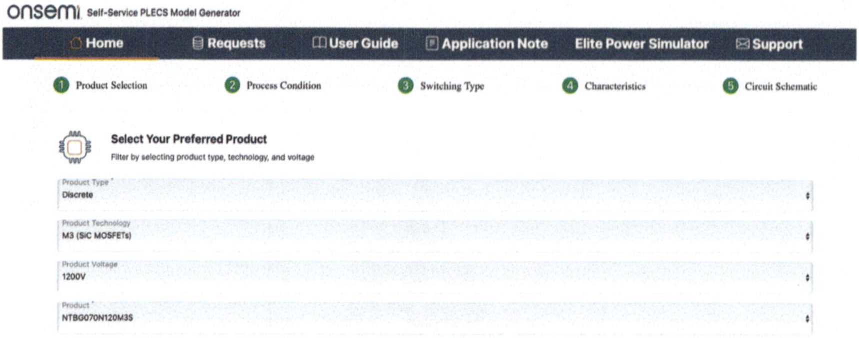

Fig. 6.46 PLECS Model Generator—step 1: select device

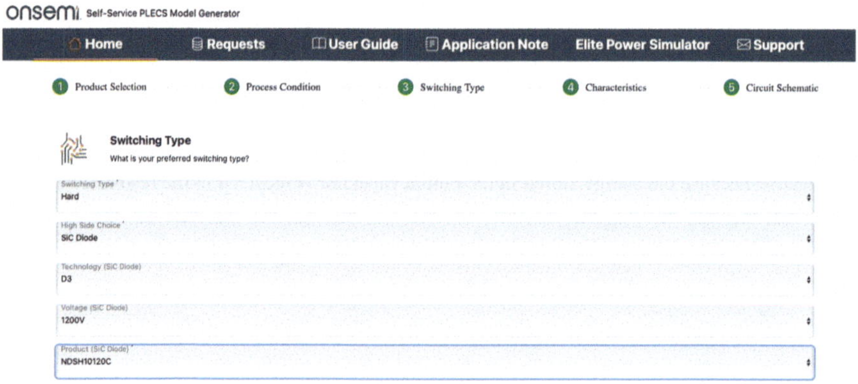

Fig. 6.47 PLECS Model Generator—step 3: select switching mode

Figures 6.46, 6.47, 6.48, 6.49, 6.50, 6.51, 6.52, 6.53, 6.54, 6.55, 6.56, and 6.57 illustrate the steps for using the PLECS online tool.

The obtained values in Figs. 6.56 and 6.57 with generated SiC MOSFET models in Figs. 6.46, 6.47, 6.48, and 6.55, including inductors with parasitic capacitors (See Sect. 6.8.9), match the SPICE simulation results on Fig. 6.44b.

6.6.9 Conclusion

Simulation can be used to easily evaluate application performances compared to measurement-based approaches. Simulation can predict waveforms and make a virtual check of the application behaviors before hardware is actually produced.

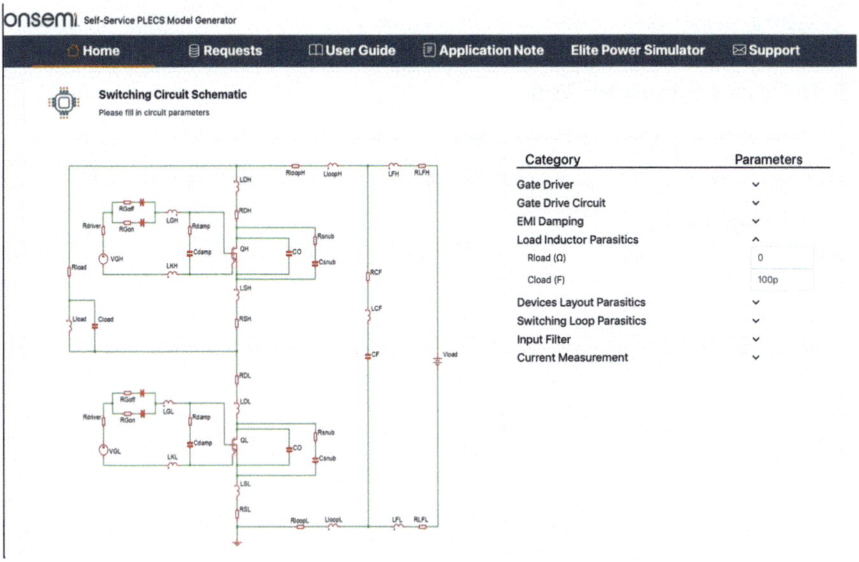

Fig. 6.48 PLECS Model Generator—step 5: include surrounding elements

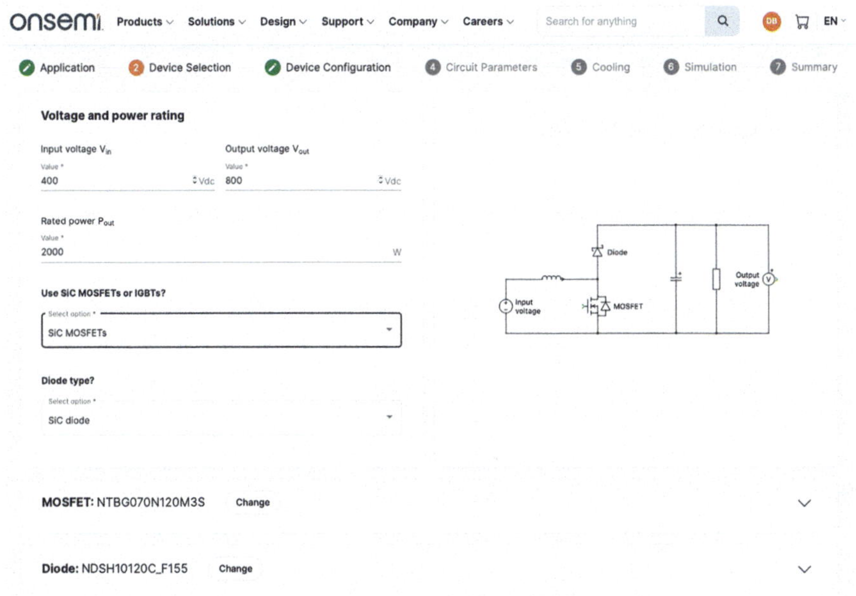

Fig. 6.49 PLECS online—steps 1 and 2: select topology, parameters, and devices

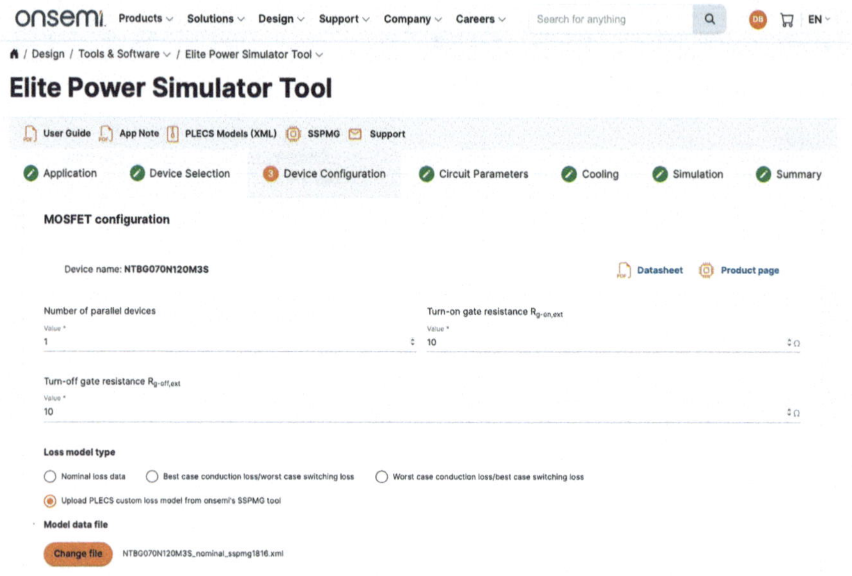

Fig. 6.50 PLECS online—step 3: define switching conditions and customized model

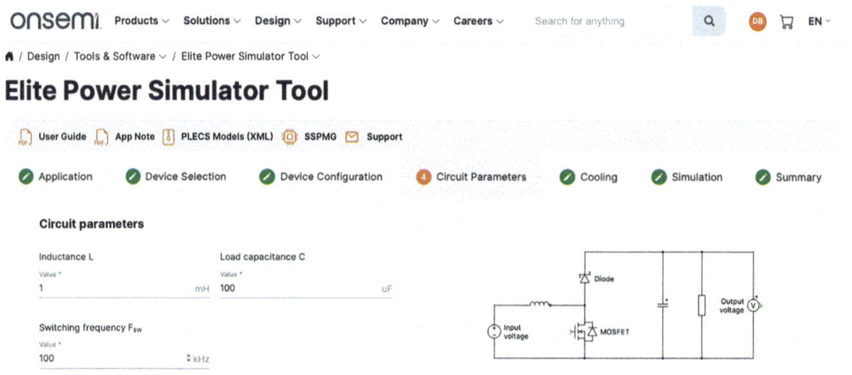

Fig. 6.51 PLECS online—step 4: define topology parameters

SPICE is the most common simulation platform that has evolved over many decades. Many options are available from a host of EDA companies and even free SPICE versions can be readily downloaded on the Internet. System-level simulations with tools such as PLECS are very useful and popular in system-level power electronic design.

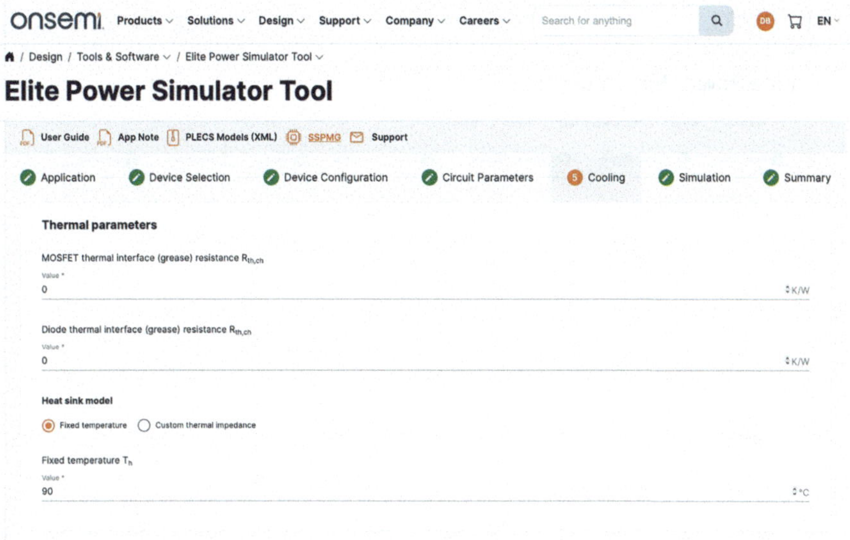

Fig. 6.52 PLECS online—step 5: define thermal elements

To obtain accurate results, it starts with accurate models for the semiconductor components. onsemi provides such accurate models through a physical and scalable modeling technique with proven results (Figures from 6.49, 6.50, 6.51, 6.52, 6.53, 6.54, 6.55, 6.56, and 6.57 for reference).

Fig. 6.53 PLECS online—step 6: simulation results. The default model does not match the simulation results on Fig. 6.43b, while the generated model matches the SPICE simulation on Fig. 6.44b

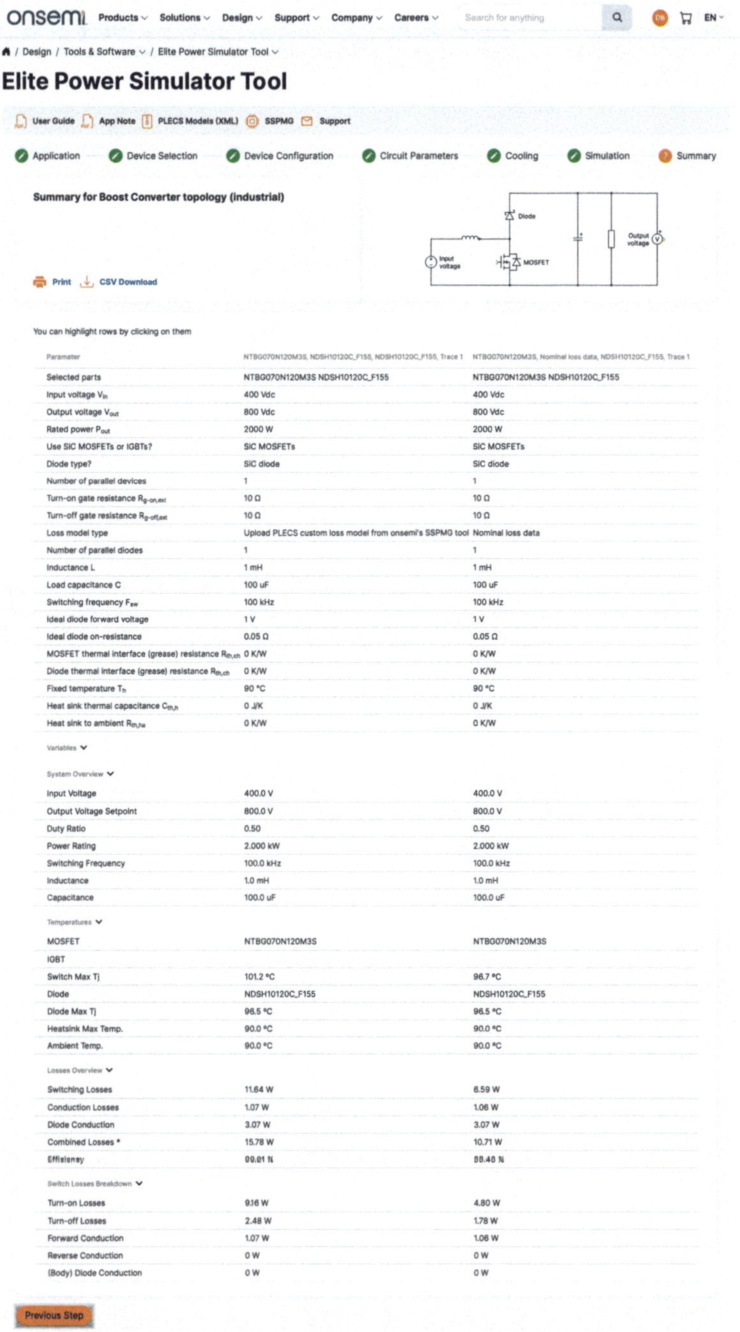

Parameter	NTBG070N120M3S, NDSH10120C_F155, NDSH10120C_F155, Trace 1	NTBG070N120M3S, Nominal loss data, NDSH10120C_F155, Trace 1
Selected parts	NTBG070N120M3S NDSH10120C_F155	NTBG070N120M3S NDSH10120C_F155
Input voltage V_{in}	400 Vdc	400 Vdc
Output voltage V_{out}	800 Vdc	800 Vdc
Rated power P_{out}	2000 W	2000 W
Use SiC MOSFETs or IGBTs?	SiC MOSFETs	SiC MOSFETs
Diode type?	SiC diode	SiC diode
Number of parallel devices	1	1
Turn-on gate resistance $R_{g\text{-}on,ext}$	10 Ω	10 Ω
Turn-off gate resistance $R_{g\text{-}off,ext}$	10 Ω	10 Ω
Loss model type	Upload PLECS custom loss model from onsemi's SSPMG tool	Nominal loss data
Number of parallel diodes	1	1
Inductance L	1 mH	1 mH
Load capacitance C	100 uF	100 uF
Switching frequency F_{sw}	100 kHz	100 kHz
Ideal diode forward voltage	1 V	1 V
Ideal diode on-resistance	0.05 Ω	0.05 Ω
MOSFET thermal interface (grease) resistance $R_{th,ch}$	0 K/W	0 K/W
Diode thermal interface (grease) resistance $R_{th,ch}$	0 K/W	0 K/W
Fixed temperature T_h	90 °C	90 °C
Heat sink thermal capacitance $C_{th,h}$	0 J/K	0 J/K
Heat sink to ambient $R_{th,ha}$	0 K/W	0 K/W
Variables		
System Overview		
Input Voltage	400.0 V	400.0 V
Output Voltage Setpoint	800.0 V	800.0 V
Duty Ratio	0.50	0.50
Power Rating	2.000 kW	2.000 kW
Switching Frequency	100.0 kHz	100.0 kHz
Inductance	1.0 mH	1.0 mH
Capacitance	100.0 uF	100.0 uF
Temperatures		
MOSFET	NTBG070N120M3S	NTBG070N120M3S
IGBT		
Switch Max Tj	101.2 °C	96.7 °C
Diode	NDSH10120C_F155	NDSH10120C_F155
Diode Max Tj	96.5 °C	96.5 °C
Heatsink Max Temp.	90.0 °C	90.0 °C
Ambient Temp.	90.0 °C	90.0 °C
Losses Overview		
Switching Losses	11.64 W	6.59 W
Conduction Losses	1.07 W	1.06 W
Diode Conduction	3.07 W	3.07 W
Combined Losses *	15.78 W	10.71 W
Efficiency	99.21 %	99.46 %
Switch Losses Breakdown		
Turn-on Losses	9.16 W	4.80 W
Turn-off Losses	2.48 W	1.78 W
Forward Conduction	1.07 W	1.06 W
Reverse Conduction	0 W	0 W
(Body) Diode Conduction	0 W	0 W

Fig. 6.54 PLECS online—Step 7: Simulations' summary. The default model does not match the simulation results on Fig. 6.43b, while the generated model matches the SPICE simulation on Fig. 6.44b

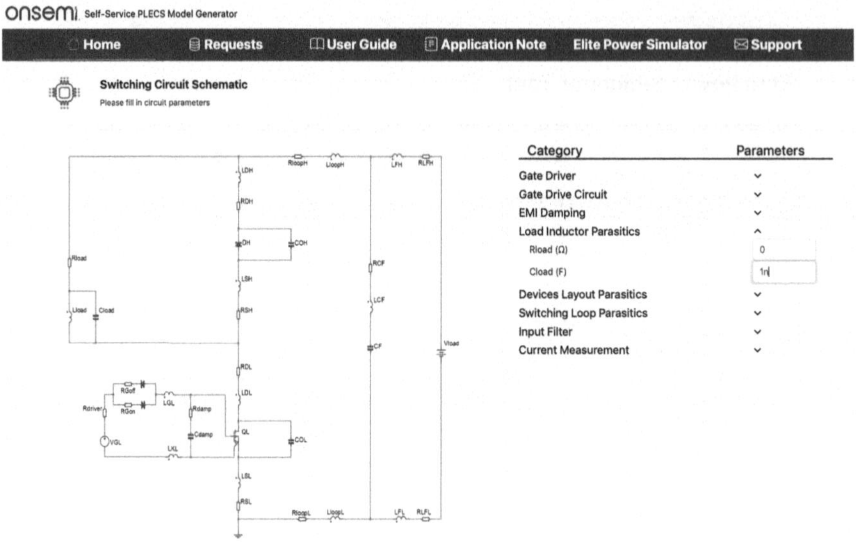

Fig. 6.55 Model Generator—Step 5: With various parasitic capacitor for various resonant frequency

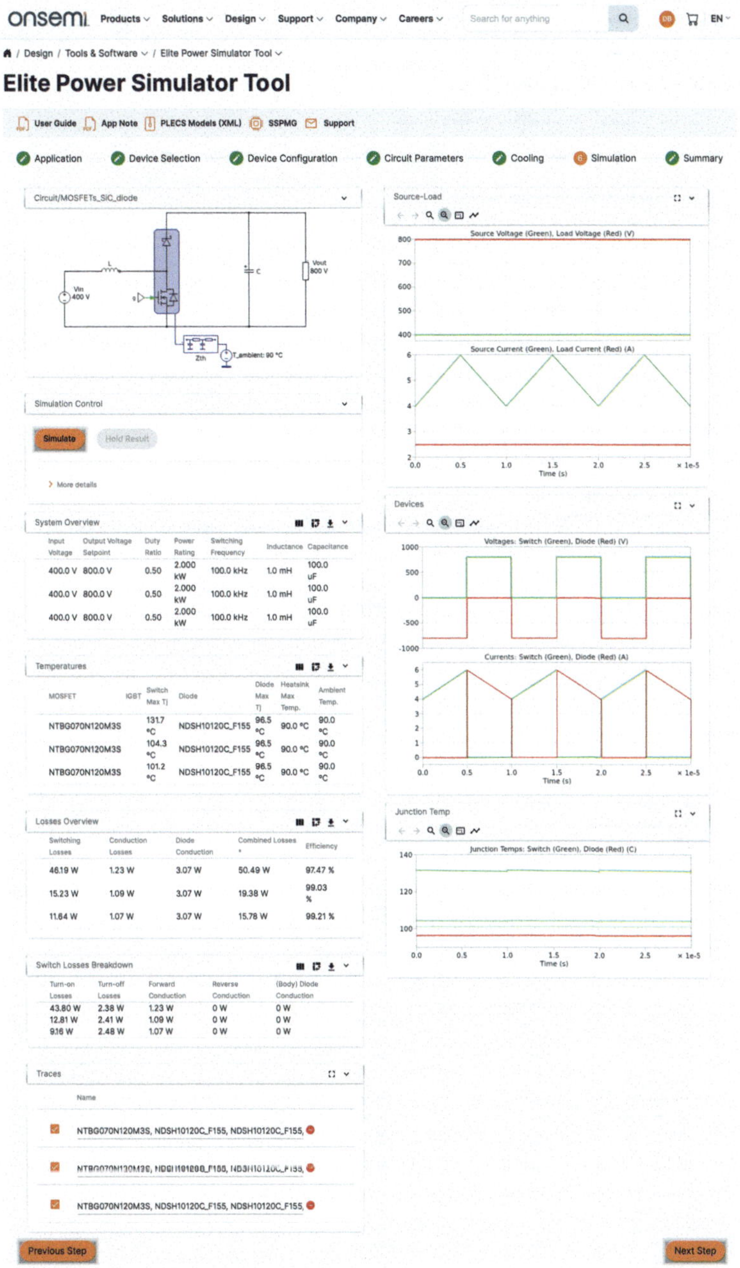

Fig. 6.56 PLECS online—Step 6: Simulation results for various resonant frequency inductors. The obtained values with generated SiC MOSFET model including inductor with parasitic capacitors match the SPICE simulation results on Fig. 6.44b with inductors having various resonant frequencies

onsemi Products Solutions Design Support Company Careers Search for anything EN

/ Design / Tools & Software / Elite Power Simulator Tool

Elite Power Simulator Tool

User Guide App Note PLECS Models (XML) SSPMG Support

Application Device Selection Device Configuration Circuit Parameters Cooling Simulation Summary

Summary for Boost Converter topology (industrial)

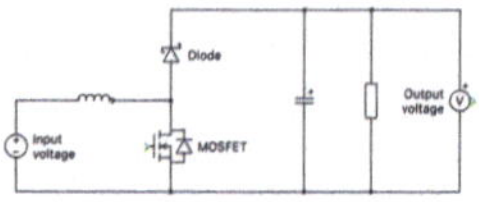

Print CSV Download

You can highlight rows by clicking on them

Parameter	NTBG070N120M3S, NDSH10120C_F155, NDSH10120C_F155, Trace 3	NTBG070N120M3S, NDSH10120C_F155, NDSH10120C_F155, Trace 1	NTBG070N120M3S, NDSH10120C_F155, NDSH10120C_F155, Trace 2
Selected parts	NTBG070N120M3S NDSH10120C_F155	NTBG070N120M3S NDSH10120C_F155	NTBG070N120M3S NDSH10120C_F155
Input voltage V_{in}	400 Vdc	400 Vdc	400 Vdc
Output voltage V_{out}	800 Vdc	800 Vdc	800 Vdc
Rated power P_{out}	2000 W	2000 W	2000 W
Use SiC MOSFETs or IGBTs?	SiC MOSFETs	SiC MOSFETs	SiC MOSFETs
Diode type?	SiC diode	SiC diode	SiC diode
Number of parallel devices	1	1	1
Turn-on gate resistance $R_{g\text{-}on,ext}$	10 Ω	10 Ω	10 Ω
Turn-off gate resistance $R_{g\text{-}off,ext}$	10 Ω	10 Ω	10 Ω
Loss model type	Upload PLECS custom loss model from onsemi's SSPMG tool	Upload PLECS custom loss model from onsemi's SSPMG tool	Upload PLECS custom loss model from onsemi's SSPMG tool
Number of parallel diodes	1	1	1
Inductance L	1 mH	1 mH	1 mH
Load capacitance C	100 uF	100 uF	100 uF
Switching frequency F_{sw}	100 kHz	100 kHz	100 kHz
Ideal diode forward voltage	1 V	1 V	1 V
Ideal diode on-resistance	0.05 Ω	0.05 Ω	0.05 Ω
MOSFET thermal interface (grease) resistance $R_{th,ch}$	0 K/W	0 K/W	0 K/W
Diode thermal interface (grease) resistance $R_{th,ch}$	0 K/W	0 K/W	0 K/W
Fixed temperature T_h	90 °C	90 °C	90 °C
Heat sink thermal capacitance $C_{th,h}$	0 J/K	0 J/K	0 J/K
Heat sink to ambient $R_{th,ha}$	0 K/W	0 K/W	0 K/W
Variables			
System Overview			
Input Voltage	400.0 V	400.0 V	400.0 V
Output Voltage Setpoint	800.0 V	800.0 V	800.0 V
Duty Ratio	0.50	0.50	0.50
Power Rating	2.000 kW	2.000 kW	2.000 kW
Switching Frequency	100.0 kHz	100.0 kHz	100.0 kHz
Inductance	1.0 mH	1.0 mH	1.0 mH
Capacitance	100.0 uF	100.0 uF	100.0 uF
Temperatures			
MOSFET	NTBG070N120M3S	NTBG070N120M3S	NTBG070N120M3S
IGBT			
Switch Max Tj	131.7 °C	101.2 °C	104.3 °C
Diode	NDSH10120C_F155	NDSH10120C_F155	NDSH10120C_F155
Diode Max Tj	96.5 °C	96.5 °C	96.5 °C
Heatsink Max Temp.	90.0 °C	90.0 °C	90.0 °C
Ambient Temp.	90.0 °C	90.0 °C	90.0 °C
Losses Overview			
Switching Losses	46.19 W	11.64 W	15.23 W
Conduction Losses	1.23 W	1.07 W	1.09 W
Diode Conduction	3.07 W	3.07 W	3.07 W
Combined Losses *	50.49 W	15.78 W	19.38 W
Efficiency	97.47 %	99.21 %	99.03 %
Switch Losses Breakdown			
Turn-on Losses	43.80 W	9.16 W	12.81 W
Turn-off Losses	2.38 W	2.48 W	2.41 W
Forward Conduction	1.23 W	1.07 W	1.09 W
Reverse Conduction	0 W	0 W	0 W
(Body) Diode Conduction	0 W	0 W	0 W

Previous Step

Fig. 6.57 PLECS online—Step 7: Simulations' summary for various resonant frequency inductors. The obtained values with generated SiC MOSFET model including inductor with parasitic capacitors match the SPICE simulation results on Fig. 6.44b with inductors having various resonant frequencies

6.7 Implementation of High-Voltage SiC in Long-Haul EV Trucking and EVTOL Aircraft Power Systems

Stephen Oliver

Wide bandgap technologies—GaN and SiC—are accelerating the transition from fossil fuels to renewable energy sources of electricity and electrified loads. In the USA alone, the forecasted adoption of passenger EVs by 2030 has increased more than 2×, from 21% in 2018 to 53% in the 2022 forecast [18]. Now, it is the turn of long-haul EV trucking—and commercial light aircraft—to move away from diesel fuels.

6.7.1 *The Road to EV Trucks*

In the early days, with only 60,000 medium- and heavy-duty BEV trucks sold in 2022, a small percentage of the approximately 5,000,000 trucks is sold every year. However, a momentum is building, with 60 medium- and 50 heavy-duty truck models introduced and international legislative pressures. In 2022, the USA and 26 other countries signed the COP27 MoU, committing to 30% ZEV sales by 2030 and 100% by 2040 [19], so a significant ramp in production is anticipated. An example "eTruck" is shown in Fig. 6.58 [20].

For power management, passenger EVs have seen a transition from silicon MOSFETs and IGBTs to SiC, with new GaN power ICs expected for some 400 V applications from 2025. For long-haul trucking, the extreme power for fast-charging large-scale batteries demands higher voltages and more advanced SiC power MOSFETs and Schottky diodes.

New infrastructure is also required; hence, schemes such as the one proposed by NREL [21] (Fig. 6.59), and built by Terawatt in Los Angeles [22], ramp to 10 MW of capacity.

6.7.2 *High-Voltage Mathematics: Why HV?*

High-voltage battery and distribution rails reduce conduction loss while increasing motor speed and torque:

$$\text{Ohm's Law}: V = \boldsymbol{IR}, \text{Power} = VI, \text{Loss} = I2R$$

$$\text{Power Loss} = \frac{\text{Power}^2 x\boldsymbol{R}}{V^2} \tag{6.1}$$

$$\text{Loss} \propto 1/V2$$

Fig. 6.58 Example "eTruck" from MAN of Germany

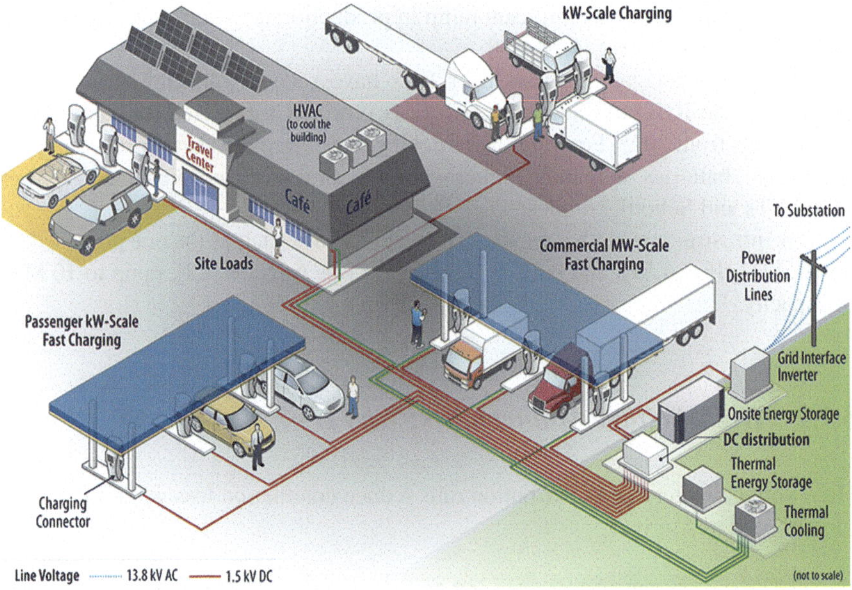

Fig. 6.59 NREL vision of multiple EV charging station

$$\text{Motor Drive}: V = \frac{T}{k}R + k\omega \qquad (6.2)$$

$$V \propto T\omega$$

where V = *voltage (V), I* = *current (A), R* = *resistance (Ω), T* = *torque (Nm), k* = *motor constant (Nm/A), and ω =rotational speed (rad/s).*

Equations (6.1 and 6.2): Ohm's law means the higher the voltage, the lower the conduction loss for a given cable bus-bar resistance (lower "I^2R" conduction loss). Motor drive equations determine that the higher the voltage, the faster the rotation and/or more torque.

The conduction loss benefit has a practical side too. Using fewer copper means lower cost and lighter, more flexible cabling.

A new "Megawatt Charging System" specification (SAE J3271 [23]) has been developed to provide up to 3.75 MW, over a 1250 V cable, as shown in Table 6.3 [24].

6.7.3 Hydrogen Versus Battery Electric?

While the majority of nonfossil fuel trucks are "pure" battery electric vehicles (BEV) that take energy in electrical form, let us take a moment to review the implications of hydrogen fuel cell electric vehicle (FCEV) schemes. First, a comparison of BEV with fuel cell trucks from Nikola is shown in Table 6.4.

While a fuel cell truck does not require the roadside charger infrastructure, it still needs a battery—though much smaller and lighter—to smooth fuel cell energy supply and traction demand. The final drive is also via an electric inverter stage, so the majority of power conversion/charging stages remain—and the benefits of high-voltage still apply—as shown in Fig. 6.60, from Eindhoven University of Technology [25].

6.7.4 The Runway to EVTOLs

Since 1978s first all-electric "power by wire" flight control aircraft (Advanced Fighter Technology Integration NF-16) [26], aircraft electrification has increased, with the 2011 Boeing 787 Dreamliner generating 1.5 MW for onboard applications [27].

Table 6.3 DC fast-charger specifications

DC fast-charger specifications	Passenger/LDV	HDV J3721 (non-cooled)	HDV J3721 (actively cooled)
Power (max, kW)	350	440	3750
Voltage (max, V)	920	1250	1250
Current (max, A)	500	350	3000
Vehicle battery (nom, V)	400/800	800, 1200	800, 1200
SiC device voltage (nom, V)	750/1200	1200/1700	1700

Table 6.4 Nikola truck comparison, BEV with fuel cell

Spec	Nikola BEV[1]	Nikola Fuel-Cell[2]
Max Speed (mph)	70	70
Range (miles)	Up to 330	Up to 500
Battery (kWh)	733	164
Fuel Cell Power Module (kWh)	-	200
Continuous Power (kW)	480	400
Instantaneous Power (kW)	797	575
Charging / Refueling Time (mins)	90 (@350 kW) *<< @ 1MW?*	20

Note the BEV charging of 90 min is prior to the introduction of MW charging stations

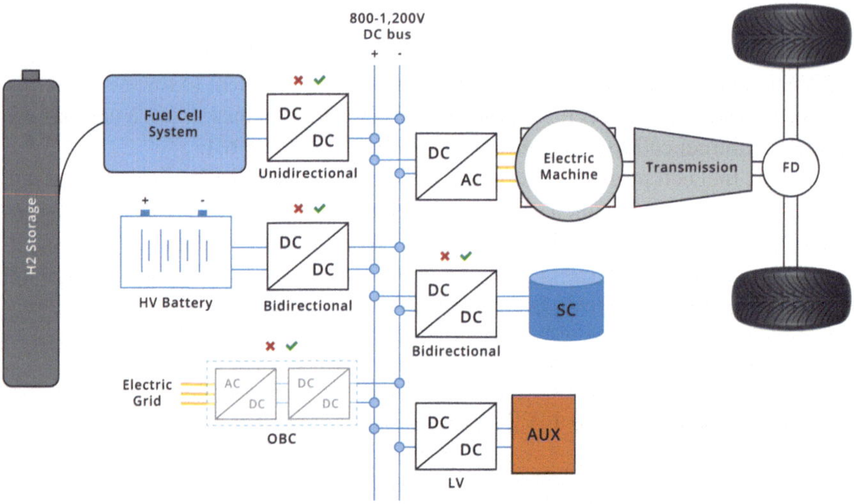

Fig. 6.60 FCEV schematic, showing multiple, high-voltage power conversion, inversion stages

The challenges faced by long-haul EV trucks are more severe in the "airborne" EV world, and the benefits of high voltage are amplified. Battery life becomes a safety-critical requirement for range, and saving weight becomes even more critical.

Recent "all-electric" designs have focused on "rotating" propelling machines, whether electric "jets" (ducted electric vectored thrust [28]) by Lilium (900 V architecture [29]), or more standard bladed propellors such as Joby 1000 V [30]) or Archer Aviation (800 V [31]), to achieve electric vertical take-off and landing (EVTOL) (Fig. 6.61).

Fig. 6.61 Archer Aviation EVTOL, using 800 V architecture

Here, we can also look at new types of electric motor architectures—enabled by high-speed, high-efficiency performance of silicon carbide. Low-frequency copper wire windings and iron "stator" assemblies are the most expensive components, and most complex manufacturing stage in BLDC motors, driving up size and cost. SiC can run at high frequency, in lightweight PCB-based stators to generate more power (inductance), in half the size and weight, at a fraction of the noise and carbon footprint created by conventional motors. Infinitum's air-core motor system is an example of next-gen 0.5–30 hp. (400 W–22 kW) motor drives for industrial/commercial applications and up to 400 hp. (300 kW) [32].

6.7.5 High-Voltage SiC for EV: Trench-Assisted Planar-Gate MOSFET Technology

SiC MOSFETs offer superior conductivity and switching performance compared to silicon due to their wide bandgap characteristics and high electric field strength. However, traditional designs using planar or trench techniques must compromise between manufacturability, performance, and/or reliability. Trench-assisted planar-gate design is a no-compromise, next-generation solution, high-yield manufacturing, fast and cool operation, and extended, long-life reliability. It enables lowest $R_{DS(ON)}$ at high temperature and lowest energy losses at high speeds. This enables unprecedented, industry-leading levels of performance, robustness, and quality.

In real-life applications, the ambient temperature in a system can be up to 80 °C (usually with liquid cooling for EV applications), and the power cycling of the device further elevates its junction temperature. Navitas' GeneSiC MOSFETs were

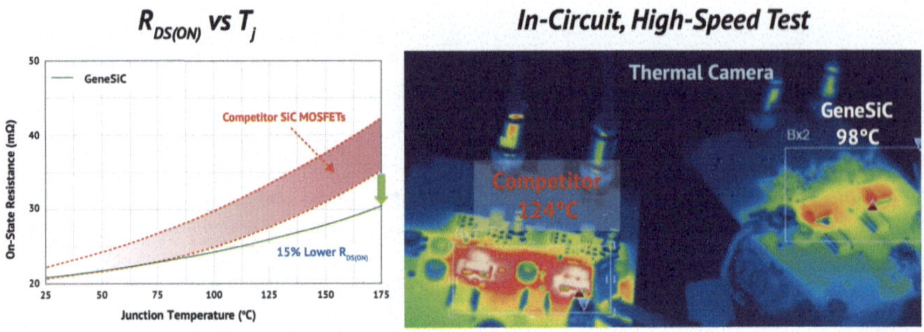

Fig. 6.62 Navitas' GeneSiC trench-assisted planar-gate FET versus industry standard (1200 V, 20 mOhm, D2PAK), showing 25 °C cooler operation at same test conditions due to lower $R_{DS(ON)}$ coefficient over temperature

designed with this in mind and support the industry's lowest $R_{DS(ON)}$ temperature coefficient, with side-by-side test results demonstrating 25 °C cooler case temperatures, which results in significantly lower losses and higher system efficiency. From a reliability standpoint, 25 °C cooler operation translates into a three times longer device lifetime (Fig. 6.62).

Navitas SiC FETs range from 650 to 6500 V—easily covering the land-based and airborne EV markets, for charging, DC-DC, and traction drive applications.

6.7.6 Summary

To enable efficient, reliable, long-haul/heavy-duty EV trucking and EVTOLs, high-voltage architectures and new high-efficiency power semiconductors are needed. High-power, high-voltage chargers are needed, and SiC shows significant benefits over Si IGBTs, with optimal trench-assisted planar-gate technology.

6.8 Advanced SiC Solutions for DC Fast Charging

Maurizio Di Paolo Emilio

To decrease the amount of time required for charging, the industry is using DC fast chargers (DCFCs) and ultrafast chargers. DCFCs and ultrafast chargers bypass the electric vehicle's onboard charger (OBC) to deliver more power straight to the battery, enabling charging at a current rating of 200–500 A, depending on the battery's capacity.

6.8.1 Overview of DCFCs

The EV industry is rapidly evolving its charging infrastructure to meet the demands of consumers and businesses. To address the critical aspect of charging time, the focus has shifted toward DCFCs and ultrafast chargers. These chargers operate at significantly higher power levels, bypassing the vehicle's onboard charger (OBC) to deliver direct power to the battery, slashing charging times considerably.

DCFCs and ultrafast chargers accommodate higher EV battery voltages, enhancing power efficiency but posing challenges for OEMs to maintain reliability and safety. These chargers integrate various components for power management, connectivity, and communication, demanding a flexible manufacturing approach to cater to diverse EV charging needs.

Contrasting AC Level 2 chargers, which rely on the OBC for charging, DC charging's direct battery approach facilitates faster speeds, ranging from 25 to 400 kW, even up to 1 MW. This speed transformation opens up new applications for EVs, allowing quick recharges during breaks and enhancing their practicality for long-distance travel.

The future of fast charging lies in dual CCS connectors and NACS connectors, with evolving standards aiming to charge EVs at rates between 100 and 500 A, delivering up to 400 kW or more. Major players such as Tesla, Ford, GM, and Rivian and charger manufacturers such as ChargePoint, EVgo, Triump, and Blink are backing these connector standards for future EVs and chargers. According to Yole, the market is anticipated to witness a rapid growth rate (CAGR >30%) in fast and ultrafast DC chargers until 2030, reflecting the industry's commitment to revolutionizing EV charging for enhanced convenience and usability.

To enhance charging speed, support increased EV battery voltages, and enhance overall power efficiency, DCFCs must operate at high voltages and power levels. This poses a problem for original equipment manufacturers, as they must provide a framework that maximizes efficiency while ensuring that both reliability and safety are not compromised.

DCFCs and ultrafast chargers incorporate several components (Fig. 6.63), including auxiliary power, sensing, power management, networking, and communications. The complexity of their design is related to the requirement of addressing the charging requirements of different EVs.

The majority of the new DCFCs are being designed with dual combined charging system connectors, equipped with 24-foot cables. These chargers can charge EVs at a rate ranging from 100 to 300 A, depending on the capacity of the EV's battery. The next DC ultrafast chargers can charge at high currents of up to 500 A and provide a power output of 400 kW.

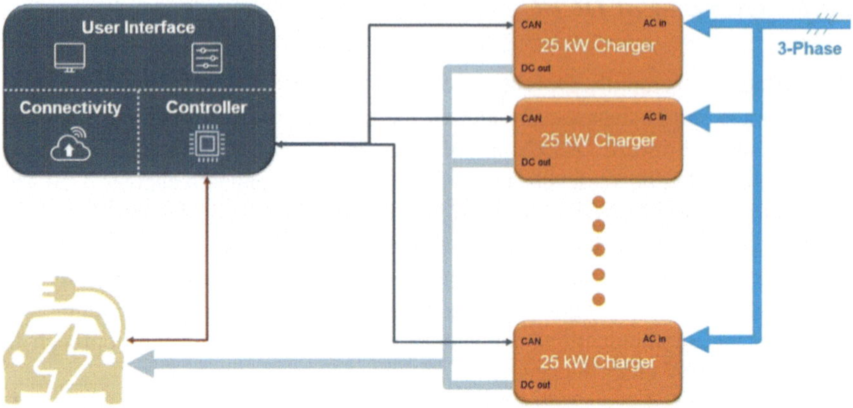

Fig. 6.63 Simplified block diagram of a DCFC. (Source: onsemi, December 2023)

6.8.2 EV Barriers

J.D. Power's 2023 Electric Vehicle Consideration Study highlights a crucial barrier for nearly half of US consumers in embracing EVs: concerns about charging accessibility and speed, crucial for a seamless driving experience akin to traditional vehicles. To meet this demand and ensure equitable access to public charging stations, the USA must quadruple charger availability by 2025 and increase it eightfold by the decade's end.

However, this surge in electric consumption poses a threat to existing electrical grids, potentially causing overloads. Bidirectional charging has emerged as a vital solution, allowing vehicles to serve as energy storage units for homes while facilitating regular battery charging and implementing vehicle-to-grid technology. This innovation not only supports a rapid DC fast-charging network but also enables swift vehicle recharging compared with lengthier methods, addressing the need for speed and accessibility in EV adoption.

6.8.3 Design Considerations

When designing a DCFC, crucial factors shape its architecture and component selection. Efficiency targets dictate the voltage- and power-range optimization, focusing on levels where efficiency profoundly impacts power delivery. The choice between discrete designs and PIMs offers flexibility versus simplified development with performance benefits such as compact packaging and enhanced reliability. Circuit topology selection (such as phase-shifted full-bridge [PSFB] and full-bridge LLC) impacts component choice for power-factor correction and DC/DC stages, involving two- or three-level systems using 650 V or 1200 V switches and diodes. These considerations drive the charger's efficiency, performance, and reliability.

The main key factors to consider when designing a DC fast charger include the following:

- Efficiency: To maximize power delivery, it is important to design the charging system for the power level at which efficiency has the most significant effect.
- Discrete or integrated solutions: Discrete designs provide optimal adaptability but necessitate a more intricate development procedure. Modules provide numerous performance advantages that are challenging to get with a discrete design, including a streamlined mechanical assembly, enhanced thermal management, increased dependability and diminished voltage spikes, and high-frequency electromagnetic interference.
- Topology: The topology, whether it is two-level or three-level, and the requirement for unidirectional or bidirectional operation will have a direct impact on the selection of components for the charger.

Numerous topologies exist for the implementation of the power factor correction (PFC) and DC/DC stages of a DC charger. This article, based on an onsemi white paper, will provide a comprehensive analysis of different standalone DC/DC topologies employed in this specific application.

6.8.4 Main Types of Topologies

Galvanic isolation is necessary between the mains grid and the battery to prevent the grounding of EV batteries. Isolation is a compulsory requirement in the DC/DC power stage of an EV DCFC.

Full-Bridge LLC

The topology comprises a primary full-bridge circuit, resonant LLC tanks, and a secondary full-bridge rectifier, as depicted. The LLC converter facilitates zero-voltage switching (ZVS) on the primary side and zero-current switching (ZCS) on the secondary side. Variable-frequency operation is required to control the output voltage. The LLC topology is suggested for applications with a small output voltage range and a set input voltage to achieve great efficiency.

Interleaved Three-Phase LLC Resonant Converter

This converter provides optimal efficiency under full-load conditions by evenly dispersing losses among three phases. Additionally, it mitigates stress on the capacitor and power switches. Nevertheless, this configuration exhibits diminished efficiency due to a decrease in charging current as the battery reaches its rated voltage.

To enhance overall efficiency under various load conditions, it is necessary to apply phase shedding. The operation of both dual-phase and single-phase resonant converters will result in excellent efficiency at both 50% load and low-load conditions. Additionally, these converters will ensure output voltage regulation even at low loads.

PSFB Converter

The phase-shifted full-bridge (PSFB) converter comprises a main full-bridge circuit, a resonant inductor, a secondary rectification bridge, and an LC filter. The system provides ZVS performance for the major switches by utilizing phase-shift control and eliminating turn-on losses.

Nevertheless, this configuration will result in substantial power dissipation during the switching-off process for the primary switches and the output rectifier. To mitigate voltage overshoot in output diodes, it is necessary to use either active or passive snubbers. PSFB converters are highly efficient for charging high-voltage batteries.

DAB Converter

Figure 6.64 shows the components of the dual active bridge (DAB) converter, including a primary full bridge, a leakage inductance from the transformer, a secondary full bridge, and a filter capacitor. The system does not require any extra resonant components. Instead, it utilizes the transformer's leakage inductance and the output capacitance of the power switches to accomplish soft-switching operation. The advent of wide bandgap semiconductors has contributed to the increasing popularity of DAB converters in applications involving bidirectional power flow.

The DAB converter accomplishes ZVS for the primary switches and ZCS for the secondary bridge switches. This converter exhibits reduced switching losses but experiences elevated conduction losses.

Utilizing low-$R_{DS(on)}$ silicon carbide MOSFETs in the secondary bridge circuit, a decrease in conduction losses is enabled. The DAB converters offer superior compactness, efficiency, reduced component strain, and smaller filter elements compared with alternative topologies.

Interleaved DAB Converter

Interleaved DAB converters can be employed to attain power levels exceeding 75 kW. This topology divides losses between two transformers and enables the utilization of cost-optimized onsemi's EliteSiC full-bridge power integrated modules.

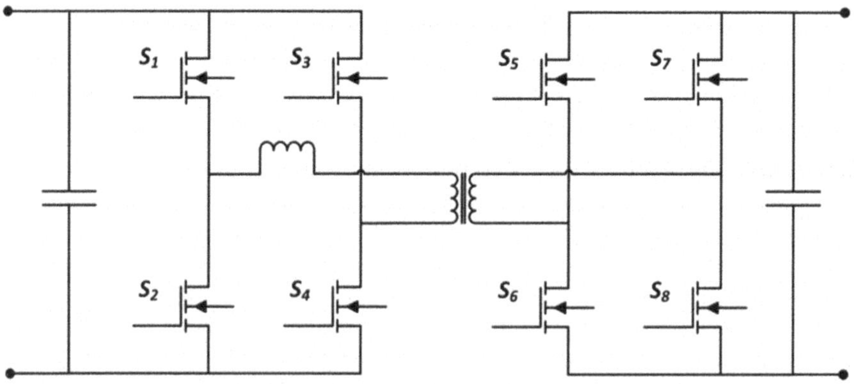

Fig. 6.64 DAB converter. (Source: onsemi, December 2023)

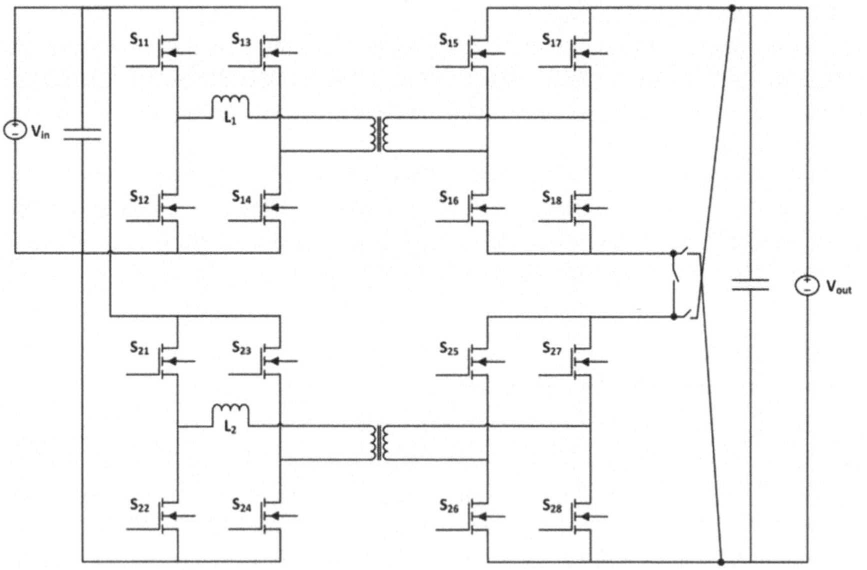

Fig. 6.65 DAB resonant converter. (Source: onsemi, December 2023)

The EliteSiC F2 full-bridge power integrated modules can reach a power output of 60 kW. To provide a 360 kW DC fast-charger system, it is possible to connect and operate six charger system modules, each with a capacity of 60 kW, in parallel.

DAB Resonant Converter

The DAB resonant converter is an alternative method to the DAB converter for achieving bidirectional power flow. To achieve the DAB resonant converter, a supplementary resonant network (LC, LLC, and CLLC) is incorporated into the DAB converter.

The CLLC DAB converter (Fig. 6.65) exhibits symmetrical circuitry, resulting in identical voltage gain characteristics for power flow in both directions. CLLC converters employ a pair of resonant capacitors positioned on either side of the transformer, hence mitigating the strain on the capacitors compared with an LLC converter.

EliteSiC power integrated modules (half-bridge or full-bridge) are recommended for high-power DAB CLLC resonant converter applications, delivering 25 kW to 120 kW (using three-phase interleaved dual active half-bridge resonant converter) output power.

Three-Level DNPC LLC Resonant Converter

This converter comprises a three-level half-bridge circuit, clamped diodes, a resonant LLC network, and a secondary full-bridge circuit. It is chosen as the principal topology for the primary side of the resonant LLC circuit due to its identical construction to the phase leg of the rectified PFC front end and two-level complete bridge.

SHB LLC Resonant Converter

This is an alternative kind of multilevel topology that can be employed as a primary side topology for LLC circuits to endure high input voltage. The series half-bridge (SHB) LLC shares the identical resonant tank and secondary full-bridge circuit with the DNPC resonant LLC design.

The primary benefit of the SHB LLC phase leg is the elimination of two clamping diodes compared with the DNPC LLC phase leg, resulting in a reduction in the overall number of components. SBH circuits have a simpler construction compared with the DNPC topology, which is its main advantage.

6.8.5 Isolated DC/DC Topologies

Galvanic isolation between EV batteries and the main grid safeguards batteries during charging. This isolation is vital in the DC/DC power stage and various fast-charger types. Unidirectional chargers employ full-bridge LLC or PSFB designs, integrating secondary full-bridge rectifiers. For bidirectional charging, the preference leans toward dual active bridge (DAB) or DAB CLLC topologies.

Ultrafast chargers, operating at 400 kW to 1 MW, commonly adopt series half-bridge (SHB) LLC or flying-capacitor three-level topologies for their efficient power delivery and management. These varied topologies ensure safe and optimized power transfer in diverse EV charging scenarios, catering to different power needs and directional charging requirements. Let us delve into the analysis of these main topologies and their key features:

- Full-bridge LLC topology: This utilizes LLC resonant tanks for ZVS/ZCS, recommended for narrow-output-voltage applications and suffers at light loads without control schemes.
- Interleaved three-phase LLC resonant converter: This enhances efficiency by distributing losses across phases, suitable for varying loads via phase shedding.
- PSFB converter: This achieves ZVS for primary switches but faces turn-off losses and requires snubbers to minimize diode voltage overshoot.
- DAB converter: This employs transformer leakage and switch capacitance for soft switching, is less sensitive to component variations, and is popular in bidirectional applications.
- Interleaved DAB converter: This scales up power using interleaving and distributes losses across multiple transformers, allowing for high-power applications.
- DAB resonant converter: This adds resonant networks for bidirectional power flow, with CLLC converters reducing capacitor stress, ensuring symmetry in voltage gain.
- Three-level DNPC LLC resonant converter: This is a complex topology using three-level half-bridge circuits that achieves ZVS but requires hybrid modulation controls.
- SHB LLC resonant converter: This is a multilevel topology handling high input voltage and shares similarities with DNPC topology.

6.8.6 SiC Devices

onsemi has honed SiC MOSFET technology, advancing from planar M1 to the latest M3E device, optimizing specific on-state resistance ($R_{DS(on)}$) and short-circuit performance. Its roadmap introduces M4T trench devices, promising enhanced performance.

Smart chip additions, such as a tunable polyresistor and temperature sensor, facilitate design flexibility and system monitoring. Module packaging innovations, such as dual-side sintering, slash thermal resistance, and boost power ratings and efficiency, are evident in onsemi's VE-Trac automotive modules. Leveraging vertical SiC manufacturing ensures rigorous quality controls, eliminating defects, while gate oxide improvements demonstrate exceptional reliability with <1 ppm failure rates and minimal Vth shift after 1e11 cycles. This comprehensive development propels onsemi's EliteSiC line for high-performance applications such as EV traction inverters.

6.8.7 PIMs

With the wide selection of PIMs that onsemi offers, designers can choose the best PIMs for their DC fast-charging or ESS applications, regardless of the market topology they serve. With the flexibility and efficiency of this extensive portfolio, designers may simplify their design cycles.

When dies from the same wafer are used in several modules, it increases consistency and dependability and removes the need to purchase discretes from different sources, which frequently results in inconsistent performance. This module portfolio has several noteworthy characteristics, such as using Gen3 M3S SiC MOSFET technology for industry-leading efficiency; supporting TNPC, half-bridge, and full-bridge topologies; and enabling scalable output power from 25 to 100 kW.

The M3S-based solution enables best-in-class $R_{DS(on)}$ and offers superior switching performance and low reverse-recovery losses. Parts with low $R_{DS(on)}$ to output charge (Qoss) facilitate short dead times, affect the resonant current settings, and enable higher-frequency operation, leading to higher efficiency, smaller form factor, and reduced cooling requirements.

These modules emphasize thermal management to minimize overheating-related system failures. They are offered in industry-standard F1 and F2 packaging with optional pre-applied thermal interface material and press-fit pins. Complete SiC modules guarantee energy conservation by reducing power losses, which saves a significant amount of money and energy. Furthermore, the focus on reliability and robustness ensures steady, dependable operations, which supports continuous performance in demanding applications.

6.8.8 Module Packaging Improvements

Improving the power module packaging for reliability while lowering thermal resistance (Rth) and stray inductance (Ls) is key to improving system performance in a difficult, high-temperature, power cycling environment, as in the EV traction inverter.

onsemi has planned a SiC power module packaging roadmap for the Rth figure of merit. Sintering on both surfaces has, for example, brought the Rth value below 0.1 K/W for the B2 sintered/sintered module compared to the 0.15 K/W achieved on the original dual-side cooling (DSC) soldered module.

The improvements in Rth and Ls enable improvements in power ratings and efficiency. Improved thermal management and loss reduction have allowed onsemi to have a range in module power ratings in their VE-Trac™ line of automotive-qualified power modules, from the cheaper 120–150 kW systems using thermal grease for the heat sink attach to the higher power 150–300 kW systems using sintered heat sinks. The substrate technology includes state of the art such as AMB silicon nitride DBC, with the Cu clip front metal bonding and Ag die attach sintering for improved reliability and thermal performance.

6.8.9 Reliability

Having a fully vertical SiC manufacturing capability from crystal growth to packaging has allowed onsemi to have greater control in screening and quality controls to ensure high reliability. In-line screening during the Fab process and customized burn-in screening on the fabricated devices serve to eliminate extrinsic defects. The planar gate oxide process flow has leveraged several decades of gate oxide improvements. The time to dielectric breakdown (TDDB) curves so obtained predict a < 1 ppm failure rate at the maximum operation gate voltage of 18 V. Threshold voltage (VTH) stability measured under a gate cycling stress test from −3 to 18 V shows a shift of only around 50 mV after 1E11 cycles.

Developing trench MOSFETs will entail recreating these high-reliability standards achieved on the planar SiC MOSFETs.

6.8.10 Hybrid Solutions

Hybrid solutions are combinations of power electronics technologies such as silicon carbide, silicon MOSFETs, and IGBTs, which meet the challenging requirements of a wide range of power applications.

In industrial applications, onsemi is using hybrid solutions to provide the following:

- Higher power density and efficiency. onsemi develops hybrid solutions that use silicon and advanced power semiconductor technologies such as SiC. These hybrid solutions provide improved efficiency and power density, allowing industrial systems to operate more effectively and with reduced energy consumption.
- Improved reliability and robustness. Industrial environments can often be harsh and demanding, with exposure to high temperatures, voltage transients, and other challenging conditions. onsemi's designs hybrid solutions with customers' end application need for ruggedness and reliability in mind. By taking advantage of the strengths of different technologies, such as SiC MOSFETs and IGBTs, onsemi enables customers to create solutions that can withstand high temperatures, handle high voltages, and provide enhanced and improved protection against short circuits and overloads.
- Better integration and system optimization. Hybrid solutions focus on integrating multiple components and functionalities into a single package or module. Combining these solutions makes it possible to simplify system design, reduce component count, and optimize performance. It also helps industrial applications save space, lower costs, and improve system efficiency.
- Customization and flexibility. onsemi offers hybrid solutions that can be customized and tailored to meet specific application needs. By working closely with customers, these solutions optimize performance, reliability, efficiency, thermal management, and protection mechanism for industrial systems in terms of technology topology and packaging requirements from the customer.

System efficiency can be increased because SiC diodes reduce switching losses, leading to higher power conversion and lower energy waste in industrial applications. IGBTs help to achieve higher switching frequencies.

An IGBT and SiC diode hybrid solution helps reduce the component size of, for example, an inductor coil, reducing the system size and weight. SiC diodes in hybrid solutions offer higher power density due to their compact size and better thermal conductivity, making them suitable for space-constrained applications. Hybrid solutions enhance reliability with superior temperature and voltage handling, ensuring robust operation in demanding industrial environments.

Other examples of industrial applications where onsemi' hybrid solutions can be implemented include EV fast-charging solutions, UPS, and energy storage solutions for renewable energy.

Since onsemi manufactures its own SiC and silicon devices, it can offer customers various solutions, from pure SiC solutions to pure Si solutions and every combination in between.

6.8.11 SiC in EVs

Range anxiety is the primary factor preventing the widespread adoption of electric vehicles. Swapping from an IGBT-based to a SiC-based inverter dramatically affects the vehicle's range. However, this can be regarded as a distinct approach to addressing another barrier to EV adoption, cost. The batteries used in EVs are the most expensive component. If using SiC extends the EV's range by 7%, the battery size can be reduced by 7% while maintaining the same range as the non-SiC baseline. A reduced battery cell will effectively reduce the overall cost of an electric vehicle. This is why SiC adoption in EVs is so robust, and it is what is driving SiC manufacturers' extremely optimistic revenue projections.

EV charging stations and the expansion of this charging infrastructure are related to EVs. The power density of EV charging stations is one of the primary considerations. SiC enables system designers to deliver more power in the same volume or maintain the same amount while reducing the volume by 300%. Using SiC is primarily motivated by the need for EV charging stations to output more power at the same rate. The aim is to charge an EV in the same time it takes to fill up a petrol tank. This can only be accomplished by increasing the electricity supplied by the charging station to the electric vehicle.

According to onsemi, a key challenge with using rare earth metals in EVs and industrial applications is their limited availability, with China owning the lion's share (estimated 70%). Additional factors include price volatility driven by supply/demand, recycling processes, and the environmental footprint.

6.8.12 Collaboration with Other Companies

onsemi has recently strengthened collaboration with other companies to meet the growing demand for SiC and sustainable energy solutions.

With Vitesco Technologies, onsemi has announced a 10-year long-term supply agreement worth $1.9 billion (€1.75 billion) for silicon carbide (SiC) products to enable Vitesco Technologies' ramp in electrification technologies. Experts from both companies will work together, leveraging onsemi's industry-leading EliteSiC die technologies packaged with innovative assembly techniques to extract the maximum performance at the highest levels of ruggedness. These advanced power device solutions will enable Vitesco's next-generation traction inverters with unprecedented power density and efficiency that can ultimately address the entire power spectrum of battery electric vehicles.

onsemi's EliteSiC silicon carbide (SiC) MOSFETs and IGBT-based high-density power integrated modules (PIMs) will be integrated by Sineng Electric into its utility-scale sustainable energy solutions, including a solar inverter and an industry-first 200 kW energy storage system (ESS). The two companies collaborate to develop solutions that maximize the efficacy of power conversion, energy storage, and solar inverter systems.

According to onsemi, the solutions resulting from this collaboration will deliver better overall system efficiency, reduced cost, and enhanced reliability, thus making renewable energy solutions more appealing and accessible and fostering greater adoption.

EliteSiC products addressing energy applications include full SiC PIMs, which support higher efficiencies by increasing switching frequency up to 40 kHz with much lower turn-on/turn-off losses than onsemi EliteSiC MOSFETs. They also help to reduce the inductor size from 0.6 to 0.2 mH. Accordingly, inverter designers achieve better system performance with fewer design efforts under the same footprint.

With Kempower, onsemi signed a strategic agreement to integrate EliteSiC D3 diodes and M3S MOSFETs into Kempower's electric vehicle (EV) charging solutions. According to onsemi, these chargers enable dynamic load balancing, ensuring optimal real-time power distribution by intelligently managing and adjusting power allocation. Moreover, they enhance efficiency by preventing power waste and ensuring even charging rates for all connected vehicles.

In the same direction goes the long-term supply agreement (LTSA) announced by onsemi with the electric mobility brand ZEEKR. onsemi will provide its EliteSiC silicon carbide (SiC) MOSFETs to enhance the powertrain efficacy of ZEEKR's smart electric vehicles (EVs), resulting in enhanced performance, quicker charging speeds, and an increased operating range. As a result of this deal, ZEEKR can integrate its silicon carbide technology into the traction engines of up to 700,000 automobiles or more per year during the next 10 years.

References

1. https://www.iea.org/energy-system/buildings/data-centres-and-data-transmission-networks
2. The growing energy footprint of artificial intelligence, Alex de Vries, Joule, October 2023
3. https://patents.google.com/patent/US20100259240A1/en
4. www.qorvo.com
5. https://www.qorvo.com/design-hub/design-tools/interactive/fet-jet-calculator/launch
6. Silicon Carbide (SiC) patents support the emergence of a complete domestic supply chain in China, KnowMade, January 2022., https://www.knowmade.com/technology-news/semiconductor-news/power-electronics-devices-news/silicon-carbide-sic-patents-support-the-emergence-of-a-complete-domestic-supply-chain-in-china/
7. Sanan Optoelectronics: a new entrant reshuffling the cards in the global SiC patent landscape and in the Chinese SiC power device market, October 2021, https://www.knowmade.com/technology-news/semiconductor-news/power-electronics-devices-news/sanan-optoelectronics-a-new-entrant-reshuffling-the-cards-in-the-global-sic-patent-landscape-and-in-the-chinese-sic-power-device-market/
8. New silicon carbide prospects emerge as market adapts to EV expansion, McKinsey & Company, October 2023, https://www.mckinsey.com/industries/semiconductors/our-insights/new-silicon-carbide-prospects-emerge-as-market-adapts-to-ev-expansion
9. Showa Denko: a patent portfolio tailored to its ambitions as a world leader in the SiC wafer market, KnowMade, August 2021, https://www.knowmade.com/technology-news/

semiconductor-news/power-electronics-devices-news/showa-denko-a-patent-portfolio-tailored-to-its-ambitions-as-a-world-leader-in-the-sic-wafer-market/
10. The vertical innovation path: from materials to power devices and their integration in electric vehicles (EV), KnowMade, August 2022, https://www.knowmade.com/technology-news/semiconductor-news/power-electronics-devices-news/the-vertical-innovation-path-from-materials-to-power-devices-and-their-integration-in-electric-vehicles-ev/
11. Power SiC patent landscape analysis: who is driving the establishment of the SiC industry in South Korea?, KnowMade, May 2022, https://www.knowmade.com/technology-news/press-release/power-sic-patent-landscape-analysis-who-is-driving-the-establishment-of-the-sic-industry-in-south-korea/
12. Silicon Carbide (SiC) Patent Landscape 2022, KnowMade, May 2022, https://www.knowmade.com/patent-analytics-services/patent-report/semiconductor-patent-landscape/power-electronics-devices-patent-landscape/silicon-carbide-sic-patent-landscape-2022/
13. SiC power semiconductor technology: A future battlefield in the US-China trade war? Wolfspeed on the front line, KnowMade, January 2023, https://www.knowmade.com/technology-news/semiconductor-news/power-electronics-devices-news/sic-power-semiconductor-technology-a-future-battlefield-in-the-us-china-trade-war-wolfspeed-on-the-front-line/
14. Trench Warfare: a SiC Battleground, PGC Consultancy, January 2024., https://www.pgcconsultancy.com/post/trench-warfare-a-sic-battleground
15. https://en.wikipedia.org/wiki/SPICE
16. https://ieeexplore.ieee.org/document/7931077/
17. https://www.onsemi.com/pub/Collateral/TND6260-D.PDF
18. Boston Consulting Group (BCG) forecasts, 2018–2022
19. Global Memorandum of Understanding on Zero-Emission Medium- and Heavy-duty Vehicles, COP27, https://globaldrivetozero.org/mou-nations/
20. https://www.man.eu/corporate/en/zero-emission/zero-emission.html
21. High-Power Medium- and Heavy-Duty Electric Vehicle Charging, NREL, https://www.nrel.gov/transportation/medium-heavy-duty-vehicle-charging.html
22. Terawatt's California heavy-duty EV charging site offers charging now, during site build-out, https://www.evearly.news/en/terawatts-california-heavy-duty-ev-charging-site-offers-charging-now-during-site-build-out/
23. SAE J3271 specification, https://standardsworks.sae.org/standards-committees/j3271-megawatt-charging-system-electric-vehicles-tf
24. US DoE/Argonne National Lab, https://www.anl.gov/reference/faq-charging-for-heavyduty-electric-trucks
25. Pardhi et al., A review of fuel cell powertrains for long-haul heavy-duty vehicles: technology, hydrogen, energy and thermal management solutions. Energies **15**(24), 9557 (2022). https://doi.org/10.3390/en15249557
26. General Dynamics NF-16A AFTI, https://www.nationalmuseum.af.mil/Visit/Museum-Exhibits/Fact-Sheets/Display/Article/195759/general-dynamics-nf-16a-afti/
27. High Voltage (HVDC) Distribution & Conversion in Datacenters, Aerospace, Microgrids & Transportation, PSMA Power technology Roadmap presentation by Stephen Oliver, Vicor, https://www.psma.com/sites/default/files/uploads/files/02%20PSMA%20PTR%202014_07_10%20HVDC%20Oliver%20Vicor.pdf
28. Lilium Jet, Ducted Electric Vectored Thrust (DEVT), https://lilium.com/newsroom-detail/technology-behind-the-lilium-jet?_gl=1*26caj4*_up*MQ..*_ga*MTA2MzM4ODM0NS4xNzA4MDQyODUx*_ga_9YC7ETNZ98*MTcwODA0Mjg1MC4xLjEuMTcwODA0Mjg1OS4wLjAuMA
29. Lilium., https://lilium.com/newsroom-detail/first-high-voltage-electrical-harnesses-roll-off-the-line-for-the-all-electric-lilium-jet
30. Joby., https://joby-site.cdn.prismic.io/joby-site/5f82ea34-645e-4468-8e3f-14a16e298941_Joby-Charging-GEACS-final.pdf
31. Archer Aviation., https://archer.com/technologies
32. Infinitum air-core motors., https://goinfinitum.com/document-library/#AE-brochures

Suresh Thangavel is working at Infineon as a global system architect for photovoltaic applications. His main focus is to understand and translate application requirements, define and analyze concepts for solution offerings, support technical go-to-market activities, and provide level 3 support at the application level. Besides having more than 16 years of experience in the area of design and development of bipolar modules, STACK, and application engineering, he holds a master's degree in engineering with a specialization in power electronics and drives.

Aparna Kumar is a product marketing manager at Qorvo based in New Jersey. She received her Bachelor of Technology Degree in Electronics Engineering from the University of Mumbai (India). She has more than 11 years of experience in the semiconductor industry with the last 5 years of it in SiC. At Qorvo, Aparna is focused on customers in the power delivery and broad industrial sectors.

Mike Zhu (Ke Zhu) is an application engineer at Qorvo, Inc. He received his MS in Electrical and Computer Engineering from The Ohio State University in 2015. He has 10 years research experience in SiC and GaN device evaluation, design of high-frequency, high-power density power electronics, and EMI solution for WBG devices.

Christoph Ihm has 23 years of experience in the semiconductor industry at Infineon and Intel. He is currently responsible for global application marketing activities for commercial and utility photovoltaic as well as wind applications at Infineon's Green Industrial Power (GIP) division. Prior to that, he was working as a senior market intelligence analyst for wireless and computing applications at Infineon and Intel. He also worked as an investor relations manager at Infineon for several years. He holds a diploma degree in business administration from the University of Leipzig, Germany, and successfully completed the "Stammhauslehre" at Siemens.

Tobias Asshauer is a global application marketing manager for energy storage and hydrogen applications at Infineon Technologies. After starting his career as an electrical designer at ABB, he worked in various management positions as an engineering manager, a business developer, an operations manager for transformers, and a head of product marketing for dry-type transformers. He graduated from the University of Applied Sciences Soest with a Dipl. Ing degree in electrical engineering, majoring in power engineering and distribution, renewable energies, and high-voltage engineering.

Rémi Comyn holds a PhD in electrical engineering and physics from the University of Sherbrooke and Côte d'Azur University. For more than 5 years, Rémi has been working with KnowMade as a patent and technology analyst, unveiling IP, R&D, and business strategies of players engaged in the development of compound semiconductors for power and RF applications. Previously, Rémi worked for several years as a research engineer, focusing on gallium nitride (GaN) technology.

Zan Huang A graduate of Tsinghua University and the University of California, Zan Huang has more than 20 years of experience in the semiconductor industry and power electronics fields. He joined Transphorm in 2009 and became one of the pioneers in promoting wide bandgap high-voltage electronic components. Since joining Wolfspeed in 2017, he has held positions in various applications, sales, and marketing and is committed to promoting Wolfspeed's industry-leading silicon carbide technology.

Didier Balocco received his engineering degree from the *École Nationale Supérieure d'Électronique et de RadioÉlectricité de Bordeaux*, France, in 1992 and his PhD degree in power electronics from the University of Bordeaux in 1997. In 1996, he joined AEG Power Solutions, formerly Alcatel Converters, as a research engineer for DC-DC and AC-DC converter design in a range of 1 W to 1 kW mainly for telecom equipment. He managed the research activities from 2000 to 2014. He published more than 10 papers on power electronics. He holds 1 patent. From 2011 to 2013, he worked 18 months on a 15 kW solar inverter (DC-AC) module for a 150 kW cabinet in Dallas, Texas, USA. His main interests during this period were switching mode power supplies, converter stability and modeling, and high-power factor rectifiers. He joined Fairchild Semiconductor in August 2014 as a field application engineer (FAE) supporting south of France, Spain, and Portugal. In 2016, *onsemi* acquired Fairchild. In 2018, he moved to a new role inside *onsemi*. He is currently a business marketing engineer for Europe focusing on power devices and silicon carbide.

James Victory is currently a fellow at onsemi, working in modeling and simulation for power discrete and module technologies. In June 2008, he co-founded Sentinel IC Technologies. Prior to that, he was the executive director of IC design enablement at Jazz Semiconductor. He started his career with Motorola in 1992 where he specialized in semiconductor device modeling for RF analog and power technologies. He received his BSEE, MSEE, and PhD in electrical engineering from Arizona State University in 1990, 1992, and 1994, respectively. He has over 50 publications, including invited papers and workshop tutorials on semiconductor device modeling. He holds six US patents.

Stephen Oliver is the VP for corporate marketing and investor relations at Navitas Semiconductor, with over 30 years of experience in the power semiconductor and power supply industries worldwide. Prior to Navitas, Stephen served as VP of the VI chip product line at Vicor, and earlier at Motorola, Philips (NXP), and International Rectifier (Infineon). He holds a B.Eng (Hons) in Electrical and Electronic Engineering from Manchester University (UK) and an MBA in Global Marketing and Strategy from the UCLA Anderson School of Management. Stephen hold several patents in power semis, is a chartered engineer, and has been the VP, president, and chairman of the Power Source Manufacturers Association (PSMA).

Maurizio Di Paolo Emilio holds a PhD in physics and is a telecommunications engineer. He has worked on various international projects in the field of gravitational wave research, designing a thermal compensation system (TCS) and data acquisition and control systems and on others about X-ray microbeams in collaboration with Columbia University, high-voltage systems, and space technologies for communications and motor control with ESA/INFN. TCS has been applied to the Virgo and LIGO experiments, which detected gravitational waves for the first time and earned the Nobel Prize in 2017. Since 2007, he has been a reviewer for scientific publications for academics such as Microelectronics Journal and IEEE journals. Moreover, he has collaborated with different electronic industry companies and several Italian and English blogs and magazines, such as *Electronics World*, *Elektor*, *Mouser*, *Automazione Industriale*, *Electronic Design*, *All About Circuits*, *Fare Elettronica*, *Elettronica Oggi*, and *PCB Magazine*, as a technical writer/editor, specializing in several topics of electronics and technology. From 2015 to 2018, he was the editor-in-chief of *Firmware* and *Elettronica Open Source*, which are technical blogs and magazines for the electronics industry. He participated in many conferences as a speaker of keynotes for different topics such as X-ray, space technologies, and power supplies. Maurizio enjoys writing and telling stories about power electronics, wide bandgap semiconductors, automotive, IoT, embedded energy, and quantum computing. Maurizio has been an AspenCore content editor since 2019. He is currently editor-in-chief of *Power Electronics News* and *EEWeb* and a correspondent for *EE Times*. He is the host of PowerUP, a podcast about power electronics, and the promoter and organizer of the PowerUP Virtual Conference, a summit where each year great speakers talk about the power electronics design trends. Moreover, he has contributed to a number of technical and scientific articles, as well as a couple of Springer books on energy harvesting and data acquisition and control systems.

Chapter 7
Innovations and Advancements

Peter Friedrichs and Victor Veliadis

In this chapter of the book on innovations and advancements in SiC power devices, Peter Friedrichs, Vice President SiC at Infineon Technologies, and Victor Veliadis, Executive Director and CTO of PowerAmerica and ECE Professor at NC State, present insightful perspectives on the transformative potential of SiC technology. Friedrichs discusses material innovations, device enhancements, and ecosystem developments driving performance and cost savings in power electronics. Meanwhile, Victor Veliadis addresses current challenges in SiC technology, including production costs, defects affecting yield, and the need for a skilled workforce. Their combined insights offer a comprehensive outlook on advancing SiC technology to meet global energy demands and environmental goals.

7.1 Emerging Trends and Innovations

Peter Friedrichs

The biggest challenge for human beings right now is the preservation of our planet, namely, the reduction of CO_2 emissions to reduce global warming. At the same time, we are facing a growth of the worldwide population with a continuously growing demand for electrical energy. Traditional sources for energy based on fossil material such as coal and oil need to be eliminated step-by-step since their contribution to the climate change is prohibitive. The only alternative is renewable energy

P. Friedrichs
Infineon Technologies, Neubiberg, Germany

V. Veliadis (✉)
PowerAmerica, Raleigh, NC, USA
e-mail: jvveliad@ncsu.edu

M. Di Paolo Emilio (ed.), *SiC Technology*,
https://doi.org/10.1007/978-3-031-63418-5_7

sources and, thus, semiconductor-based power electronics will be required more and more. Wide-bandgap-based components such as SiC MOSFETs are preferred since they offer new levels of efficiency and power density. Both benefits meet the global trends and support a lot of emerging applications and new trends for power electronic systems.

Besides the known function of power semiconductors in the energy flow, namely, the conversion of electrical energy from the available source toward the consumer-friendly version, new elements become visible due to the nature of the modern energy sources. Since wind and solar as the most prominent examples are not in line with the consumption demands from a generation logistics point of view, storage elements will be essential components to manage the grid of the future. They will also play a substantial role in the worldwide charging infrastructure since a direct grid connection of high-power fast-charging farms cannot be realized in many cases by using the pure grid power capability. Storage is mostly known in the form of batteries that are stationary concepts. Here, the efficiency of the power semiconductors counts double since they both feed the battery and extract the energy. But also other kinds of green storage, e.g., the generation of hydrogen or ammonia by renewable energies in periods with low demand, will grow in the future.

Hydrogen can be transported and act as green fuel for mobile applications. An efficient generation of hydrogen via electrolyzer systems once more requires powerful semiconductors, ideally based on high-power wide-bandgap devices, due to their performance and efficiency. They will replace today's thyristor-based systems, as with the growing connection of electrolyzers to the net grid codes most probably will inhibit the future use of thyristor-based solutions. Furthermore, when connected to solar generation, a direct DC-DC conversion, for example, will take place, favoring the use of fast unipolar high-power switches such as SiC MOSFETs. Another trend, although requiring more substantial changes and efforts, e.g., on the regulation side, can be the step-by-step move from AC- to DC-dominated energy grids. Already today, long-distance energy transmission is favorably realized by HVDC systems. Solar and hydrogen sources originally generate DC, and many consumers require DC power, which is served by a DC link, such as inverterized motors. Thus, large-scale DC power grids might come up step-by-step, enabling significant material savings, e.g., for cables. For a large-scale rollout, however, technical hurdles such as safe disconnect principles need to be solved. Yet, a major roadblock will be the legacy and reluctance of the established grid. At a smaller scale, for example, in moderate household installations, this trend can probably be realized earlier. It will be strongly connected to the new scheme of so-called prosumers instead of pure consumers of electrical energy in the private sector. Solar panels, heat pumps, storage (via battery or hydrogen), and bidirectional charging are key elements already present in many households today. Efficiency and secondary features such as noise, space, and lifetime will serve as selling arguments in this sector.

Another trend, already evident in innovative designs today, is the ongoing reduction of cooling efforts, for example, by spending more semiconductor area than traditionally used, thereby avoiding bulky coolers. The availability of MOSFETs, with their linear I-V characteristic, opens up this opportunity for high-voltage applications as well. The benefits of such systems are clear: efficiency increases substantially, nearly no heat is generated that requires management by bulky heat sinks, the material used for such components is conserved, and the total system size shrinks. Furthermore, reliability increases significantly, as typical long-term degradations caused by temperature swings and gradients during operation are nearly eliminated, and other sources of failure, such as failed forced cooling elements (e.g., fans) or degrading thermal pastes between power semiconductors and heat sinks are no longer present. Finally, this shift will also help address a severe challenge regarding the thermal handling capability of typical packages that come along with further shrinks of the area used for a power chip with a given on-resistance (defining efficiency to a large extent). Keeping today's design principles in the form of utilizing the maximum power handling capability defined either by the power cycling performance or the maximum junction temperature pushes package technologies to their limits or requires expensive new solutions. By increasing the semiconductor area and thus keeping total losses and the connected loss density small, existing packages can still be used easily.

In addition, the features and capabilities of new wide-bandgap power devices can enable or trigger completely new system solutions. A current prominent example is the iron-less motor invented by Infinitum. The concept is characterized once more by a significant saving potential regarding scarce materials—in this case, iron and magnetics for motors. Furthermore, the concept reduces energy consumption and has a smaller footprint compared to traditional motors, representing a perfect match to the challenges we have for our future. However, it comes along with a very small inductance and thus required fast switching power semiconductors that are now available due to the rollout of SiC MOSFETs. Another example can be the replacement of mechanical breakers/fuses by semiconductor-based solutions.

Such components can act much faster than typical relay-like components, reducing damage to systems in case of failures. Furthermore, they can be combined with intelligence and communication features, which are prerequisites, for example, for the switch from a consumer at the end of an energy grid toward the future prosumer, who partially generates their own energy but also serves as a grid supplier. These components will differ substantially from the devices we know today, as, in contrast to the usual power semiconductors, they are not designed to switch permanently. Thus, dynamic performance will be of secondary importance, and the designer can fully concentrate on the lowest resistance in on-state. On the other hand, a robust overcurrent handling capability and the ability to turn off such currents fast and safely are key requirements for these new devices (Figs. 7.1, 7.2, and 7.3).

As a continuous trend, services around the pure power component will gain importance in the future. Health monitoring or, in general, smart and intelligent driver solutions will bring more and more functionalities into the power semiconductor world. Managing critical operating modes by circuitries around the power switch—mainly short circuits—seems to be an inevitable strategy in the future to further increase efficiency and power density of energy conversion systems.

Finally, the discussion about very high-voltage devices remains ongoing. SiC has proven multiple times the ability to serve as single-stage high-voltage solutions,

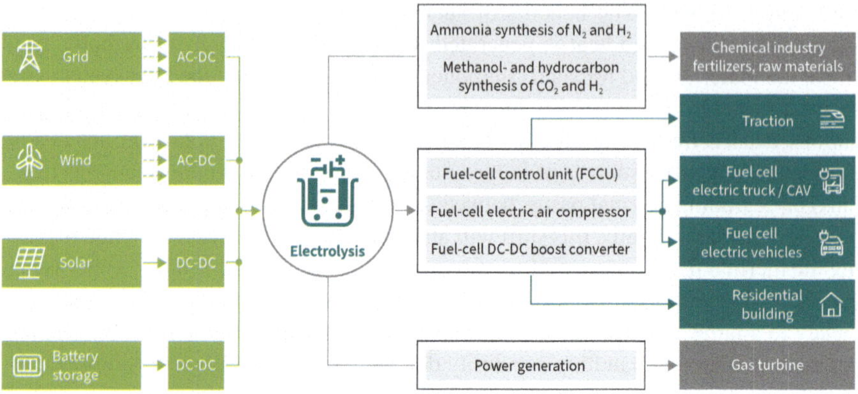

Fig. 7.1 Hydrogen as a storage and transportation medium for renewable energy

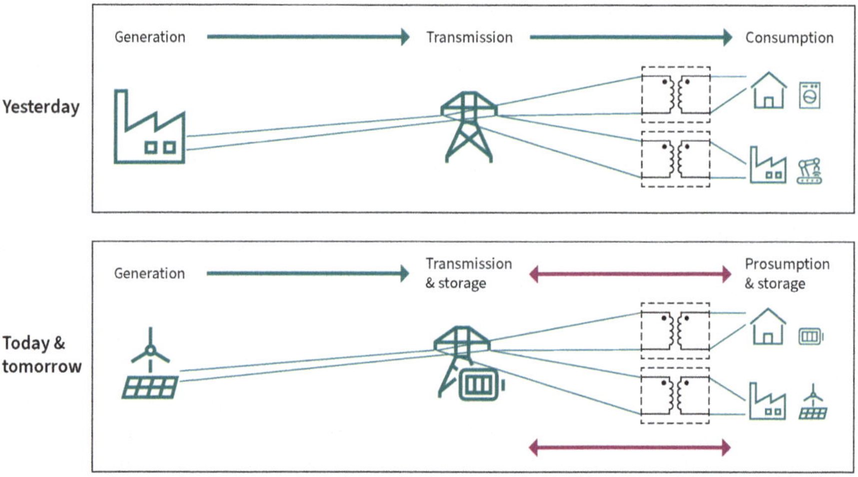

Fig. 7.2 Prosumer versus consumer—the energy chain of the future

such as in the form of MOSFETs or IGBTs with blocking capabilities well above 10 kV. However, these devices remain on lab scale. In fact, a strong pull from the industry is missing, even if the overall storyline with reduced numbers of stacks in HVDC applications seems straightforward. The major hurdles today can be seen in three areas. Firstly, the device itself—going beyond the rolled-out voltage levels such as 3.3 kV will likely require new device concepts, such as charge-compensated devices, and complex technology steps, mainly for growing the active layer by epitaxy. Commercially, it would work only if there were a common understanding of a building block definition in the form of a certain blocking voltage in this region, but such feedback is missing from the application side. Additional severe efforts will be required to develop a suitable housing technology capable of managing high voltages and higher switching frequencies as expected from such switches. It is a matter

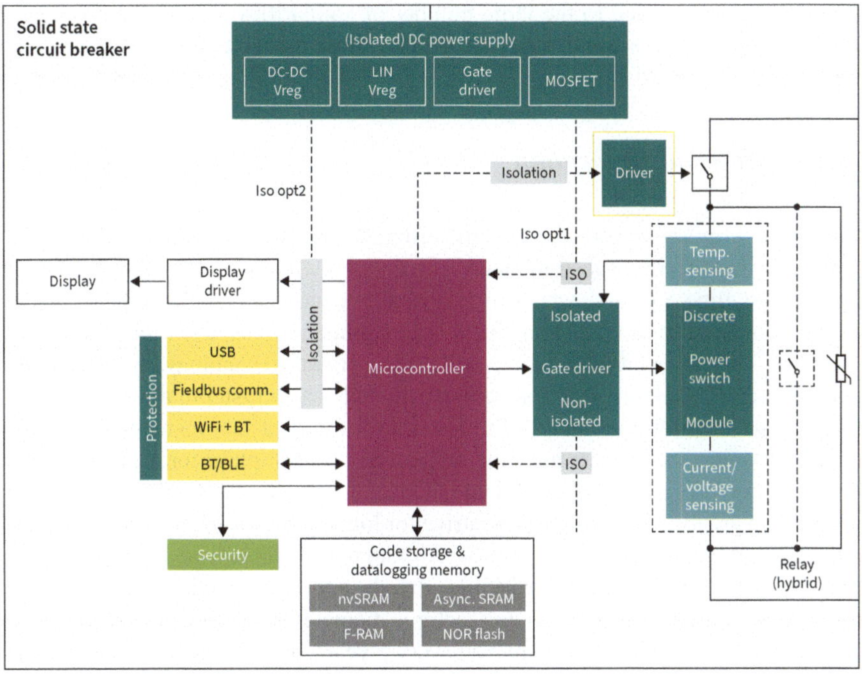

Fig. 7.3 Intelligent solid-state breaker principle

of fact stacking more components with lower voltage capabilities which will bring substantial advantages regarding loss distribution and thermal handling and will therefore provide better efficiency. Secondly, for the intended system solution, other high-voltage elements besides the semiconductor are required, mainly capacitors. As long as the complete picture is not visible, a fast implementation is questionable. Finally, the targeted industry is characterized by extremely long design cycles and small innovation steps to reduce risks in the end application. Thus, the payback of enormous development efforts will be visible on a long-term basis, impacting the priority to spend development efforts in this direction.

7.2 Potential Advancements and Their Impact

Peter Friedrichs

SiC power devices are still in the emerging phase compared to their silicon-based counterparts. Consequently, there is immense potential for substantial improvements in performance and related cost savings. Some aspects of the SiC development process mirror the learning curves seen in the silicon world, such as the innovation of cell concepts for switching devices, which can often be transferred from silicon to SiC due to the large number of similarities in basic device ideas. However, other aspects are unique, particularly those related to material.

7.2.1 Material Innovation

One of the primary reasons for the cost difference between silicon and SiC-based devices is the different substrate material technology and maturity. The growth of single-crystalline SiC is slower, which typically achieves crystal lengths of below 10 cm compared to three to four times longer silicon ingots, even for 300 mm in diameter. This is mainly due to the underlying growth technology, as silicon can only be pulled from a melted source using a small seed crystal in the case of SiC, whereas sublimation growth (PVT) is currently used for commercially available material. Major cost drivers are time and energy consumption for a given crystal length. Therefore, leading players in the field are using advanced PVT technologies to speed up growth (more mm/h) and strive for longer crystals in one run to increase efficiency and reduce consumable costs. Although this data is typically not publicly accessible, it appears that this path has already added significant productivity in recent years. Another measure in this sector is diameter enlargement. Historically, a new step in diameter has been achieved every 5 years, using the same milestones as for silicon wafers, except that 125 mm wafers have been skipped for SiC. However, the impact of a larger diameter only slowly translates into a measurable benefit for the end user, as initially, yield and defect density are lower compared to the previous diameter, and thus, the cost per cm^2 might temporarily even increase during the transition phase. A larger diameter does offer higher productivity in the form of more capacity, but this scaling is much less than using the pure area factor due to the nature of many processing steps.

In addition to advancements in PVT, new growth technologies are regularly investigated. Currently, solution growth, which features significantly reduced numbers of certain critical defects such as basal plane dislocations (BPD), is in the focus of several startups. However, the size of such material is currently still limited. Additionally, the high-temperature CVD technology (HTCVD) is gaining new attention after being discarded decades ago due to severe limitations regarding the achievable crystal length. To evaluate the commercial potential of such alternatives, more data points need to be collected.

Other approaches are based on layer transfer technologies such as SMARTSiC by Soitec or SiCKREST by SiCOXS. These technologies use just a small high-quality seed layer for epitaxy, which is transferred from a reusable monocrystalline donor to a potentially low-cost carrier. The potential of this technology still needs to be completely evaluated. While it can address high-quality material bottlenecks due to the multiplication factor, the transfer process itself and the cost of the carrier need to be low enough to be competitive with standard material in the long run.

Finally, besides the growth, the wafer production process can also be optimized, as a significant material loss impacts cost of wafers produced by the abrasive and yield risky multi-wire sawing process. Currently, many groups are working on virtually loss-free separation technologies, mostly laser-assisted, such as the KABRA process or cold split technology.

7.2.2 Device Innovation

The development focus is primarily on chip design and technology, which is currently most visible in power transistors, while diode technology has a secondary priority. In future device generations, advancements are expected to be developed in three key areas that are crucial for the performance and eventual cost position of new power switches: cell design, gate concept, and drift zone design. These three segments contribute to essential performance indicators such as area-specific RDS_{on} or switching performance, as well as final reliability and ruggedness under extreme operating conditions.

The initial focus is still heavily on the cell concept and layout, with cell shrink being the primary target. As a result, trench-based devices with critical dimensions in the vertical direction will increasingly become mainstream. Despite differences in design principles between silicon and silicon carbide power switches, driven by the need to shield the gate oxide more effectively in blocking mode in SiC MOSFETs compared to silicon, it is anticipated that successful concepts developed for silicon devices can be reused in SiC. Examples include FINFET-like concepts with 3D gate arrangements.

Currently, work on the gate structure is directed in two directions. In the short term, further improvements based on the silicon dioxide structure are a target for innovative research. However, the achievable channel mobility, a key parameter defining the final channel resistance, is well below the bulk mobility (max around 10% while silicon achieves more than 90%). Breakthroughs are not ruled out, and history has shown leapfrog-like steps, such as when NO-based treatments were introduced. A second path is the use of alternative dielectrics such as high *k* materials. The advantage in case of success if obvious—much lower electric fields in blocking mode and better threshold voltage control. However, like in the silicon world, current drawbacks such as high leakage currents and lack of data regarding long-term stability need to be addressed. Finally, when it comes to improvements in the drift zone design, learnings from the silicon technology are being investigated, such as co-doping effects to enhance mobilities (such as SiGe concepts in silicon) or the switch to charge-compensated drift zone design. The latter approach promises significant steps forward regarding the $RDS_{on} \times A$, but the technological implementation is very challenging due to lack of diffusion in SiC and limited implantation depth when standard tools are considered. Thus, the final cost performance achievable with such an approach will decide its commercial success.

In addition to the core elements of the actual power switch, more functionalities can be implemented into the chip, mainly sensing and protection elements. Many examples from the silicon world can be reused, for instance, current sense designs using current mirror technologies, temperature sensor based on well-designed resistors on-chip, or gate protection elements to avoid breakdowns under spikes at the gate.

7.2.3 Innovation Impacting the Ecosystem Around the Chip: Package and Control

To leverage the full potential of silicon carbide-based power devices today and even more important in the future, substantial advancements are needed in the components around the chip. Ongoing chip shrinks will result in increasing power densities to handle at the package level—unless the trend to replace cooling efforts by chip area becomes mainstream. Higher power densities at the package level require new die attach methods, such as those already implemented by Infineon in its discrete device under the label. XT.

In this case, the typical backside solder is replaced by a joint formed directly between the die and the lead frame, creating a perfect thermal contact between the two thermally highly conductive materials SiC and copper. This technology not only improves thermal impedance substantially but also extends the lifetime under cycling conditions. Regarding the backside connection, further improvements can be expected from heat spreading layers, which will be enabling elements for very small dies in discrete housings, for example. Although today, through-hole package platforms such as TO247 still cover a large portion of SiC-based products, it is expected that more and more SMD packages will be introduced due to the easier assembly process and the meanwhile significantly improved thermal handling capabilities. This trend is combined with increasing confidence at the user side regarding cooling via well-designed power PCBs and/or the utilization of top-side cooling approaches, recently introduced for high-power SMD packages as well. Foreseeable trends in this segment include higher peak temperature capabilities to cover overload conditions and new mold materials that can improve the ruggedness under extreme conditions measurably.

Furthermore, the integration of small systems, such as half-bridge circuits, partially with drives in the same package, will become an interesting solution for customers in applications with short design cycles. The border between classical discrete mold packages and so-called IPMs will become fuzzier in the future, also driven by the possibility of having sensing and protection functions on-chip, as mentioned in the previous paragraph. For classical frame-based power modules, addressing power handling capabilities of 50 kW and above, it is expected to see more solutions with reduced stray inductance, advanced ceramics for better thermal handling capability, and improved die attach to enhance power cycling capabilities. Front-side die attach trends are clearly alternatives to bond wires, mainly for high-power densities. Clips for double-side cooled setups based on soldering processes

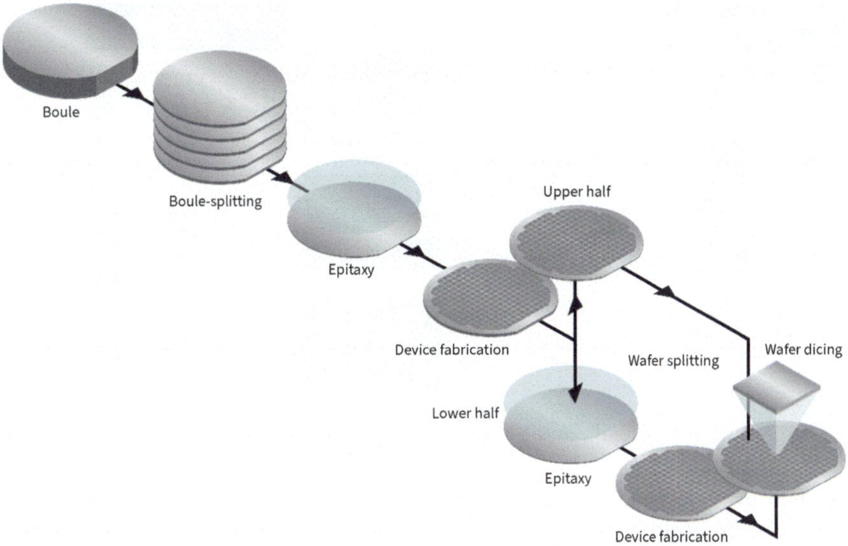

Fig. 7.4 Material innovations enabled by cold split

Fig. 7.5 Device structure, for example, trench MOSFET, with labels for the different regions

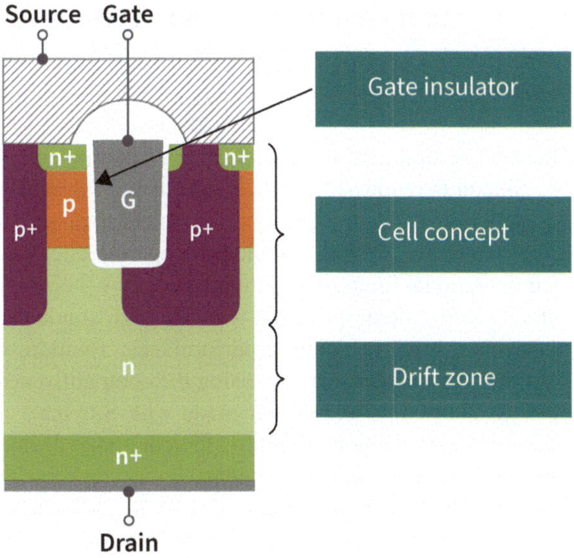

are expected to gain attention in the future for both modules and discrete housings (Figs. 7.4, 7.5, and 7.6).

The next level in the ecosystem of a power device is the control element connected to the gate. Future innovation will target taking over more and more protection functionalities into the driver and out of the chip in order to utilize the full

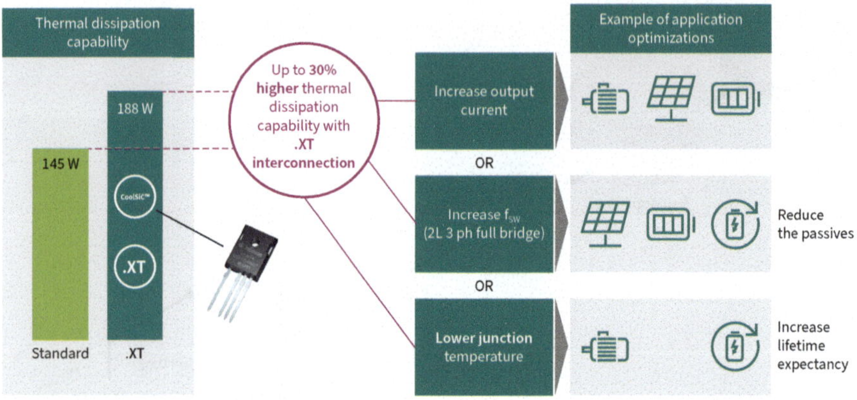

Fig. 7.6 XT technology for die attach improvement—potential in the application

potential of wide-bandgap transistors. Examples include smart short-circuit handling, gate-controlled overcurrent protection, or safe turn-off from high currents without critical spikes.

7.3 Current Challenges in SiC Technology

Victor Veliadis

There are several reasons behind silicon's dominance of the power electronics market. Silicon is renowned for its excellent starting material quality, ease of processing, opportunity for low-cost mass production, proven reliability, and circuit design legacy. However, despite significant progress, silicon devices are now approaching their operational limits. They are held back by their relatively low bandgap and low critical electric field, traits that result in high conduction and switching losses and substandard high-temperature performance. To address these shortcomings, much effort has been directed at increasing the competitiveness of commercial SiC power devices. Transistors and diodes made with SiC have superior material properties, enabling the production of highly efficient power devices with a smaller form factor and simplified cooling management.

It has been over two decades since Infineon launched the world's first commercial SiC Schottky diode in 2001. The first commercial SiC power MOSFET was released by Cree in 2011. As SiC continues to grow to a projected US$10 billion market by 2029 [1], the industry is lifting the last barriers to mass commercialization, which primarily come down to three challenges. First, compared to mass-produced silicon, costs are higher—chiefly due to the complexity of synthesizing SiC substrate and to how much more labor-intensive SiC fab manufacturing and modules are. Defects are the second significant issue, as they limit chip yield and area and compromise reliability and ruggedness. The third issue is that the workforce lacks expertise in integrating SiC technologies into systems—like many industries, the power electronics industry is traditionally slow to change and adapt to new technologies.

7.3.1 Cost Analysis

Today, the SiC wafer represents 50–65% of the overall SiC device cost [1], a consequence of its unique complex fabrication specifics. Conventional SiC substrates are primarily grown through seeded sublimation, at temperatures of 2300–2500 °C, which creates process control challenges. Because crystal expansion is limited, requiring the use of large high-material quality seeds, sublimation growth rates can be relatively low, in the order of 0.5–2 mm/h. Furthermore, the hardness of SiC, which is comparable to that of diamond, makes sawing and polishing SiC substrates slow and costly relative to Si.

The epitaxial layers, where SiC devices are fabricated, are grown by chemical vapor deposition in horizontal or planetary reactors at 1500–1650 °C. Pressure typically ranges from 30 to 90 Torr, and growth rates can be as high as 46 mm/hr. Epitaxial growth is done on 4-degree off-cut substrates to maintain the polytype stability of the substrate. Since defects in SiC wafers limit large-area device yields, and numerous devices are paralleled in modules to increase current output, tight epitaxial doping and thickness uniformities are required—particularly as the wafer area increases.

For mass SiC commercialization, high-yielding fabrication processes are required. Numerous well-established processes from silicon technology have been successfully transferred to SiC. However, SiC material properties necessitate specific processes not available from silicon, including wafer thinning, etching, heated implantation and anneal, and low-resistivity Ohmic contact formation. SiC is inert against chemical solvents, and only dry etching is practical. Furthermore, the hardness of SiC results in low photoresist selectivity, and a "hard" mask, usually composed of metal or dielectrics, is required for SiC photolithographic patterning and etching. Conventional thermal diffusion is not practical in doping SiC due to its high melting point and the low diffusion constant of dopants within SiC—600–900 °C heated ion-implantation is typically performed. After ion implantation, a 1600–1800 °C anneal assists in lattice damage recovery and dopant electrical activation. During the anneal, a protective cap layer protects the wafer surface from degradation due to the higher than carbon desorption and migration of silicon surface atoms. The high value of the SiC/metal barrier results in rectifying metal contacts, and post-metal deposition anneal is required for Ohmic contact formation. Typically, a 50–100 nm Ni layer is blanket deposited and patterned on the wafer for the simultaneous Ohmic contact formation on the n-type- and p-type-doped regions. Unlike Si wafers, SiC wafers are transparent. This complicates the use of "silicon" tools for inspection and metrology measurements, as the focal plane is determined with the use of an optical microscope. SiC-specific wavelength metrology/inspection tools are now available from multiple vendors. Another issue is the SiC wafers' relative lack of flatness, compared to that of Si, which can complicate photolithography. In addition, the high temperatures of SiC processing can further degrade wafer flatness, occasionally rendering wafers unusable. This is particularly problematic with the thick epitaxy wafers used in +3.3 kV device fabrication. Lastly, the poor SiC/SiO_2 gate interface quality reduces inversion layer mobility and causes threshold voltage instability. Thus, passivation techniques, including annealing in nitrides, are utilized to improve the SiC/SiO_2 interface quality, similar to what is done in silicon [2].

Overall, SiC wafer fabrication and fab processing are more complex, more time-consuming, and more labor-intensive than those in silicon. The result is more expensive manufacturing, contributing to higher device costs. A key part of the vertical integration occurring in today's SiC industry is securing internal substrate and epitaxy wafer capabilities to eliminate "outside purchasing" profit margins. In addition, investments in disruptive SiC wafering, such as engineered substrates and synthesis, boule slicing, and sawing/polishing with less wasted material, have high returns and are actively being sought.

SiC wafer processing savings can be introduced by shifting production from today's 150- to 200-mm wafers. Presently, the majority of SiC device production is on 150-mm wafers. The 200-mm SiC wafers were demonstrated in 2015 and are in the process of becoming commercially available. For wide adoption, it is highly desirable that the 200-mm wafer defect density and cost of material, per square area, are equal or lower than those of 150-mm wafers. In addition, wafer planarity should not be of lower quality in 200-mm wafers. Due to the large fab overhead costs, the price of a fully processed wafer is to a first approximation independent of its area. Thus, the cost of processing a 150-mm wafer is similar to that of a 200-mm wafer, with the latter producing about 1.8 times more devices. Of course, a 200-mm SiC wafer will be more expensive than a 150-mm wafer, and that needs to be factored into the overall cost equation. Several silicon fabs are starting to process SiC wafers, and given the plethora of mature 200-mm fabs with fully depreciated equipment, there are many companies waiting on the sidelines to enter SiC production when 200-mm wafers become widely available. These are companies that have established 200-mm silicon wafer production and do not want to retool to fabricate the 150-mm SiC wafers that are currently commercially available. To illustrate the economic benefits of moving to larger SiC wafer area, let us assume cost and defect-density parity, per cm^2, between 150- and 200-mm wafers. Further assuming a \$1500 cost for a fully processed 150-mm SiC wafer—60% of that coming from the "starting" wafer and thus 40% from fab manufacturing cost—a back-of-the-envelope calculation points to a 17% device cost reduction when switching to 200 mm. The same fabrication cost scenario, with 50% of the overall cost coming from the "starting" wafer, allows for a 22% device cost reduction when switching to 200-mm production. These calculations do not include the additional processing cost reductions of streamlined mass production using the newer equipment of 200-mm volume fabs, nor the higher yield of "edge" devices in larger-area wafers.

Device manufacturers have developed IP for several SiC process steps and compete on both design and processing. Although SiC is not fully "silicon fab" compatible, the SiC industry has leveraged Si technology processes and infrastructure by making the relatively small financial investments required to adapt existing fabs. Today, SiC manufacturing has matured, and its fab infrastructure mirrors that of Si. Integrated SiC device manufactures coexist with foundries and fabless companies, and design houses provide know-how and IP that can be leveraged to accelerate entry to market, as shown in Fig. 7.7. Securing a supply of high-quality substrates as the industry rapidly ramps up production to meet demand is of particular importance in SiC device manufacturing.

With respect to fabrication, the SiC industry is successfully leveraging the fully depreciated legacy silicon fab infrastructure and is making the relatively small financial investments that allow mature silicon fabs to process SiC. Consequently, SiC chip fabrication alongside silicon has emerged as a cost-effective model that exploits silicon manufacturing economies of scale. Through repurposing mature, fully depreciated 150-mm (and soon 200-mm) Si fabs, SiC power devices can be manufactured with the relatively small investments necessary to support the unique SiC processing steps. Minimizing fabrication cost by exploiting the mature Si volume production assumes the fab is loaded close to capacity with standard Si and SiC processes running in the same lines. In addition, aggregating the demand for SiC

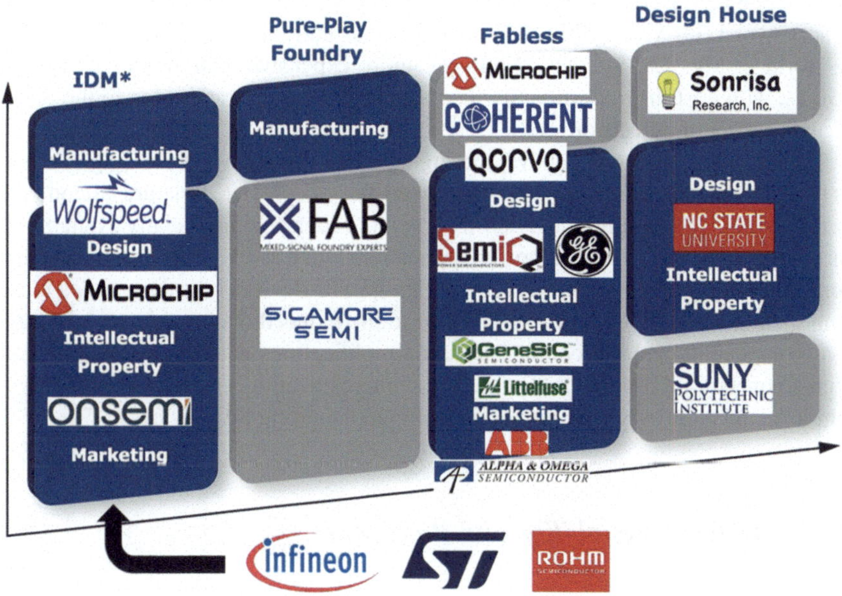

Fig. 7.7 The US mature SiC fab infrastructure mirrors that of Si (not an inclusive list). It consists of *integrated device manufacturers, foundries, fabless companies, and design houses. ST Microelectronics, Infineon, and ROHM are large non-US-based IDMs with a big share of SiC volume

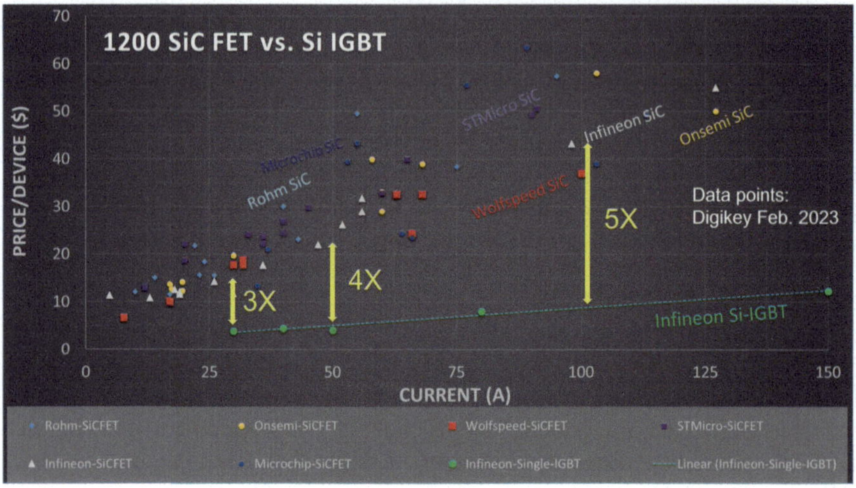

Fig. 7.8 1200-V SiC MOSFET pricing versus current rating. The SiC MOSFETs are more expensive than similarly rated Si IGBTs, with the cost disproportionally increasing with current rating due to the detrimental impact of defects on yield

substrates and epilayers in volume fabs contributes to lower material costs. Lower fabrication costs in a fully depreciated "Si + SiC capacity-loaded" fab, coupled with decreased material costs, can lead to significant price reductions for SiC devices [3]. A survey of 1200-V SiC MOSFET pricing versus current rating is given in graphical form in Fig. 7.8. SiC MOSFETs are more expensive than similarly rated Si IGBTs, with cost disproportionally increasing with current rating due to the detrimental impact of defects on yield.

Based on the discussion above, the higher cost of SiC devices is being lowered as mass production brings economies of scale, fully depreciated mature silicon fabs introduce SiC in their lines, and 200-mm wafers become commercially available. It should be noted that in several power applications, such as PV, price parity with silicon is achieved at the system level even though SiC devices cost three times more than their silicon counterparts. This is due to the reduced mass/volume of the passive components and the cooling system simplifications enabled by the efficient high-frequency SiC operation. For mass commercialization, cost savings at the system level must outweigh the higher cost of manufacturing SiC devices.

7.3.2 Defects, Reliability, and Ruggedness

"Killer" defects have been virtually eliminated in modern SiC wafers. Basal plane dislocations (BPDs) are the last major defect degrading device performance [4]. When bipolar current flows through a SiC device, electron-hole pair recombination at BPDs provides the energy to activate dislocation glides, which give rise to stacking faults that degrade electric performance. BPDs can propagate from the wafer substrate through the thickness of the epitaxial layers where devices are fabricated. Luckily, in commercial wafers, more than 95% of substrate BPDs propagate as relatively "benign" threading edge dislocations in epitaxial layers grown off-axis by chemical vapor deposition [5]. BPDs can also be generated during epitaxy and during the high-temperature ion-implantation fabrication process. BPD-related degradation can be readily identified in post-fabrication production testing and is thus a yield issue, not a reliability one. Defects impact yield, disproportionally increasing the cost of large-area devices (see Fig. 7.8).

Threshold voltage instability is the main remaining reliability concern in SiC MOSFETs, which are the dominant transistors in SiC-based power electronics, and is primarily due to the high density of oxide traps at the SiC/gate-oxide interface. A positive shift in the MOSFET's threshold voltage increases resistance and associated conduction losses, while a negative shift is undesirable as it makes the device susceptible to unintentional turn-on. In addition, oxide traps reduce electron mobility, leading to higher channel resistance. SiC MOSFETs exhibit short-circuit withstand times well below the 8–10 μs of silicon IGBTs. Short-circuit withstand time increases with longer MOSFET channel length and device area, but both increase resistance. SiC devices can be made more rugged through design trade-off optimizations. Fast gate drives, with prognostic and diagnostic functions, have been developed and provide adequate short-circuit protection within the typical 3–4 μs SiC MOSFET short-circuit capability [6]. Extensive reliability data have been accumulated over many years of field operation and are building confidence in the insertion of SiC devices in power electronic systems.

7.3.3 Education and Workforce Training

A highly skilled workforce is the third challenge to creating a substantial demand for wide bandgap devices, which in turn spurs SiC mass manufacturing and lowers overall costs. The high dV/dt and dI/dt of power SiC devices can lead to severe overshoot and ringing system issues, impacting reliability, safety, protection, efficiency, and performance. Circuit insertion of high-voltage SiC packages and modules must address the high dV/dt and related exacerbated EMI/EMC issues. Entities such as PowerAmerica carry out university–/industry-applied collaborative projects, offer industry-driven WBG short courses and tutorials at conferences, and match students with industrial internship opportunities. These activities train the existing workforce and prepare the next generation of SiC technologists, facilitating accelerated deployment of SiC power electronics. Additional educational initiatives run by PowerAmerica include free, available-to-all monthly technical webinars delivered by member experts; WBG lectures, which are being added to power electronics courses across member universities; and tutorials offered at the winter and summer PowerAmerica member meetings, which provide an informal setting for learning and networking.

References

1. Yole Group market research., https://www.yolegroup.com/
2. S. Wolf, R.N. Tauber, *Silicon Processing for the VLSI Era*, vol 1 (Lattice Press, 1986), pp. 222–223
3. V. Veliadis, Compd. Semicond. Mag. **25**(36), 36–42 (2019)
4. V. Veliadis, H. Hearne, E.J. Stewart, M. Snook, W. Chang, J.D. Caldwell, H.C. Ha, N. El-Hinnawy, P. Borodulin, R.S. Howell, D. Urciuoli, C. Scozzie, Degradation and full recovery in high-voltage implanted-gate SiC JFETs subjected to bipolar-current stress. IEEE Electron. Dev. Lett. **33**(7), 952–954 (2012)
5. T. Kimoto, Jpn. J. Appl. Phys. **54** (2015), 040103-1-27
6. A. Kumar, S. Parashar, S. Sabri, E. Van Brunt, S. Bhattacharya, V. Veliadis, IEEE 30th Interernational Symposium on Power Semiconductor Devices and ICs (ISPSD), (2018), pp. 423–426

Peter Friedrichs is the vice president SiC at Infineon Technologies. He has been with the company since 2011. Prior to that, he was with the Corporate Research and Development of Siemens AG and the Fraunhofer Institute for Integrated Circuits in Erlangen. Peter Friedrichs has received the Ing. diploma in electrical engineering from the Slovak Technical University of Bratislava in 1993 and the Dr.-Ing. degree from the Friedrich-Alexander University of Erlangen.

Victor Veliadis is the executive director and CTO of PowerAmerica, a member-driven consortium of industry, universities, and national labs accelerating the commercialization of energy-efficient silicon carbide and gallium nitride power semiconductor chips and electronics. At PowerAmerica, he has managed a budget of $156 million that he strategically allocated to over 210 industrial and university projects to catalyze SiC and GaN semiconductor and power electronics manufacturing, workforce development, and job creation. His PowerAmerica educational activities have trained 430 full-time students in collaborative industry/university WBG projects and engaged over 7000 attendees in tutorials, short courses, and webinars. Dr. Veliadis is currently negotiating a $64 M US Department of Energy PowerAmerica renewal to further catalyze WBG power technologies.

Victor Veliadis is an ECE professor at NC State and an IEEE fellow and EDS distinguished lecturer. He has 27 issued US patents, 6 book chapters, and over 150 peer-reviewed publications to his credit. He has given over 150 keynote/tutorial/invited presentations including keynotes at ICSCRM, APEC, ECCE, and WiPDA. Prior to entering academia and taking an executive position at Power America in 2016, Dr. Veliadis spent 21 years post-PhD in the semiconductor industry where his work included design, fabrication, and testing of SiC devices, GaN devices for military radar systems, and financial and operations management of a commercial semiconductor fab. Dr. Veliadis has a Ph.D. degree in electrical engineering from Johns Hopkins University (1995).

Chapter 8
SiC Market

Aleksei Cherkasov

The semiconductor market has been dominated by silicon-based power components for a very long time. Silicon metal–oxide–semiconductor field-effect transistors (MOSFETs) and insulated gate bipolar transistors (IGBTs) have taken the bigger part of power electronics market, and only a few could think of any new semiconductors to shatter the first-choice product of the majority of engineers worldwide. If you would ask any of semiconductor market experts 10 or 15 years ago about the prospects for silicon carbide (SiC) to break into massive semiconductor industry and successfully compete with much cheaper silicon, they would fail to predict the current situation.

Silicon carbide as a semiconductor material has been familiar to engineers for over a century. Offering many technical benefits compared to silicon, many companies in this field have been involved in research and development to get closer to commercialization of SiC. And it was not until 2001 when CREE introduced the first SiC Schottky diode chips and 2011 when they brought to life the first SiC power MOSFETs. Since then, what seemed impossible, or extremely hard to introduce to the semiconductor mass market, step-by-step began to bring value to power electronics engineers and gain its place among the other electronic components.

Today, even those who are not involved in semiconductors business know well about the new wide bandgap (WBG) material. Widely adopted by the electric car manufacturers, silicon carbide displaces silicon from the automotive power electronics market.

Right now, companies have various approaches to vehicle electrification such as mild-hybrid electric vehicles (MHEV), full hybrid electric vehicles (HEV), plug-in hybrid electric vehicles (PHEV), battery electric vehicles (BEV), and fuel cell electric vehicle (FCEV). Some car manufacturers completely stopped production of

A. Cherkasov (✉)
Infineon Technologies, Villach, Austria

M. Di Paolo Emilio (ed.), *SiC Technology*,
https://doi.org/10.1007/978-3-031-63418-5_8

internal combustion engines and now focus on xEVs only. What 20 years ago seemed like a big step forward is a reality we live in now.

And to make this dream come true became possible with the help of power semiconductors. For a long time, semiconductors were used in automotive industry, and the evolution of power semiconductor materials pushed the transition to electrification of vehicles. Electric vehicle performance and cost depend on the technical level of the motor control system. Previously, silicon (Si) IGBT modules served as a heart of electronic control system with their relatively high switching speed and low conduction loss. But with the growth of silicon carbide (SiC) technology, xEVs step into the new era of electrification.

The real revolution for SiC power devices happened when Elon Musk's Tesla Model 3 became the first mass-produced commercial passenger electric car to use silicon carbide MOSFETs from the two leading semiconductor companies—Infineon Technologies and STMicroelectronics. This made Tesla the first company to implement SiC MOSFETs in the main traction inverter of the electric propulsion system.

Being one of the pioneers in the electric vehicle market and the one who accelerated the shift from the internal combustion engine cars to the electric ones, Elon Musk and his Tesla cleared the way for the semiconductor companies to concentrate their new investments on bringing more silicon carbide semiconductors qualified for the automotive power electronics.

Thus, according to the leading company in production of the silicon carbide wafers Wolfspeed, the cost of SiC-based semiconductors in the electric vehicles will multiply almost five times by 2027 and reach 5000 US dollars per electric passenger car. This forecast in turn demonstrates great prospects for suppliers of SiC power discretes and modules. An additional incentive is the announcement by the leading automakers of more than 500 billion US dollars of investments in the development of electric vehicles by 2030 [1].

The success of SiC power devices in the automotive and electric vehicle industry, including the commercial, public, agricultural, marine, and other vehicles, encourages and continues to motivate silicon bound industries to introduce new products and projects using silicon carbide semiconductors. Among them are the renewable energy companies adopting SiC solutions in the converters and inverters, industrial organizations building general-purpose SiC-based motor drives, uninterruptible power supplies (UPS), energy storage systems (ESS), and, of course, electric vehicle charger manufacturers.

During the past 5 years, the SiC-based power semiconductor market has grown from hundreds to billions of US dollars showing an incredible year-to-year growth. Every year, the leading market analysts predicted market growth, but no one could imagine that the growth would be so high and so fast. In 2021, the SiC power device market was predicted to reach six billion US dollars by 2027. As shown in Fig. 8.1, today, this number almost doubled, and the new estimates tell us to expect that penetration of the silicon carbide into power electronics will reach ten billion US dollar share by 2027–2028, and by 2030, the silicon carbide power device market may reach unbelievable 14 billion US dollars [2].

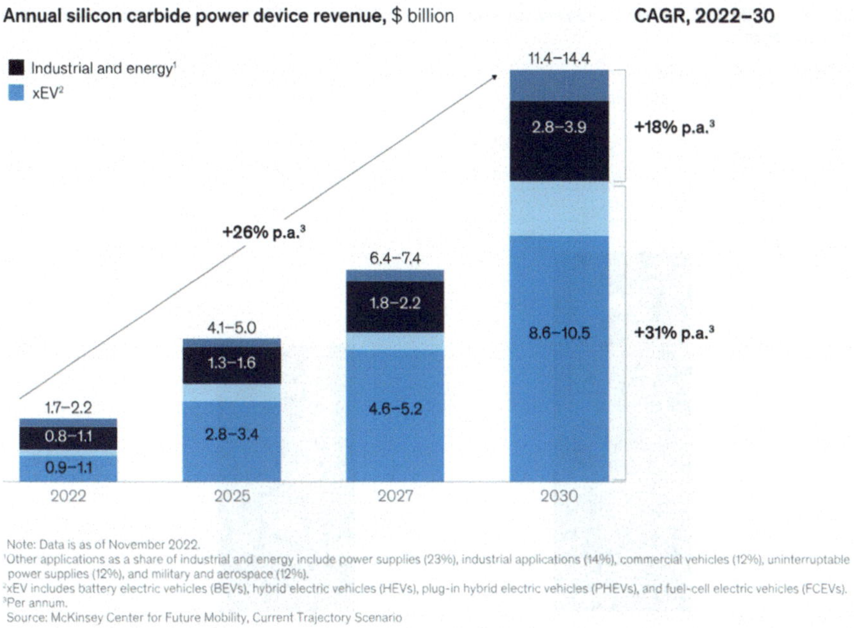

Fig. 8.1 Annual silicon carbide power device revenue

Nowadays, the silicon carbide market can be divided into two major categories—companies that produce SiC wafers and companies that produce SiC-based devices—power discretes and power modules. The SiC power device market is growing rapidly with many new companies going into this field across Asia, especially in China. On the other hand, the SiC wafer market is much more difficult to enter, and 90% of it is controlled by a few leading companies.

The SiC-based power semiconductor industry is still dominated by companies from the EU, the USA, and Japan due to their earlier entry into the market. STMicroelectronics has been leading SiC device revenue over the past few years. The company generated around $700 million in 2022 and expected to exceed $1 billion in SiC sales revenue in 2023 [3]. Followed by the leading global power electronics company Infineon Technologies, it showed significant revenue growth in the past several years. STMicroelectronics introduced its first SiC diodes back in 2004. The SiC MOSFET was introduced in 2009 and launched into mass production in 2014.

As shown in Fig. 8.2, in 2022, the following companies accounted for 96% of all SiC-based power semiconductor devices produced in the world:

STMicroelectronics
Infineon Technologies
Wolfspeed
ROHM
Onsemi
Mitsubishi Electric

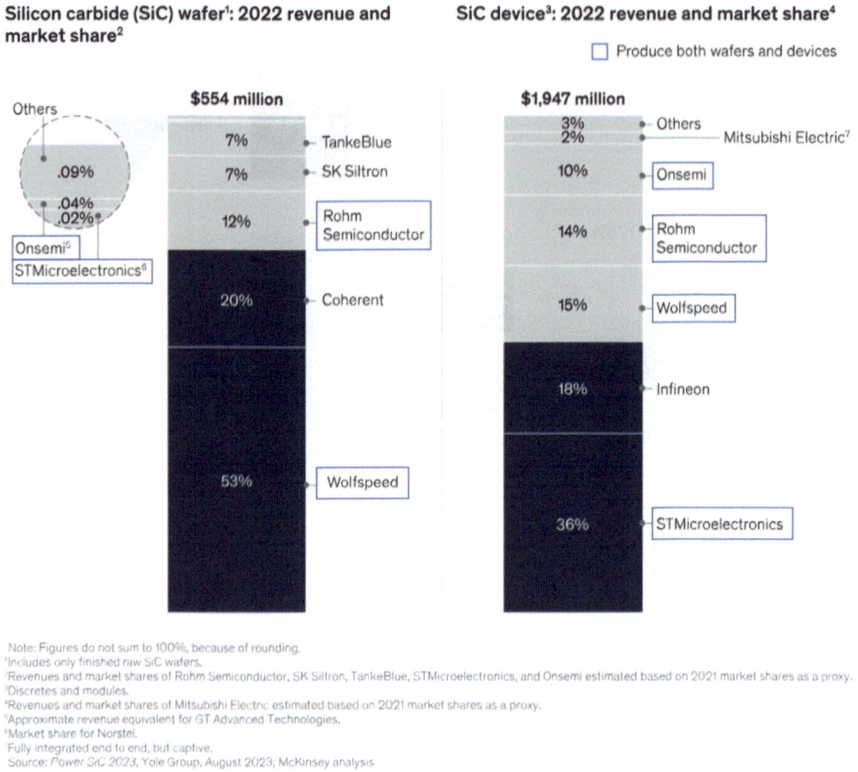

Fig. 8.2 Silicon carbide wafer and device market split by manufacturers

Even though silicon carbide has been known for a long time, the production of the first SiC-based power devices began only recently, in the early 2000s, when wafers with a diameter of 4 inches were introduced. A few years ago, manufacturers such as Resonac (former Showa Denko), Coherent (former II-VI), and Wolfspeed completed the transition to 6 inches, while large-scale production of 8 inches wafers is planned to be launched in the next 1–2 years.

Given the forecast of the research company Canaccord Genuity shown in Fig. 8.3, demand for SiC wafers will continue to exceed supply, and the gap will only increase by 2030. This means the further growth of investments in SiC both from existing and the emerging manufacturers of silicon carbide wafers and power devices.

In 2022, as it is shown in Fig. 8.2, the leading position in production of SiC wafers belonged to Wolfspeed. Considering their investments announced from 2021, they will continue to hold their leadership position at least till the end of this decade.

At the moment, there are more than 200 companies globally that produce power semiconductor devices, as well as more than a hundred companies from the

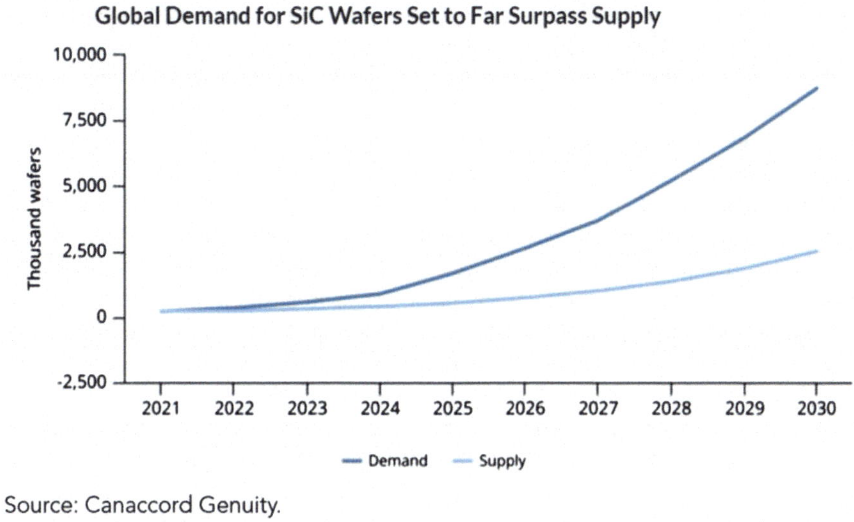

Fig. 8.3 Silicon carbide wafer demand versus supply trend

automotive and other industries that have their own R&D and in-house semiconductor assembly, and they as well are direct consumers of semiconductor wafers and chips.

It is promised that in the near future, the production of 300 mm silicon wafers will meet the market demand; however, if we speak of 200 mm silicon and 150 mm and 200 mm silicon carbide wafers, the gap between supply and demand will continue to grow.

The transition from the 4 inch SiC wafers to 6 inch SiC wafers caused many problems to guarantee the same level of quality and yield. The reason is hidden within silicon carbide properties. SiC hardness is close to diamond, and to grow it, you need much higher temperatures compared to growing silicon, more energy, and what is even more important—more time. Besides, for the 4H-SiC structure, which is prevailing in the SiC market, it is difficult to analyze the surface defects, and this in turn may lead to the problems during the epitaxial growth and lower yield of chips per wafer [4].

The common defects occurring during the manufacturing of silicon carbide wafers are crystalline stacking faults, micropipes, pits, scratches, stains, and surface particles. All of these affect the characteristics and performance of the SiC-based power devices and the equipment they are being installed in. And this is more common with 6 inch wafers rather than with 4 inch wafers. Silicon carbide is a very fragile material and to produce it with the required properties and in required volume takes a lot of time and many challenges to overcome.

Taking into account that it took much time and effort to make a step from 4 inch to 6 inch SiC wafer production, it is obvious that further transition to wider diameters of 8 inches, and later to 12 inches, would face more challenges, even having the background knowledge gained in the past decades.

In addition to the trend of switching to larger wafer diameters, silicon carbide market leaders are working on various technological aspects to improve the quality and yield and what is of high importance to reduce the cost of SiC-based power semiconductors.

Thus, Kiselkarbid i Stockholm, a semiconductor company from Sweden, is working on a new and innovative technology for growing silicon carbide, which will allow the production of practically defect-free SiC wafers. Kiselkarbid i Stockholm's ability to cost-effectively grow defect-free SiC wafers will be providing power component manufacturers with a drastically improved way to increase yield and build larger components at higher voltages. The company promised to ship defect-free 8 inch SiC wafers in the nearest future [5].

In 2018, Infineon Technologies acquired a German-based Siltectra GmbH, a start-up company from Dresden, which has developed a cold split innovative technology to process crystal material efficiently with extremely low loss of material. Since then, Infineon Technologies has been working on further development of SILTECTRA™ COLD SPLIT technology to successfully use it to split silicon carbide wafers. According to the results received, SILTECTRA™ COLD SPLIT allows to double the yield of chips per wafer [6].

Another approach to improving silicon carbide chips yield is offered by Soitec, a company from Bernin in the southeast of France. They call it SmartSiC™. According to Soitec, the SmartSiC™ offers improved manufacturing yield and device performance, unleashing new levels of system compactness. Soitec's proprietary Smart Cut™ technology combined with ultrahigh-conductive handle substrates are the keystones of creating this new generation of engineered substrate [7].

For a very long time until the present, the semiconductor market has not been a central topic on a global level, and for the first time in history, it attracts so many investments into further growth and development. It is clear that the global economic and environmental trends provide incredible opportunities for the silicon carbide market to grow further and faster.

8.1 Opportunity

Seeing the increasing demand in silicon carbide wafers and power devices, such companies such as SK siltron, Wolfspeed, onsemi, Infineon Technologies, ROHM, Mitsubishi Electric, and Fuji Electric have announced their projects to increase the production of semiconductor wafers in order to partially compensate for the growing demand. But even with the announced projects, the global semiconductor industry will experience a shortage of wafers, primarily based on SiC, which opens additional opportunities for new market players with a sufficient level of investment and technology.

With government support, companies from China and South Korea are rapidly joining this race, where plans have been announced for the development of wide bandgap semiconductors through subsidies, loans with a low or almost zero rate, a reduction in the tax base, and a reduction in export/import duties on required materials and equipment.

With some delay, Japanese companies realized the potential of the silicon carbide market, and during 2023, companies such as Mitsubishi Electric, ROHM, and Toshiba announced their investment plans to support the growing global demand for SiC power devices.

As a semiconductor material, silicon carbide is still much younger than silicon when it comes to power electronics application. SiC has not been studied completely yet, and it can offer much in the near future. Today, we can only speak of the applications listed above, but in 5–10 years from now, we can expect silicon carbide to open the road to the new applications as its theoretical properties have more to offer.

Thus, we can expect SiC-based semiconductor devices to be used in mission critical applications in space. Recently, a team of scientists from GE Research demonstrated SiC MOSFETs that can withstand temperatures above 800 °C [8]. It proved the theoretical calculations done by many semiconductor physicists. This achievement can open the road to a whole new level of sensing, actuation, and control applications for space exploration and hypersonic vehicles using SiC MOSFETs.

Another point to consider for the future growth of the silicon carbide market is the approach to design. Today, we can see that most of the companies pursue the planar MOSFET design with a few who work with trench MOSFET design [9].

At present, the SiC-based drivetrain inverter for the electric vehicles is a planar-dominated market, and with the growing market feedback and eventual approval of silicon carbide as the reliable technology for the next generations of the inverters to come, it is clear that soon the end market will push for lower price product. This will further add competitiveness to the market and the rivalry between the companies that use either planar or trench design.

A major step in semiconductor industry growth, including silicon carbide, has been made several years ago when pushed by the global pandemic outcomes and continuing geopolitical uncertainties, many governments decided to launch the chip initiatives to support the development, growth, and supply chain stability. Thus, hundreds of billions of US dollars have been confirmed to be invested into semiconductor industry worldwide during the next 10 years in the USA, China, Japan, India, South Korea, the UK, and the European Union.

Thus, the US CHIPS and Science Act provides 52.7 billion US dollars for American semiconductor research, development, manufacturing, and workforce development. This includes 39 billion US dollars in manufacturing incentives, including 2 billion US dollars for the legacy chips used in automobiles and defense systems, 13.2 billion US dollars in R&D and workforce development, and 500 million US dollars to provide for international information communications technology security and semiconductor supply chain activities [10].

As for the European Union, with the European Chips Act, the EU plans to address semiconductor shortages and strengthen Europe's technological leadership. It will mobilize more than 43 billion Euro of public and private investments till 2030 and set measures to prepare, anticipate, and swiftly respond to any future supply chain disruptions, together with Member States and international partners [11].

Unlike the USA and EU, China has been heavily investing in the semiconductor industry during the past 10 years. In 2021–2022, over 290 billion US dollars was allocated to semiconductor-related investment, in which one-third of the fund was for semiconductor equipment and materials [12]. It is clear that the Chinese government will continue to accelerate the advancements in semiconductor industry by investing more in the next 10 years to reach their goal of semiconductor supremacy after 2030.

Government support is stimulating the industry to grow; however, major companies in this field, such as Infineon Technologies, STMicroelectronics, Wolfspeed, onsemi, Microchip Technology, ROHM, Mitsubishi Electric, Fuji Electric, Renesas Electronics, and Toshiba, have been investing into their future as well.

As shown in Fig. 8.4, from 2021 and up to the present time, over 80 projects have been launched globally to build new factories, R&D centers, and joint ventures or simply expand the production capacity. The total amount of investment into these projects surpasses 70 billion US dollars [13]. Many market leaders are also pursuing vertical integration by acquiring companies to fill in the gaps in the production chain of silicon carbide power devices and systems.

If we take a closer look at the semiconductor projects, of silicon and gallium nitride (GaN), SiC is the leading technology with over 60% of total investment. Over 25 market leaders announced their plans to invest in silicon carbide.

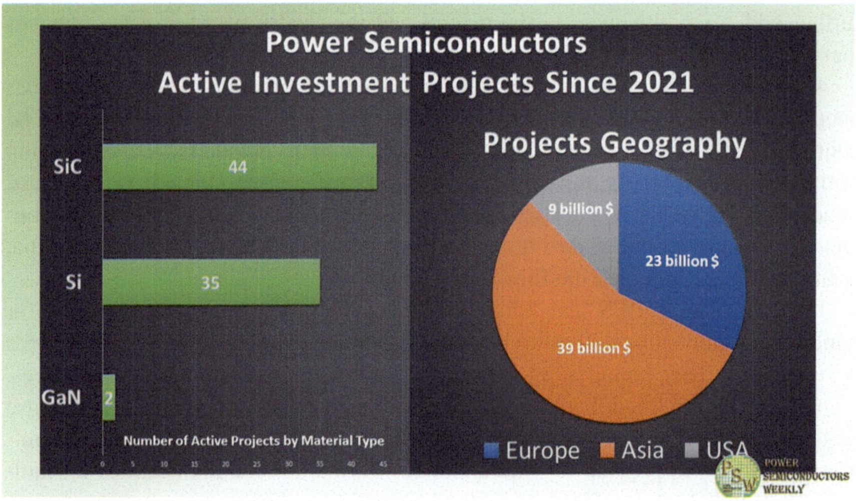

Fig. 8.4 Power semiconductor active investment projects since 2021

Another growth driver for silicon carbide and semiconductor market in general is the lesson learned from the pandemic, when many economies and industries suffered much due to the problems they faced with the supply chain. Even today, 4 years later, many companies have not recovered to the pre-pandemic level, and many went bankrupt. The situation triggered many industries, including semiconductors, to rethink their approach to managing their supply chain.

For many decades, the US and EU semiconductor companies have been expanding their ecosystem network in Asia—China, Malaysia, Indonesia, Singapore, Thailand, and Vietnam. To stay competitive in the market, many companies were seeking ways to optimize their production and logistic costs. Thus, it was one of the obvious choices, until they started to be very dependent on their network in this region. No one could predict the scale of the pandemic, and with the sophistication of the supply chain network came the consequences.

Considering the lesson learned and to avoid any future similar situations, today, we can see the shift from completely relying on so to say outsourcing to investing more into domestic or close to the headquarter semiconductor and semiconductor-related hubs. The outsourcing trend will continue to grow, but clearly, the supply chain will become shorter and closer to the market strategic spots.

Another element of the growth opportunity pyramid for the semiconductor industry would be human resources.

For a decade, human resources issue has remained one of the most important in the semiconductor industry, and it concerns absolutely all companies in the production chain—from suppliers of materials and equipment to manufacturers and companies that assemble power semiconductor devices. And this is not a local issue, but a global one.

The semiconductor industry is aging in terms of human resources. So, many companies employ engineers with more than 20–30 years of experience, and often they have no one to transfer their experience, knowledge, and area of responsibility to. In the past, many companies lacked a well-developed succession system. But the situation is changing with more companies adapting to the current market circumstances.

Another problem that the semiconductor industry faced and now tries to overcome is that for many it is associated with something unattractive, not prestigious, repulsive. Everyone in the industry understands that with the growing demand for semiconductors and silicon carbide-based in particular, the race cannot be won only with monetary support. Qualified staff have always been crucial to the overall success of any company. The leading companies are already creating and using many internal and external programs to attract the semiconductor bright minds to guarantee their own and industry progress in the future.

Another issue that was contributing to the human resources problem is the public awareness of the value of the semiconductor industry and often the lack of information at all. The market leaders have already started to make a change, and we can see the result today, that more and more people are aware and realize what benefits silicon carbide provides when we speak of the automotive industry and residential renewables.

Last but at the same time one of the most important factors contributing to the lack of qualified personnel in the industry is the insufficient number of necessary courses and educational institutions for training specialists in this area. That is why many companies in the industry have taken the following actions in the last 3 years:

Investments and joint programs of manufacturers with technical educational institutions
Creation of nonprofit organizations aimed at developing training and popularization of the semiconductor industry
Increased marketing activity of industry leaders to highlight the semiconductor industry and its value

Given the measures being implemented, the situation with the shortage of qualified personnel in the industry can be changed in the next 5–7 years to correspond to the market growth trends.

Despite all the present and possible obstacles, there is no doubt that the semiconductor industry, and silicon carbide part of it in particular, will continue to grow driven by the global megatrends of electrification and decarbonization and the opportunities it provides to the greener and safer future for humanity.

References

1. Wolfspeed Inc., Wolfspeed Investor Day 2022
2. A. Brothers, O. Burkacky, J. Dragon, J. Kakarwada, A. Mahindroo, J. Patel, A. Suryanarayanan, New silicon carbide prospects emerge as market adapts to EV expansion. https://www.mckinsey.com/industries/semiconductors/our-insights/new-silicon-carbide-prospects-emerge-as-market-adapts-to-ev-expansion
3. S. Rothrock, J.-C. Eloy, P. Chiu, Silicon carbide: from gold rush to commodity? https://www.yolegroup.com/strategy-insights/silicon-carbide-from-gold-rush-to-commodity/
4. S. Lovati, SiC Technology: challenges and future perspectives. https://www.powerelectronics-news.com/sic-technology-challenges-and-future-perspectives/
5. Kiselkarbid i Stockholm AB, KISAB raises a €7.5 m round led by Fairpoint Capital in order to expand production of its defect free silicon carbide wafers. https://kisabsemi.com/kisab-raises-a-e7-5-m-round-led-by-fairpoint-capital-in-order-to-expand-production-of-its-defect-free-silicon-carbide-wafers/
6. Infineon Technologies AG, SILTECTRA™—innovative splitting Technologies. https://www.infineon.com/cms/en/about-infineon/company/siltectra/
7. Soitec SA, Auto SmartSiC™. https://www.soitec.com/en/products/auto-smartsic
8. GE Aerospace, GE Scientists demonstrate ultra-high temperature SiC MOSFET electronics. https://www.ge.com/news/press-releases/ge-scientists-demonstrate-ultra-high-temperature-sic-mosfet-electronics
9. P. Gammon, Trench warfare: A SiC battleground. https://www.pgcconsultancy.com/post/trench-warfare-a-sic-battleground
10. The White House, FACT SHEET: CHIPS and science act will lower costs, create jobs, strengthen supply chains, and counter China. https://www.whitehouse.gov/briefing-room/statements-releases/2022/08/09/fact-sheet-chips-and-science-act-will-lower-costs-create-jobs-strengthen-supply-chains-and-counter-china/

11. European Commission. European Chips Act. https://commission.europa.eu/strategy-and-policy/priorities-2019-2024/europe-fit-digital-age/european-chips-act_en
12. J. Lin, China invested US$290.8 billion in semiconductor projects between 2021–2022. https://www.digitimes.com/news/a20230627VL205/china-ic-manufacturing-semiconductor-chips+components.html
13. A. Cherkasov, Power semiconductors investment projects surpass 70 billion USD. https://www.powersemiconductorsweekly.com/2023/07/19/power-semiconductors-investment-projects-surpass-70-billion-usd/

Aleksei Cherkasov is a business development, marketing, and sales professional with 15 years of experience in the power semiconductor industry and holds an MBA degree in marketing and brand management. He is the founder of Power Semiconductors Weekly platform. Currently, Aleksei works as a senior manager product marketing at Infineon Technologies Austria AG in Villach.

Chapter 9
Conclusion

Maurizio Di Paolo Emilio

In recent years, silicon carbide (SiC) has emerged as a transformative force in the power electronics industry, promising unparalleled efficiency and performance improvements. As SiC IDMs (integrated device manufacturers) and companies within the supply chain continuously pursue efficiency gains and cost reductions, its dominance continues to grow, particularly in the automotive sector. Moreover, cost reductions will inevitably expand the reach of SiC into other critical applications, including solar inverters, energy storage systems, shipping, and haulage. Advancements in power device development are expected to broaden the product range of SiC, enabling previously unexplored applications due to limited voltage capabilities. These developments not only highlight the potential for the continued growth of SiC but also support the disruptive impact of emerging technologies throughout the supply chain. The insights provided here are guided by the expertise of industry speakers Peter Gammon, PGC Consultancy and Professor at University of Warwick, who offers invaluable perspectives on the future trajectory of SiC innovation, and Vernon Rogers, Executive Vice President of Sales and Marketing at Aehr Test Systems, who expands on the advancements in wafer-level test and burn-in, enabling a future for high quality and reliable SiC MOSFET.

9.1 Substrate Innovation

In the ever-evolving landscape of semiconductor technology, disruptive innovations are constantly reshaping the industry's capabilities and potential. Among these disruptive technologies, the emergence of innovative SmartSiC substrates by Soitec

M. Di Paolo Emilio (✉)
Power Electronics News, Pescara, Italy

M. Di Paolo Emilio (ed.), *SiC Technology*,
https://doi.org/10.1007/978-3-031-63418-5_9

offers the potential to revolutionize various sectors. Building on their successful layer transfer technology known as SmartCut, which propelled SOI (silicon-on-insulator) technologies to the forefront in the 1990s, Soitec is now pioneering a similar process for silicon carbide substrates.

The SmartSiC substrates leverage a groundbreaking approach where a single SiC substrate is transformed into up to ten individual substrates. This process involves transferring only the top micron of a conventional single-crystal SiC to a polycrystalline SiC handle wafer. Through meticulous polishing, the original single-crystal SiC substrate is rejuvenated, enabling multiple repetitions of this process. Not only does this innovation promise to address the pressing need to increase wafer supply, but it also offers the additional benefit of the polycrystalline substrate lowering resistance, thereby enhancing device efficiency.

Qualification of these groundbreaking substrates is currently underway in collaboration between Soitec and STMicroelectronics ("ST"), as part of ST's vertical integration strategy to secure full supply chain control, from SiC wafer production to device design and manufacture. This partnership underscores the industry's collective effort to harness disruptive technologies for real-world applications.

In parallel, other substrate innovations are also making waves in the semiconductor realm. KISAB, a Swedish start-up, has achieved a remarkable feat by producing 150 mm SiC substrates entirely devoid of basal plane dislocation defects that have been posing significant challenges in SiC device manufacturing. Eliminating these "killer" defects is pivotal for the advancement of bipolar devices such as IGBTs (insulated gate bipolar transistors) in SiC, promising a leap forward in performance and reliability.

Furthermore, advancements in laser processing technologies by companies such as Disco, Halo, and Infineon's Siltectra are revolutionizing the wafering process of SiC boules. By minimizing wasted material (known as kerf losses), these technologies significantly enhance yield and efficiency. Infineon's innovative approach, in particular, aims to reuse each single-crystal SiC wafer twice through laser splitting, further optimizing resource utilization and reducing production costs.

The industry's transition to 200 mm substrates is gaining momentum, driven by the substantial increase in die yield compared to traditional 150 mm wafers. This transition not only offers greater economies of scale but also aligns with the prevalent technology node in modern silicon power device manufacturing. Companies such as Wolfspeed, with the commissioning of the first functional 200 mm fab in Mohawk Valley and ongoing expansion projects by Coherent, onsemi, STMicroelectronics, and Infineon, are spearheading this transformative shift toward larger substrates.

In conclusion, the semiconductor industry is witnessing a paradigm shift fueled by disruptive technologies across substrate innovation, wafering processes, and transition to larger wafer sizes. These advancements hold immense promise for addressing critical challenges, unlocking new possibilities, and driving unprecedented progress in various application domains, from electric vehicles to renewable energy systems and beyond. As industry players continue to push the boundaries of technological innovation, the future of semiconductor devices appears brighter than ever before.

9.2 Device Innovation

Advancements in semiconductor technology are at the forefront of innovation in various industries, particularly in the field of power electronics. Manufacturers are constantly striving to enhance the performance of their devices while simultaneously reducing costs and improving efficiency. One of the primary strategies employed by these manufacturers is the reduction of die size, which not only maximizes yield but also lowers production costs.

In power MOSFET designs, a split exists between companies opting for trench MOSFET designs and those sticking with planar MOSFET layouts. ROHM and Infineon are leading the charge with trench MOSFET designs, while Wolfspeed, ST, and onsemi continue to utilize planar layouts. Both approaches aim to reduce cell pitch, thereby increasing current and power density within the die and ultimately reducing die size.

Traditionally, a cell pitch of around 6 microns has been standard across the industry. However, recent advancements have pushed the boundaries even further. ST's latest Gen 3 planar designs have been reduced to under 5 microns, while ROHM's Gen 4 trench designs are now as narrow as 2 microns. Impressively, despite this reduction in size, ROHM's technology has actually increased the short-circuit resistance time, indicating improved robustness.

While conservative planar designs currently dominate the electric vehicle (EV) drivetrain market, the potential cost savings associated with trench designs may entice other players to adopt them. Onsemi, for example, announced their plans to release a line of 1200 V trench MOSFET products alongside their planar products in 2024.

Reducing resistances within MOSFETs is another avenue for enhancing device efficiency and die shrinkage. One crucial resistance is that of the substrate, which can be addressed through innovative substrates such as Soitec's SmartSiC substrates. These substrates offer the potential to lower substrate resistance, thereby improving overall device performance.

Additionally, the resistance within the channel region, a critical part of the MOS interface, poses a challenge for device optimization. Research into oxide regions processed using atomic layer deposition has emerged as a promising solution. Hitachi Energy, for instance, has led the way in incorporating high-k dielectrics, such as aluminum oxide (Al_2O_3) or hafnium oxide (HfO_2), into their devices. These high-k dielectrics have been shown to boost channel mobility while maintaining reliability in standard characterization methods such as time-dependent dielectric breakdown.

In conclusion, advancements in MOSFET technology are driving significant improvements in power electronics. The competition between trench and planar designs, coupled with innovations in reducing resistances, is reshaping the landscape of the semiconductor industry. As manufacturers continue to push the boundaries of what is possible, we can expect further advancements that will revolutionize power electronics and drive progress in various sectors, including electric vehicles and renewable energy systems.

9.3 Packaging and Modules

In the pursuit of scaling down silicon carbide (SiC) dice, overcoming thermal challenges becomes paramount. The ability to effectively dissipate heat from increasingly smaller die sizes is critical to maintaining reliability and performance. Without advancements in packaging, module design, and cooling techniques, the persistent heat trapped within these diminutive structures poses a significant obstacle.

Innovations in packaging must prioritize efficient heat transfer mechanisms, such as advanced thermal interface materials and optimized interconnect designs. Module designs need to consider not only the physical constraints of smaller form factors but also the integration of effective cooling solutions. This may involve new approaches such as microfluidic cooling or advanced heat sinks tailored for compact configurations.

Moreover, cooling methodologies must evolve to match the pace of miniaturization. Traditional cooling methods such as air or liquid cooling may need enhancements or entirely new approaches to effectively manage heat dissipation in ever-shrinking SiC dice.

Failure to address these thermal challenges would not only compromise reliability but also necessitate derating, undermining the very purpose of scaling down SiC die. As the industry pushes the boundaries of miniaturization, collaborative efforts across packaging, module design, and cooling technologies are imperative to sustain ongoing progress while ensuring the reliability and performance of future SiC-based systems.

9.4 Wafer-Level Test and Burn-in

The rise of silicon carbide (SiC) MOSFETs in multi-die power modules for e-mobility and power transformation has brought a surge in demand for wafer-level test and burn-in. This approach ensures high yield by screening for reliable die, such as known good die (KGD), before assembly into an expensive multi-die module. However, traditional burn-in techniques faced hurdles when applied to SiC technology and wafer form factor. This chapter explores the innovative solutions tackling these challenges and solidifying wafer-level test and burn-in as the future of SiC MOSFET product quality and reliability. SiC MOSFETs operate at significantly higher voltages compared to their silicon counterparts. This presents a significant obstacle for wafer-level test and burn-in, where the common cathode connects all the dies together creating a challenge for high parallelism and high-voltage testing which can lead to arcing between closely spaced conductors as examples. The arcing of several thousand volts can damage the devices and compromise test results as well as damage the system electronics. Innovations in arc-mitigation techniques have addressed this issue. One approach involves using inert gases and other special gases during high-voltage testing. These gases have a higher breakdown voltage compared to air, effectively preventing arcing while maintaining the integrity of the test environment.

Another challenge in wafer-level test and burn-in for SiC MOSFETs is the high current capabilities inherent to the technology. Traditional test and wafer sort equipment often struggle to handle the high current and high thermal power demands when testing a large number of dies in parallel. Advancements in test electronics, power supply design, and wafer interfacing to deliver thermal performance are overcoming this hurdle. Specialized test equipment with innovative power supply design has been developed to deliver the high currents associated with SiC MOSFETs while maintaining accurate measurement capabilities. Wafer-level test and burn-in offers a significant cost advantage compared to traditional packaged device burn-in. By testing the entire wafer in a single operation, the cost of test (COT) is dramatically reduced. This "one-touch" approach eliminates the need for multiple test cycles on individual packaged devices. Furthermore, advancements in equipment design allow for parallel testing of multiple wafers within the same system. Some equipment such as Aehr test system's FOX-XP can accommodate up to 18 wafers simultaneously, within the footprint of a single tester, further multiplying the efficiency and cost savings associated with wafer-level test and burn-in.

The success of wafer-level test and burn-in relies heavily on automation. This automation minimizes human error and ensures consistent, reliable test results throughout the manufacturing process. Automation of the wafer-level test and burn-in enables Industry 4.0, smart manufacturing where the digital information allows for real-time decision-making and integration of artificial intelligence (AI), and machine learning in production facilities.

By overcoming the challenges of COT through high parallelism at the wafer and system level and high-voltage testing utilizing innovative wafer interfacing and arc suppression techniques at wafer-level test and burn-in has established itself as the leading solution for SiC MOSFET reliability testing. With its high throughput, automation capabilities, and cost-effectiveness, wafer-level test and burn-in ensures the quality and reliability of SiC MOSFET power devices, paving the way for the widespread adoption of this revolutionary technology.

Maurizio Di Paolo Emilio holds a PhD in physics and is a telecommunications engineer. He has worked on various international projects in the field of gravitational wave research, designing a thermal compensation system (TCS) and data acquisition and control systems, and on others about X-ray microbeams in collaboration with Columbia University, high voltage systems and space technologies for communications, and motor control with ESA/INFN. TCS has been applied to the Virgo and LIGO experiments, which detected gravitational waves for the first time and earned the Nobel Prize in 2017. Since 2007, he has been a reviewer for scientific publications for academics such as Microelectronics Journal and IEEE journals. Moreover, he has collaborated with different electronic industry companies and several Italian and English blogs and magazines, such as *Electronics World*, *Elektor*, *Mouser*, *Automazione Industriale*, *Electronic Design*, *All About Circuits*, *Fare Elettronica*, *Elettronica Oggi*, and *PCB Magazine*, as a technical writer/editor, specializing in several topics of electronics and technology. From 2015 to 2018, he was the editor-in-chief of *Firmware* and *Elettronica Open Source*, which are technical blogs and magazines for the electronics industry. He participated in many conferences as a speaker of keynotes for different topics such as X-ray, space technologies, and power supplies. Maurizio enjoys writing and telling stories about power electronics, wide bandgap semiconductors, automotive, IoT, embedded energy, and quantum computing. Maurizio has been an *AspenCore* content editor since 2019. He is currently the editor-in-chief of *Power Electronics News* and *EEWeb* and a correspondent for *EE Times*. He is the host of PowerUP, a podcast about power electronics, and the promoter and organizer of the PowerUP Virtual Conference, a summit where each year great speakers talk about the power electronics design trends. Moreover, he has contributed to a number of technical and scientific articles, as well as a couple of *Springer* books on energy harvesting and data acquisition and control systems.

Peter Gammon is a professor of Power Semiconductor Devices in the School of Engineering at the University of Warwick (www.warwick.ac.uk/pmgammon) and the founder of PGC Consultancy (www.pgcconsultancy.com). Peter leads a team of researchers and PhD students, in the development of next-generation SiC power electronic devices. Besides helping improve today's automotive-grade SiC MOSFETs, his focus is on developing ultrahigh-voltage (10 kV+) power devices for the future power grid and radiation hard power electronics for operation in space. He is the principal investigator on six major research projects; has published over 100 journal and conference papers, with over 900 citations; and is the author of three patents. He is a senior member of the IEEE, a member of the IET, and a Senior Fellow of the Higher Education Academy (SFHEA).

Vernon Rogers Executive Vice President of Sales and Marketing at Aehr Test Systems. Vernon Rogers is the Executive Vice President of Sales and Marketing since 2018. Prior to joining Aehr, Vernon served as head of sales and marketing at GES preceding its acquisition by Kimball Electronics. Mr. Rogers, over the past 22 years, has held numerous senior executive sales, marketing and operations management positions, successfully building multi-channel sales distribution and support organizations at both public and start-up technology companies in the semiconductor and system level test space. His career has focused on nanoscale inspection, manufacturing, and test technologies with such leaders as LitePoint (acquired by Teradyne), Credence Systems Corporation (acquired by Xcerra), NPTest (acquired by Credence Systems) and Schlumberger Automated Test Equipment. He also oversaw sales for FlexStar Technology, a Bay Area leader in hard disk drives (HDD) and solid-state disk drives (SSD) test and burn-in. Mr. Rogers received a B.S. in Electrical Engineering from North Dakota State University.

Index

M. Di Paolo Emilio (ed.), *SiC Technology*,
https://doi.org/10.1007/978-3-031-63418-5

The manufacturer's authorised representative in the EU is Springer Nature Customer Service Centre GmbH, Europaplatz 3, 69115 Heidelberg, Germany. If you have any concerns regarding our products, please contact ProductSafety@springernature.com

Printed and bound by CPI Group (UK) Ltd, Croydon, CR0 4YY
01/10/2026
02225904-0009